Systems Biology in Animal Science

Systems Biology in Animal Science

Edited by Jessica Marshall

SYRAWOOD
PUBLISHING HOUSE
New York

Published by Syrawood Publishing House,
750 Third Avenue, 9th Floor,
New York, NY 10017, USA
www.syrawoodpublishinghouse.com

Systems Biology in Animal Science
Edited by Jessica Marshall

International Standard Book Number: 978-1-64740-410-9 (Hardback)

Cataloging-in-publication Data

Systems biology in animal science / edited by Jessica Marshall.
p. cm.
Includes bibliographical references and index.
ISBN 978-1-64740-410-9
1. Zoology. 2. Systems biology. I. Marshall, Jessica.
QL45.2 .S97 2023
590--dc23

TABLE OF CONTENTS

PREFACE

In my initial years as a student, I used to run to the library at every possible instance to grab a book and learn something new. Books were my primary source of knowledge and I would not have come such a long way without all that I learnt from them. Thus, when I was approached to edit this book; I became understandably nostalgic. It was an absolute honor to be considered worthy of guiding the current generation as well as those to come. I put all my knowledge and hard work into making this book most beneficial for its readers.

Systems biology refers to the study of the behavior and interactions of the components of biological entities such as cells, organisms, molecules and organs. It seeks to comprehend entire biological systems functioning as a unit instead of only focusing on their individual components. Systems biology combines mathematical modeling, statistics and bioinformatics to analyze and integrate large amounts of data. This analysis helps in better understanding the biology of the given biological systems and helps in predicting the behavior of biological systems. There are three basic concepts that are critical for comprehending complex biological systems, namely, their emergence, robustness and modularity. Systems biology utilizes a variety of quantitative experimental and computational methods for deriving information flow from proteins, genes and other subcellular components. These methods help in determining the signaling, regulatory and functional pathways that control organ, cell, tissue and organismal level functions. This book aims to shed light on the role of systems biology in animal science. Its extensive content will provide the readers with a thorough understanding of the subject.

I wish to thank my publisher for supporting me at every step. I would also like to thank all the authors who have contributed their researches in this book. I hope this book will be a valuable contribution to the progress of the field.

Editor

Integrated Analysis of miRNA-mRNA Network Reveals Different Regulatory Patterns in the Endometrium of Meishan and Duroc Sows during Mid-Late Gestation

Kaijie Yang [1,†], Jue Wang [1,†], Kejun Wang [1,2], Yabiao Luo [1], Qiguo Tang [1], Ximing Liu [1] and Meiying Fang [1,*]

1 Department of Animal Genetics and Breeding, National Engineering Laboratory for Animal Breeding, MOA Laboratory of Animal Genetics and Breeding, Beijing Key Laboratory for Animal Genetic Improvement, College of Animal Science and Technology, China Agricultural University, Beijing 100193, China; kjyang2014@163.com (K.Y.); tctcttc@hotmail.com (J.W.); wangkejun@163.com (K.W.); mothluo@foxmail.com (Y.L.); tango@cau.edu.cn (Q.T.); Lximing2018@163.com (X.L.)

2 College of Animal Science and Veterinary Medicine, Henan Agricultural University, Zhengzhou 450002, China

* Correspondence: meiying@cau.edu.cn

† These authors are contributed equally.

Simple Summary: Meishan pigs have a lower fetal loss rate during mid-late gestation compared to Duroc pigs. Differentially expressed mRNAs and miRNAs detected in endometrial tissue from Meishan and Duroc sows at mid-late gestation are involved in regulating hormone and oxygen levels, blood vessel development, and developmental processes affecting reproduction. In addition, ssc-miR-503 and ssc-miR-671-5p were shown to target the *EGF* and *ESR1* genes, respectively. These findings provided an important resource for studying embryonic mortality during mid-late gestation in pigs.

Abstract: Embryo loss is a major factor affecting profitability in the pig industry. Embryonic mortality occurs during peri-implantation and mid-late gestation in pigs. Previous investigations have shown that the embryo loss rate in Meishan pigs is significantly lower than in commercial breeds. Most studies have focused on embryonic mortality during early gestation, but little is known about losses during mid-late gestation. In this study, we performed a transcriptome analysis of endometrial tissue in mid-late gestation sows (gestation days 49 and 72) sampled from two breeds (Meishan (MS) and Duroc (DU)) that have different embryo loss rates. We identified 411, 1113, 697, and 327 differentially expressed genes, and 14, 36, 57, and 43 differentially expressed miRNAs in four comparisons (DU49 vs. DU72, DU49 vs. MS49, DU72 vs. MS72, and MS49 vs. MS72), respectively. Subsequently; seven differentially expressed mRNAs and miRNAs were validated using qPCR. Functional analysis suggested the differentially expressed genes and miRNAs target genes mainly involved in regulation of hormone levels, blood vessel development, developmental process involved in reproduction, embryonic placenta development, and the immune system. A network analysis of potential miRNA-gene interactions revealed that differentially expressed miRNAs in Meishan pigs are involved in the response to estradiol and oxygen levels, and affect angiogenesis and blood vessel development. The binding site on ssc-miR-503 for epidermal growth factor (*EGF*) and the binding site on ssc-miR-671-5p for estrogen receptor α (*ESR1*) were identified using a dual luciferase assay. The results of this study will enable further exploration of miRNA-mRNA interactions important in pig pregnancy and will help to uncover molecular mechanisms affecting embryonic mortality in pigs during mid-late gestation.

Keywords: RNA-seq; miRNA-mRNA interaction; endometrium; Meishan and Duroc sows

1. Introduction

Litter size, which depends on ovulation rate, fertilization rate, and conceptus survival to term, is one of the most important economic traits affecting production efficiency in pig the industry [1]. Although fertility in the sow is directly linked to the ovulation rate, the main barrier to increasing litter size in pigs is prenatal mortality [2]. The Chinese Meishan breed is one of the most prolific, bearing three to five piglets more per litter than European and North American commercial breeds despite having a similar ovulation rate [3]. In contrast, the Duroc is considered to be a low-prolificacy breed. Meishan embryos develop more slowly and produce less estrogen, resulting in increased embryo survival [4,5]. Meishan pigs also possess higher uterine capacity and placental efficiency than other commercial breeds, due to greater vascular density in the developing placenta and endometrium [6,7].

Between 20–45% of embryos are lost in pig pregnancy, which is a major challenge to the commercial swine industry [8]. In pigs, embryonic losses occur during two critical developmental stages; 20–30% occur during the peri-implantation (day 10–30 of gestation (GD10–30)), and an additional 10–15% of embryos are lost at mid-late gestation (GD50–90) [1,8]. Although investigators have identified genetics [9], uterine capacity [10], and angiogenesis at the maternal-fetal interface [11] as important factors, the underlying reasons responsible for spontaneous conceptus loss are still not well understood. After GD30, embryo losses in swine are thought to result from intrauterine crowding caused by limited uterine capacity [12].

miRNAs are a class of non-coding RNAs, approximately 22 nucleotides in length, which negatively regulate the translation of target genes by binding to complementary sequences in the 3′UTR (untranslated region) [13,14]. In mammals, miRNAs are predicted to control the activity of about 50% of all protein-coding genes and regulate gene expression in every cellular pathway [15]. In pigs, miRNA-mRNA interactions play an important role in pregnancy. Two important miRNAs, miR-26a and miR-125b, have been identified in extracellular vesicles isolated from uterine luminal flushings during early pregnancy. miR-125b regulates the expression of genes involved in embryo development and implantation [16]. In addition, miR-181a and miR-181c are thought to have important roles in embryo implantation and placentation in pigs by targeting *SPP1* (secreted phosphoprotein 1), *ESR1* (estrogen receptor 1), and *ITGB3* (integrin beta3) [17]. Finally, miR-130b appears to facilitate the expression of *HPSE* (heparanase), which changes placental folding and increases placental efficiency by suppressing the expression of *PPARG* (peroxisome proliferator activated receptor gamma) [18].

Previous studies show that Meishan pigs have less embryo loss than other commercial breeds, not only during early gestation but also mid-late gestation [4,5,19]. The studies conducted using microarrays and RNA-seq to identify and characterize the genes and miRNAs expressed in the uterine endometrium during early gestation [20–23] are an important step toward understanding the complex processes that govern the establishment and maintenance of pregnancy. However, little is known about genes and miRNAs expressed in the uterine endometrium during mid-late gestation.

Here, we used RNA-sequencing (RNA-seq and miRNA-seq) to compare the uterine endometrium transcriptomes from Chinese Meishan and Duroc sows during mid-late gestation. The data and our analysis may provide deeper insight into the molecular mechanisms affecting embryonic mortality in pigs.

2. Materials and Methods

2.1. Animal Sample Collection and RNA Extraction

Six unrelated Meishan pigs (no common grandparents) and six unrelated Duroc pigs were selected at two different stages of gestation (GD49 and GD72). The sows were divided into four experimental groups, each of which contained three sows of the same breed (MS49, MS72, DU49, DU72). The pigs were raised in identical conditions by the Shanghai Zhu Zhuang Yuan Company (Shanghai, China). All pigs had previously given birth to three litters and were in their fourth pregnancy. Pigs at the same day of gestation were slaughtered in the same week. After slaughter, endometrial samples were collected from each pig and stored until use at −80 °C or in liquid nitrogen. Animal treatment and sample collection were approved by the Animal Welfare Committee of the State Key Laboratory of Agricultural Biotechnology, China Agricultural University (approval No. XK257), and in accordance with the Regulations on Administration of Affairs Concerning Experimental Animals, revised June 2004 (Ministry of Science and Technology).

Total RNA was extracted from endometrial tissues using TRIzol® Reagent (Invitrogen, San Diego, CA, USA), using the manufacturer's protocol. RNA degradation and contamination were evaluated on a 1% agarose gel. A Qubit 2.0 Fluorometer and Qubit RNA analysis kit (Life Technologies, Carlsbad, CA, USA) were used to measure RNA concentration. RNA purity and integrity were checked using the RNA Nano 6000 Assay Kit for the Bioanalyzer 2100 system (Agilent Technologies, Foster, CA, USA) and a NanoPhotometer® spectrophotometer (IMPLEN, Westlake Village, CA, USA), respectively.

2.2. Library Preparation for mRNA Sequencing and Data Analysis

A quantity of 3 μg RNA from each sample was used for sample preparation. Sequencing libraries were constructed according to the manufacturer's recommendations using the NEBNext® Ultra™ RNA Library Prep Kit for Illumina® (NEB, Ipswich, MA, USA), and index adapters were added to each sample. Briefly, poly-T oligo-attached magnetic beads were used to purify mRNA from total RNA. Divalent cations were used to fragment RNA at high temperature in NEBNext First Strand Synthesis Reaction Buffer (5×). The first cDNA strand was synthesized using random hexamer primers and M-MuLV Reverse Transcriptase (RNase H-). The second strand was synthesized using DNA polymerase I and RNase H. Overhangs were converted into blunt ends by the exonuclease/polymerase activity of PolI. After the 3′ ends of the DNA fragments were adenylated, they were ligated to EBNext Adaptors. To enrich for cDNA fragments around 250–300 bp, library fragments were purified using the AMPure XP system (Beckman Coulter, Beverly, MA, USA). The size-selected adaptor-ligated cDNA was incubated with 3 μL USER Enzyme (NEB, Ipswich, MA, USA) at 37 °C for 15 min, followed by 5 min at 95 °C. PCR was then conducted with Phusion High-Fidelity DNA polymerase, Universal PCR primers, and Index (X) Primer. Finally, the PCR products were purified using the AMPure XP system, and library quality was evaluated with the Agilent Bioanalyzer 2100 system. A HiSeq 2000 platform (Illumina, San Diego, CA, USA) was used to generate 150 bp paired-end reads. Raw reads in FASTQ format were initially processed using custom Perl scripts. During this step, clean data were obtained by removing reads with adapter sequences, poly-N sequences, or low-quality scores. Q20, Q30, and GC content were calculated to evaluate the quality of the clean data, and only high-quality clean data were used in subsequent analyses. Reads were mapped to the pig reference genome (Sus scrofa 11.1, Ensembl, ftp://ftp.ensembl.org/pub/release-95/fasta/sus_scrofa/dna/) with HiSAT2 (http://ccb.jhu.edu/software/hisat2/index.shtml) [24]. The number of reads for each gene was determined by featureCounts v1.5.0-p3 (http://subread.sourceforge.net) [25], and FPKM (fragments per kilobase of transcript per million fragments mapped) for each gene was calculated according to gene length and read count. Differences in mRNA expression between Meishan and Duroc sows were analyzed using the DESeq2 R package (1.16.1) (http://www.bioconductor.org/packages/release/bioc/html/DESeq2.html) [26]. mRNAs with an adjusted p-value < 0.05 were classified as differentially expressed genes (DEGs).

2.3. Library Preparation for MicroRNA Sequencing and Data Analysis

A quantity of 3 μg RNA from each sample was used to construct a small RNA library. Sequencing libraries were generated according to the manufacturer's protocol accompanying the NEBNext® Multiplex Small RNA Library Prep Set for Illumina® (NEB, Ipswich, MA, USA). Index adapters were added to each sample. Briefly, NEB 3′ SR Adapters were directly ligated to the 3′ ends of the RNA.

After the 3′ ligation reaction, SR RT Primers were hybridized to the free 3′ SR adaptors (i.e., unligated adaptors remaining after the ligation reaction) to transform the single strand DNA adaptors into double-strand DNA molecules. The first cDNA strand was then synthesized using M-MuLV reverse transcriptase (RNase H -). PCR amplification was conducted using LongAmp Taq 2X Master Mix, Index (X) Primer, and SR Primer from Illumina. After purification on an 8% polyacrylamide gel (100 V, 80 min), DNA fragments between 140 and160 bp (the length of small non-coding RNA plus 3′ and 5′ adaptors) were recovered and dissolved in 8 μL elution buffer. Finally, the quality of the library was evaluated with an Agilent Bioanalyzer 2100 system using DNA High Sensitivity Chips. Following the manufacturer's instructions, a TruSeq SR Cluster Kit v3-cBot-HS (Illumina, NEB, Ipswich, MA, USA) was used to cluster the index-coded samples on a cBot Cluster Generation System (Illumina, NEB, Ipswich, MA, USA). After clustering, the library was sequenced on the Illumina Hiseq 2500/2000 platform and 50 bp single-end reads were generated. Raw reads were collected in FASTQ format and initially processed using custom Perl and Python scripts. In this step, clean data were generated by removing reads with low-quality scores and reads containing poly-n, 5′ adapters, poly A or T or G or C, or that did not contain 3′ adapters or insert tags. Q20, Q30, and GC content were calculated to evaluate clean data quality. Clean reads, 18–35 nucleotides in length, were selected for all subsequent analyses. Bowtie [27] was used to align small RNAs to the porcine reference genome (Sus scrofa 11.1), and Bedtools (https://bedtools.readthedocs.io/) was used to identify known miRNAs by matching them to entries in miRBase20.0 (http://www.mirbase.org/). After excluding matching reads, the remaining reads were reanalyzed to predict novel miRNAs using miRDeep2 [28]. miRNA expression was estimated by RPM (reads per million total reads). Differential expression of miRNA between samples was analyzed with the DEGseq R package (1.8.3) (http://bioinfo.au.tsinghua.edu.cn/software/degseq) [29]. miRNAs with significantly different levels of expression (p-value < 0.05) were classified as differentially expressed (DE) miRNAs.

2.4. Gene Ontology and Kyoto Encyclopedia of Genes and Genomes Enrichment Analyses

Metascape (http://metascape.org) [30] was used to conduct gene ontology (GO) enrichment analysis of the DEGs and target genes of DE miRNAs. Pig gene ensemble IDs were first converted into human gene IDs by the Ensembl BioMart tool (http://www.ensseml.org/biomart/martview) because the functional annotation for the pig is incomplete. The corresponding human gene ensemble IDs were then submitted to the Metascape database for functional annotation. GO items with adjusted p-values less than 0.01 were considered to be significantly enriched by DEGs or target genes of DE miRNAs. We used KOBAS V3.0 (http://kobas.cbi.pku.edu.cn) [31] to test the statistical significance of the enrichment of DEGs or target genes of DE miRNAs in Kyoto Encyclopedia of Genes and Genomes (KEGG) pathways. Pathways with p-values lower than 0.05 were regarded as significantly enriched.

2.5. Integrated miRNA-mRNA Analysis

To construct the miRNA-target gene network, BLASTN (https://blast.ncbi.nlm.nih.gov) was first used to recognize and remove pre-microRNAs based on high levels of similarity. Then, target relationships with miRNAs were predicted by miRanda [32], requiring an alignment score N > 140 and energy < −10 kcal/mol. Further analyses of miRNA-gene pairs were conducted based on the common miRNA-binding sites. The miRNA-mRNA interaction network was constructed and visualized using Cytoscape v3.7.2 (https://cytoscape.org) [33].

2.6. Quantitative Polymerase Chain Reaction

To detect DEGs and DE miRNAs, 1 g of RNA from endometrium tissue was transcribed into cDNA using a FastQuant RT Kit (with gDNase) (Tiangen Biotech Co., Ltd., Beijing, China) and a miRcute Plus miRNA First-Strand cDNA Kit (Tiangen Biotech Co., Ltd., Beijing, China), following the manufacturer's instructions. The expression levels of seven genes and seven miRNAs were detected by quantitative PCR with a miRcute Plus miRNA qPCR Kit (Tiangen Biotech Co., Ltd., Beijing, China) and SuperReal PreMix Plus (SYBR Green) (Tiangen Biotech Co., Ltd., Beijing, China), respectively. Primer Premier 5.0 (Premier Biosoft International, Palo Alto, CA, USA) was used to design primers for qPCR of genes and miRNAs, and the primers were synthesized by SANGON biotechnology company (Beijing, China). Gene and miRNA primer sequences are listed in Supplementary Material Table S1. The cycling parameters used for qPCR amplification of genes were as follows: initial heat denaturation at 95 °C for 15 min; 40 cycles at 95 °C for 30 s, 60 °C for 30 s, and 72 °C for 30 s; and a final extension at 72 °C for 5 min. The cycling parameters used for qPCR amplification of miRNAs were as follows: initial heat denaturation at 95 °C for 15 min; 5 cycles at 94 °C for 20 s, 65 °C for 30 s, and 72 °C for 34 s; 40 cycles at 94 °C for 20 s and 60 °C for 34 s. A melting curve analysis was conducted to eliminate DNA contamination and confirm primer specificities. The $2^{-\Delta\Delta Ct}$ method with *β-actin* was used as the endogenous control for the normalization of gene expression levels. Relative miRNA expression was also normalized using the $2^{-\Delta\Delta Ct}$ method with the U6 small nuclear RNA as an internal standard. Forward primers of U6 were designed and synthesized by RiboBio (Guangzhou, China). Each biological replica consists of three technical replicates.

2.7. Cell Culture and Transfection

HEK293T cells were obtained from the Institute of Biochemistry and Cell Biology, Chinese Academy of Science, P. R. China. HEK293T cells were cultured in DMEM with 10% FBS and 1% penicillin/streptomycin. The cells were cultured at 37 °C with 5% CO_2. All reagents for cell culture were purchased from Gibco (Langley, OK, USA) and Invitrogen (Carlsbad, CA, USA).

2.8. Dual Luciferase Assay

Potential binding sites for ssc-miR-503 and ssc-miR-671-5p were predicted using miRanda as described above. The target sequence for ssc-miR-503 (within the CDS of the *EGF* gene) and the target for ssc-mir-671-5p binding (within the 3′-UTR region of *ESR1*) were amplified from pig genomic DNA and cloned into the psiCHECK-2 plasmid (Promega, Madison, WI, USA) using the XhoI and NotI restriction sites downstream from the Renilla luciferase gene. The two primer pairs are shown in Supplementary Material Table S2. Mutant plasmids with altered putative binding sites were synthesized by BGI (Beijing), and Geneprama (Suzhou, China) synthesized ssc-miR-503 and ssc-miR-671-5p mimics and negative control mimics. Plasmids containing mutant or wild-type binding sites were co-transfected into HEK293T cells with either negative control mimics or miRNA mimics at a concentration of 100 nM using Lipofectamine™ 2000 (Invitrogen, Carlsbad, CA, USA). Experiments were conducted in triplicate. Cells were collected 36 hours after transfection and luciferase activity was measured using the Dual-Glo® Luciferase Assay System (Promega, Madison, WI, USA) in accordance with the manufacturer's instructions. The data were first normalized by the ratio of Renilla luminescence (CDS or 3′UTR region) to firefly luminescence (transfection control). Relative luciferase activity was defined by calculating the ratio of the mean of the three biological replicates in each miRNA transfection group to the mean of the corresponding control siRNA transfection group.

2.9. Statistical Analysis

Data are shown as means ± standard deviation (SD). A one-way ANOVA was used to test for significance. Levene's test was used to assess homogeneity of variance, followed by Student's t-test. Data analysis was conducted using SAS version 9.0 (SAS, Cary, NC, USA). Differences were regarded as statistically significant for p-values < 0.05.

3. Results

3.1. Overview of Sequencing Data

Between 51.29 to 66.25 million raw reads were acquired for each sample. After quality control, 48.76 to 61.19 million clean reads remained and were aligned to the pig reference genome (Sus scrofa

11.1) (Supplementary Material Table S3). Approximately 90% of total reads could be mapped, and 87% mapped to unique genomic positions (Table S3). As expected, most reads mapped to coding regions (84.13–90.84%), whereas 5.16–9.63% mapped to introns and 4.00–6.24% mapped to intergenic regions (Supplementary Material Table S4). Expression of 21,333 genes was detected in the pig endometrial tissues, and 17,779 genes exhibited expression in all four experimental groups. The miRNA libraries generated an average of 15.73 million raw reads. After quality control, about 97.8% of raw reads were retained for further analysis (Supplementary Material Table S5). Most reads were 20–24 nt, with a modal length of 22 nt. An average of 13.76 million (93.40%) of the total sRNAs were aligned to the pig reference genome (Sus scrofa 11.1) (Table S5). 484 mature miRNAs were identified, including 355 annotated mature miRNAs from 321 precursors (Supplementary Material Table S6) and 129 novel mature miRNAs from 134 precursors (Supplementary Material Table S7).

3.2. *Identifying Differentially Expressed mRNAs and miRNAs*

Using RNA samples from two points of gestation, we performed four comparisons (DU49 vs. DU72, DU49 vs. MS49, DU72 vs. MS72 and MS49 vs. MS72). The analyses revealed, 411, 1113, 697, and 327 genes that met criteria for differential expression (adjusted p-value < 0.05), respectively (Table 1 and Supplementary Material Table S8). Several differentially expressed genes have previously been identified as important for fetal loss or litter size (Table 2).

Table 1. Number of differentially expressed mRNAs and miRNAs.

Comparison	DE mRNAs			DE miRNAs		
	Total	Up	Down	Total	Up	Down
DU49 vs. DU72	411	273	138	14	10	4
DU49 vs. MS49	1113	678	435	36	24	8
DU72 vs. MS72	697	397	300	57	26	31
MS49 vs. MS72	327	185	142	43	21	22

Table 2. Differentially expressed genes associated with fetal loss or litter size based on previous studies.

Comparison Group	Ensembl ID	Gene Name	Log_2 Fold Change	Adjusted p-Value
DU49 vs. DU72	ENSSSCG00000030484	*AHR* [34]	−0.62	0.0247
	ENSSSCG00000004283	*DPPA5* [35]	4.68	0.0098
	ENSSSCG00000010698	*FGFR2* [36]	0.96	0.0446
	ENSSSCG0000012229	*GP91-PHOX* [37]	−0.74	0.0467
	ENSSSCG00000015579	*PTGS2* [38]	1.85	0.0031
	ENSSSCG00000037754	*SLC39A11* [39]	1.08	0.0235
DU49 vs. MS49	ENSSSCG00000013303	*ABTB2* [40]	−0.98	0.0497
	ENSSSCG00000017101	*ADCY2* [41]	−1.35	0.0137
	ENSSSCG00000011437	*ALAS1* [42]	1.33	0.0283
	ENSSSCG00000025578	*ALDH1A2* [43]	−1.24	0.0422
	ENSSSCG00000004283	*DPPA5*	3.85	3.66×10^{-6}
	ENSSSCG00000012229	*GP91-PHOX*	−1.05	0.0174
	ENSSSCG00000016204	*IHH* [44]	−1.50	0.0049
	ENSSSCG00000040575	*ISG15* [45]	−1.23	0.0171
	ENSSSCG00000015579	*PTGS2*	1.87	0.0419
	ENSSSCG00000007836	*SCNN1G* [38]	1.44	0.0412

Table 2. *Cont.*

Comparison Group	Ensembl ID	Gene Name	Log_2 Fold Change	Adjusted *p*-Value
DU72 vs. MS72	ENSSSCG00000009134	*EGF* [46]	−1.57	0.0012
	ENSSSCG00000014336	*EGR1* [47]	−1.44	3.75×10^{-6}
	ENSSSCG00000002383	*FOS* [47]	−0.79	0.0040
	ENSSSCG00000010639	*HABP2* [41]	2.48	0.0248
	ENSSSCG00000016204	*IHH*	2.20	5.36×10^{-8}
	ENSSSCG00000040575	*ISG15*	−1.01	0.0059
	ENSSSCG00000000455	*LRIG3* [48]	2.14	1.37×10^{-10}
MS49 vs. MS72	ENSSSCG00000014336	*EGR1*	−1.59	0.0002
	ENSSSCG00000025777	*ESR1* [49]	0.85	0.0036
	ENSSSCG00000002383	*FOS*	−1.34	9.03×10^{-8}
	ENSSSCG00000015144	*GRAMD1B* [40]	−1.39	0.0294
	ENSSSCG00000016204	*IHH*	4.13	6.46×10^{-19}

We also identified 14, 36, 57, and 43 DE miRNAs (*p*-value < 0.05) in the four comparisons described above (Supplementary Material Table S9). Among these, miR-19a [50], miR-323 [51], miR-27a [51,52], miR-30a-5p [53] may be involved in fetal loss during pregnancy. Venn diagrams showing the differentially expressed genes and miRNAs from the four groups are presented in Figure 1A (mRNAs) and Figure 1B (miRNAs).

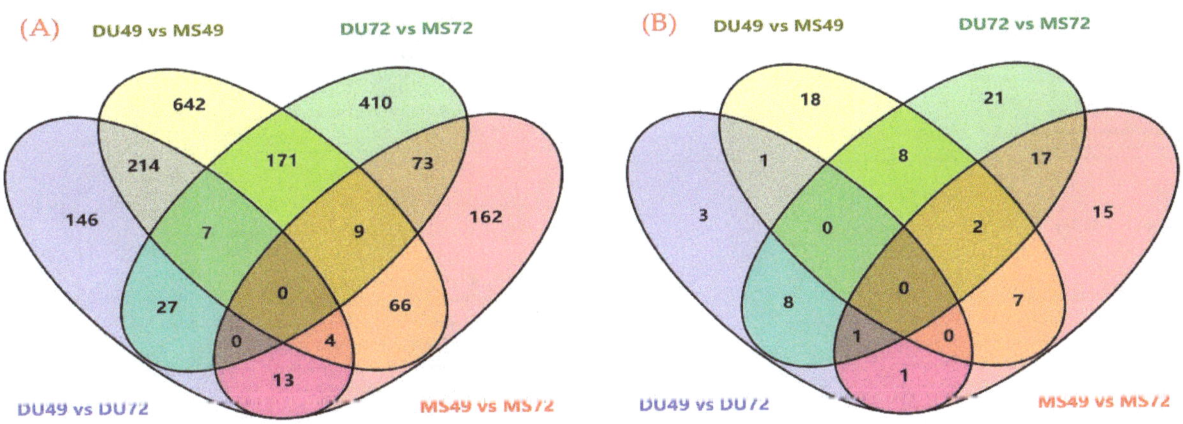

Figure 1. Venn diagram displaying the overlap among different groups of differentially expressed genes (DEGs) (**A**), differentially expressed (DE) miRNAs (**B**).

3.3. Functional Analysis of DEGs and DE miRNA Target Genes

GO and KEGG pathway analyses were performed to evaluate the potential functions of DEGs. The most enriched terms and pathways associated with DEGs identified in the comparison of GD49 and GD72 are shown in Figure 2. Similarly, terms and pathways identified in the comparison of Meishan and Duroc pigs are shown in Figure 3. Pregnancy-associated GO biological process (BP) terms were significantly enriched for DEGs from all four comparisons, such as "regulation of hormone levels", "blood vessel development", "developmental process involved in reproduction", "embryonic placenta development", and "immune" (Supplementary Material Table S10). KEGG pathways related to pregnancy were also enriched significantly in the four comparison groups, such as "steroid hormone biosynthesis", "ovarian steroidogenesis", and "estrogen signaling pathway" (Supplementary Material Table S11).

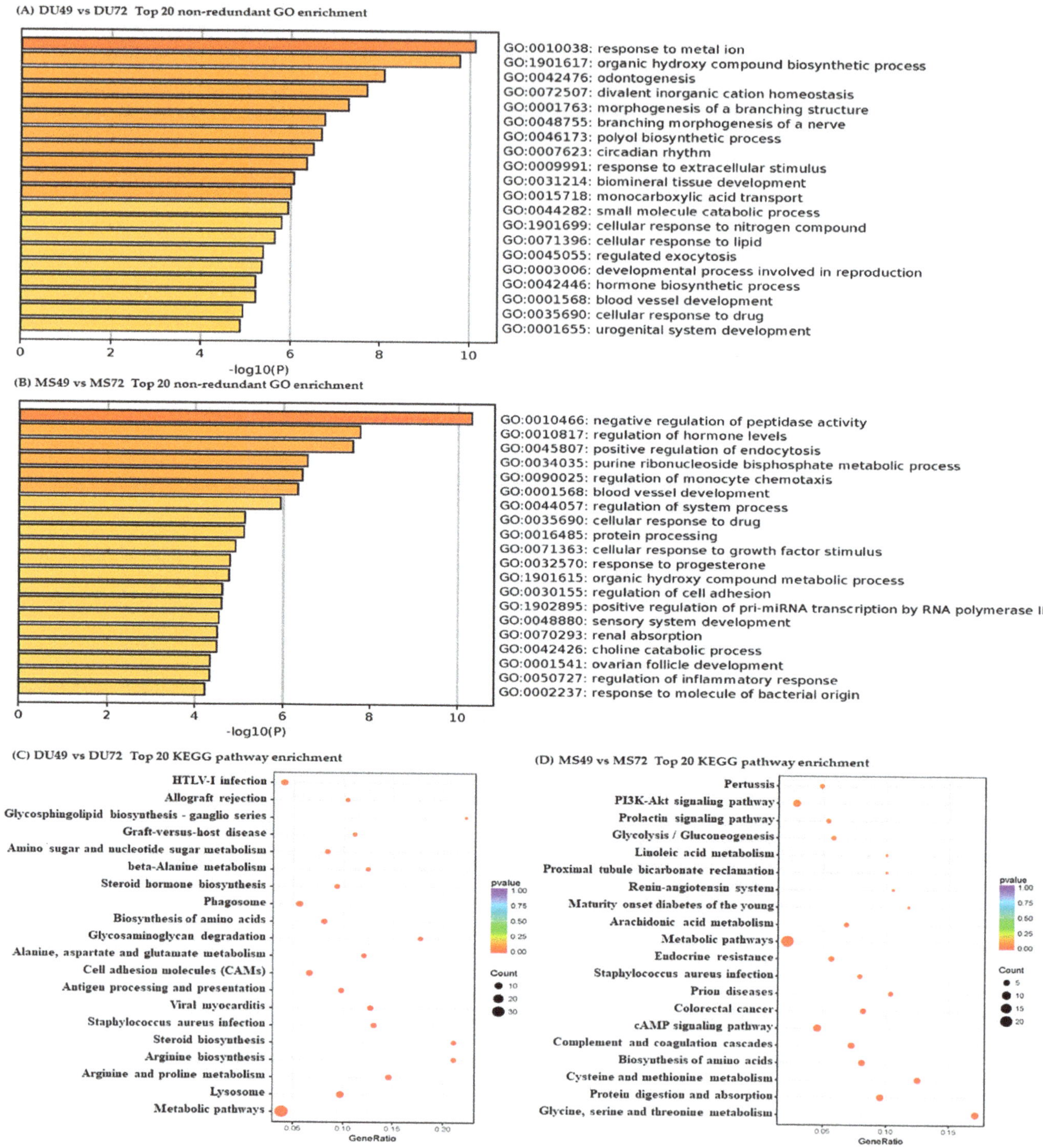

Figure 2. Pathways clustered from DEGs detected between GD49 and GD72. Top 20 non-redundant enriched gene ontology (GO) terms (**A**,**B**) and top 20 enriched KEGG pathways (**C**,**D**).

miRanda was used to predict DE miRNA target genes. A GO analysis suggests that DE miRNAs target genes primarily associated with the terms "metabolic processes", "biological regulation", "developmental processes", and "immune" (Supplementary Material Figure S1A–D). "Rap1 signaling pathway", "ras signaling pathway", "ovarian steroidogenesis", and "vascular smooth muscle contraction" were significantly enriched in the KEGG pathway analysis (Supplementary Material Figure S1E–H).

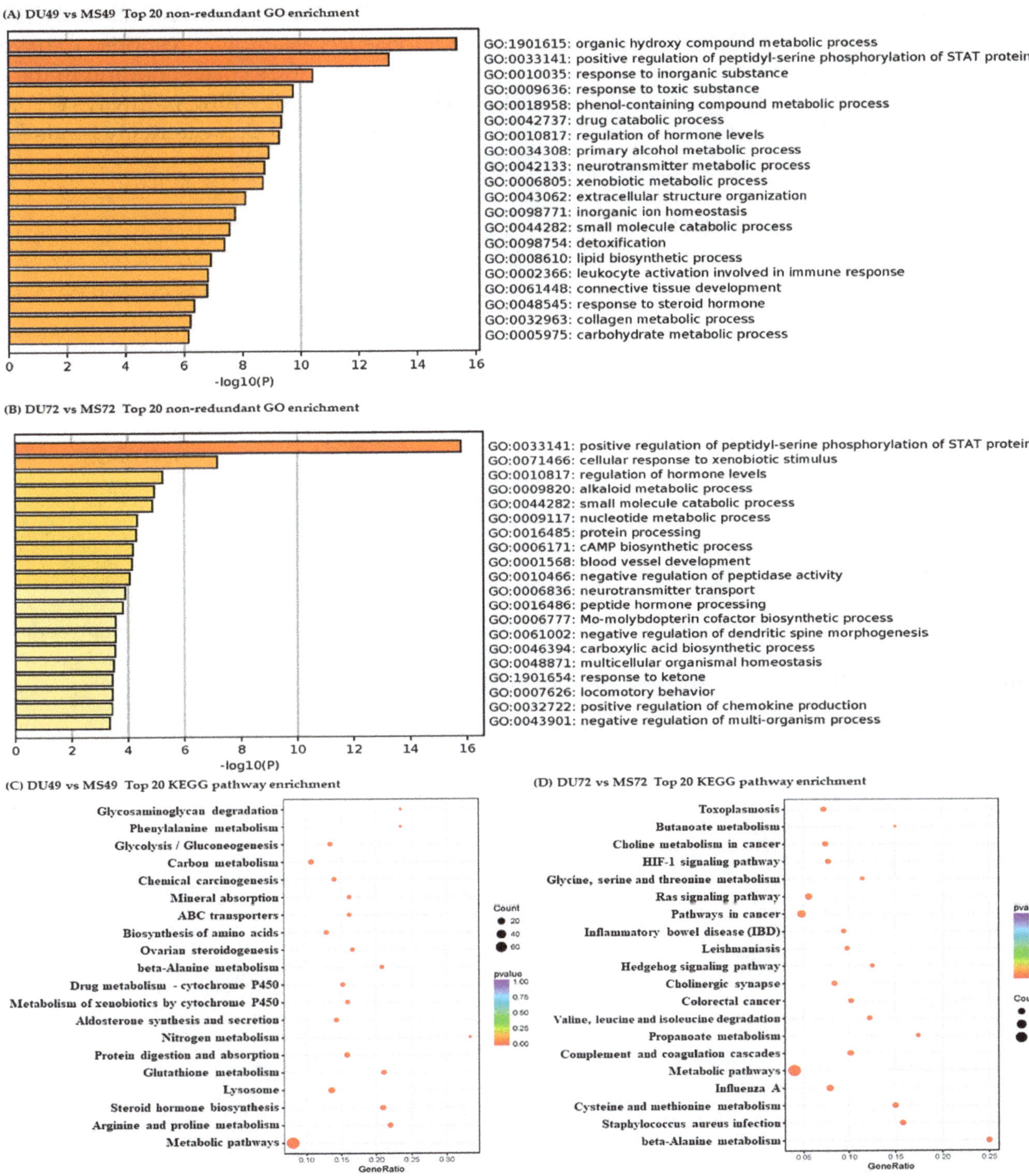

Figure 3. Pathways clustered from DEGs detected between Duroc and Meishan sows. Top 20 non-redundant enriched GO terms (**A**,**B**) and top 20 KEGG pathways (**C**,**D**).

3.4. miRNA-mRNA Interaction Analysis

To characterize the regulatory roles of miRNAs in the endometrium during mid-late gestation, we used miRanda to predict potential target relationship between DE miRNAs and DEGs. We identified 3 (DU49 vs. DU72), 18 (DU49 vs. MS49), 22 (DU72 vs. MS72), and 30 (MS49 vs. MS72) potential mRNA targets for 12, 131, 250, and 129 miRNAs, respectively. The predictions were used to construct four miRNA-gene interaction networks (Figures 4–7 and Supplementary Material Table S12), which show possible interactions amongst these genes and miRNAs. GO and KEGG pathway analyses were performed for DEGs in the networks. Several significantly enriched GO terms were shared by DEGs from MS49 vs. MS72 and DU72 vs. MS72, including "response to estradiol", "response to oxygen levels", "angiogenesis", and "blood vessel development" (Table 3). DEGs corresponding to the GO terms are significantly enriched in the Rap1, Ras, PI3K-Akt, and FoxO signaling pathways (Table 3). Interestingly, ssc-miR-503 potentially binds *EGF* mRNA and ssc-miR-671-5p potentially binds *ESR1* mRNAs. Both genes are involved in litter size in pregnancy (Table 2).

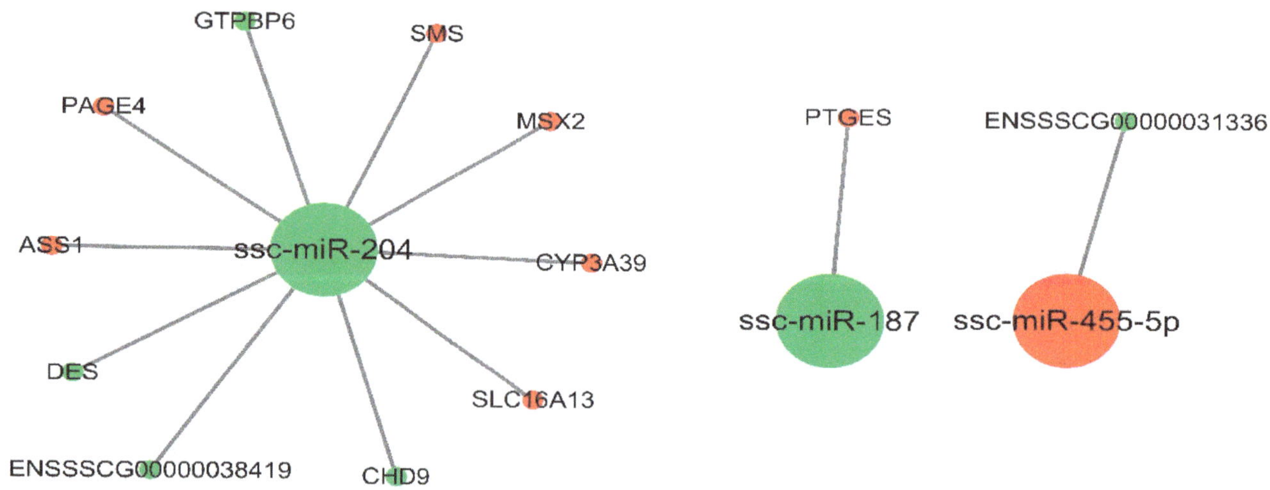

Figure 4. miRNA-gene interaction network (DU49 vs. DU72). Large and small circles represent miRNAs and genes, respectively. Red indicates that the RNA is relatively more abundant, and green indicates that the RNA is relatively less abundant in DU49 than in DU72.

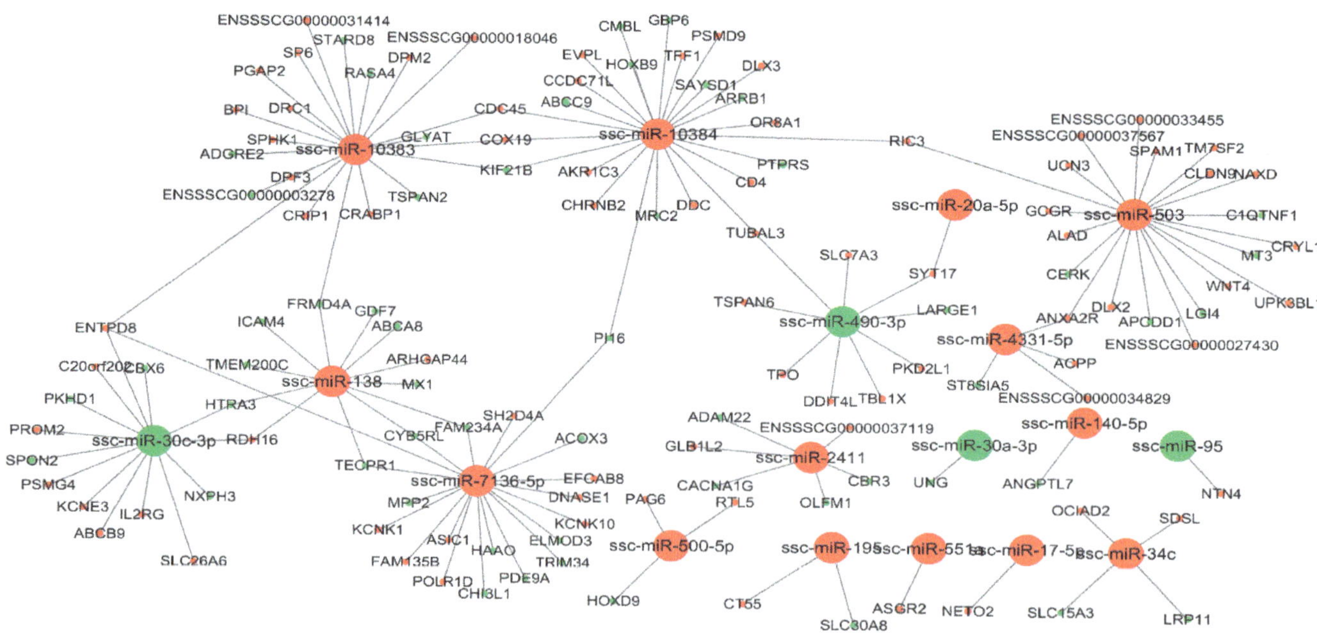

Figure 5. miRNA-gene interaction network (DU49 vs. MS49). Large and small circles represent miRNAs and genes, respectively. Red indicates that the RNA is relatively more abundant, and green indicates that the RNA is relatively less abundant in DU49 than in MS49.

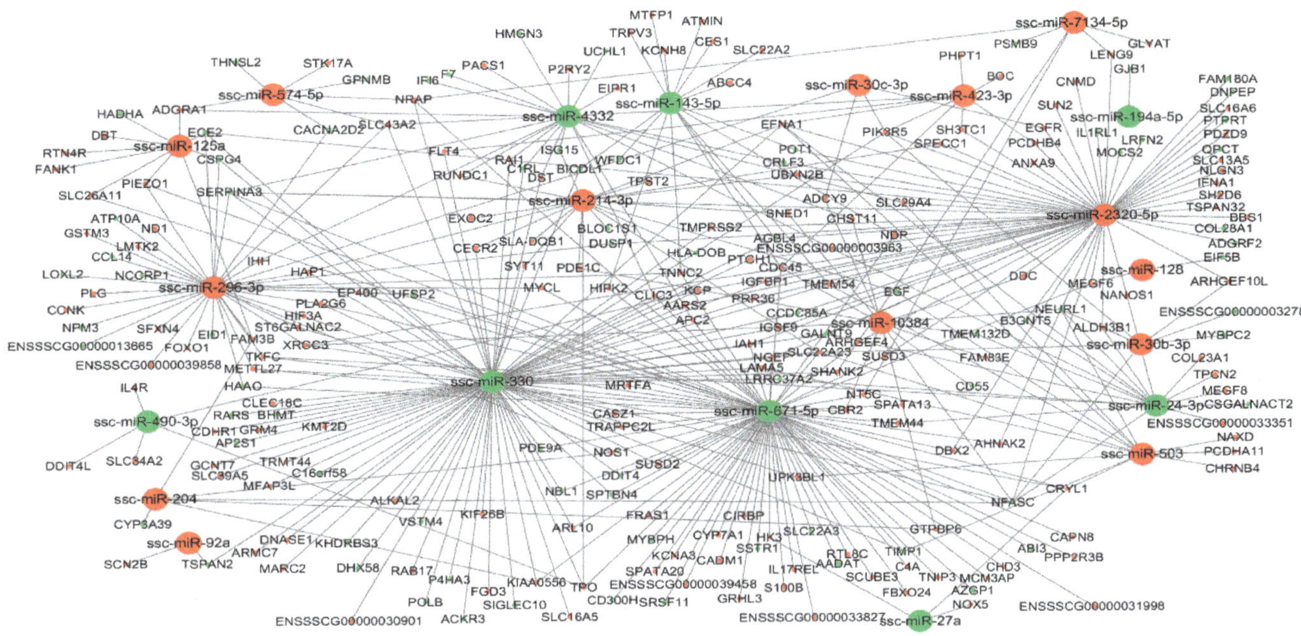

Figure 6. miRNA-gene interaction network (DU72 vs. MS72). Large and small circles represent miRNAs and genes, respectively. Red indicates that the RNA is relatively more abundant, and green indicates that the RNA is relatively less abundant in DU72 than in MS72.

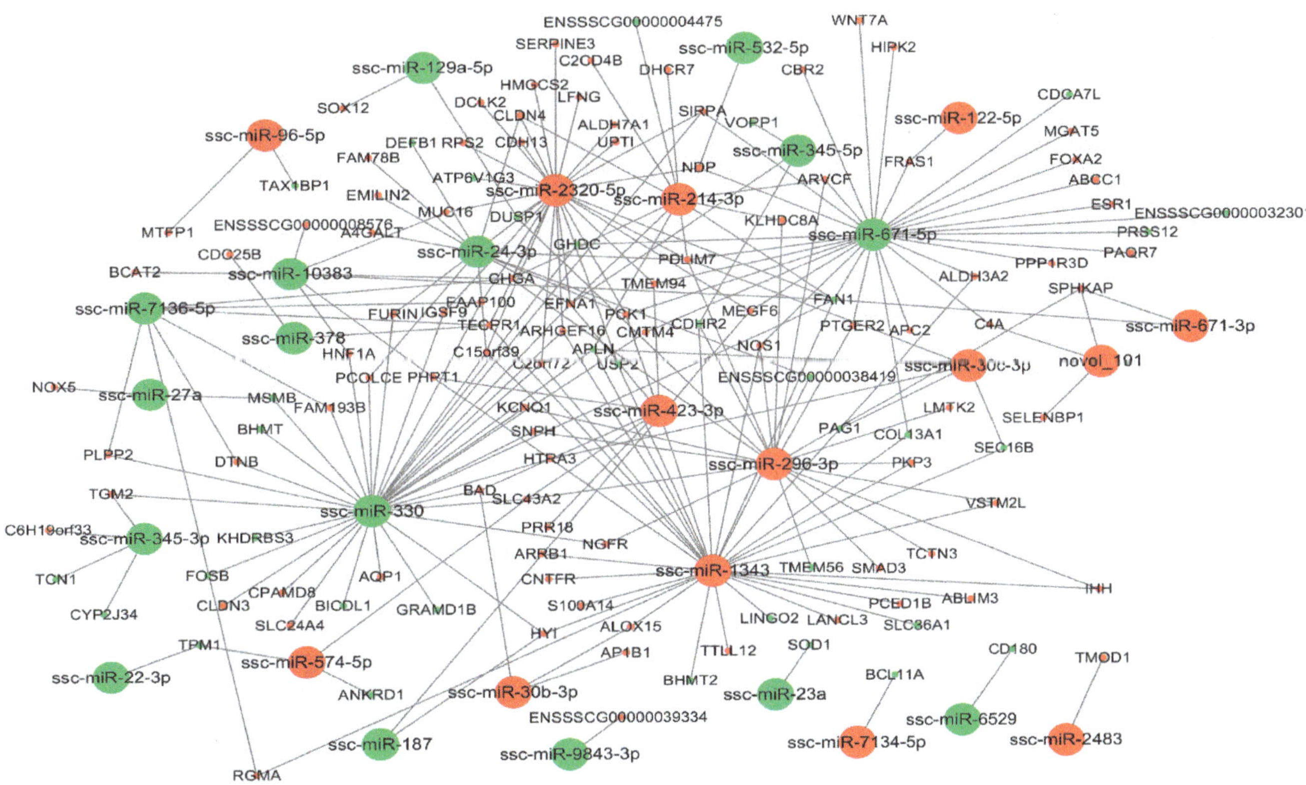

Figure 7. miRNA-gene interaction network (MS49 vs. MS72). Large and small circles represent miRNAs and genes, respectively. Red indicates that the RNA is relatively more abundant, and green indicates that the RNA is relatively less abundant in MS49 than in MS72.

Table 3. Significantly enriched GO terms and associated KEGG pathways shared between DU72 vs. MS72 and MS49 vs. MS72 in miRNA-mRNA interaction networks.

Comparison Group	Terms	DEGs No.	*p*-Value	Genes
DU72 vs. MS72	GO:0032355 ~ Response to estradiol	7	0.00025	*DUSP1, EGFR, F7, IHH, PTCH1, SSTR1, WFDC1*
	GO:0070482 ~ Response to oxygen levels	10	0.00379	*F7, FOXO1, LOXL2, NOS1, POLB, PSMB9, S100B, HIPK2, DDIT4, HIF3A*
	GO:0001525 ~ Angiogenesis	13	0.00337	*CSPG4, EFNA1, EGF, FLT4, IHH, LOXL2, GPNMB, CNMD, HIPK2, ACKR3, HIF3A, NOX5, COL23A1*
	GO:0001568 ~ Blood vessel development	15	0.00497	*CSPG4, EFNA1, EGF, MEGF8, FOXO1, FLT4, IHH, LOXL2, GPNMB, CNMD, HIPK2, ACKR3, HIF3A, NOX5, COL23A1*
	ssc04015 ~ Rap1 signaling pathway	4	3.52×10^{-5}	*FLT4, EGF, EGFR, EFNA1*
	ssc04014 ~ Ras signaling pathway	4	5.41×10^{-5}	*FLT4, EGF, EGFR, EFNA1*
	ssc04151 ~ PI3K-Akt signaling pathway	3	0.00019	*FLT4, EGF, EGFR, EFNA1*
	ssc04068 ~ FoxO signaling pathway	4	0.00031	*EGF, EGFR, FOXO1*
MS49 vs. MS72	GO:0032355 ~ Response to estradiol	5	0.00037	*BAD, DUSP1, ESR1, IHH, WNT7A*
	GO:0070482 ~ Response to oxygen levels	8	0.00045	*AQP1, BAD, CLDN3, SMAD3, NOS1, PCK1, ANKRD1, HIPK2*
	GO:0001525 ~ Angiogenesis	8	0.00550	*AQP1, CDH13, EFNA1, IHH, NGFR, WNT7A, HIPK2, NOX5*
	GO:0001568 ~ Blood vessel development	9	0.00809	*AQP1, CDH13, DHCR7, EFNA1, IHH, NGFR, WNT7A, HIPK2, NOX5*
	ssc04015 ~ Rap1 signaling pathway	2	0.00580	*EFNA1, NGFR*
	ssc04014 ~ Ras signaling pathway	3	0.00026	*BAD, EFNA1, NGFR*
	ssc04151 ~ PI3K-Akt signaling pathway	4	2.26×10^{-5}	*BAD, EFNA1, PCK1, NGFR*
	ssc04068 ~ FoxO signaling pathway	2	0.00291	*PCK1, SMAD3*

3.5. qPCR Validation for DEGs and DE miRNAs

Seven differentially expressed mRNAs (Table S8) and seven differentially expressed microRNAs (Table S9) were selected to validate the RNA-sequencing results. Among them, *EGF*, *ESR1*, ssc-miR-503, and ssc-miR-671-5p were selected because they appear to be involved in the determination of litter size based on the miRNA-gene interaction networks we constructed. The remaining five DE mRNAs and DE microRNAs were randomly selected from the DEGs and DE miRNAs identified in the four comparison groups. The expression patterns for all 14 RNAs were examined in Meishan and Duroc endometrium using quantitative RT-PCR. The results were consistent with the RNA-sequencing data (Figure 8A–N).

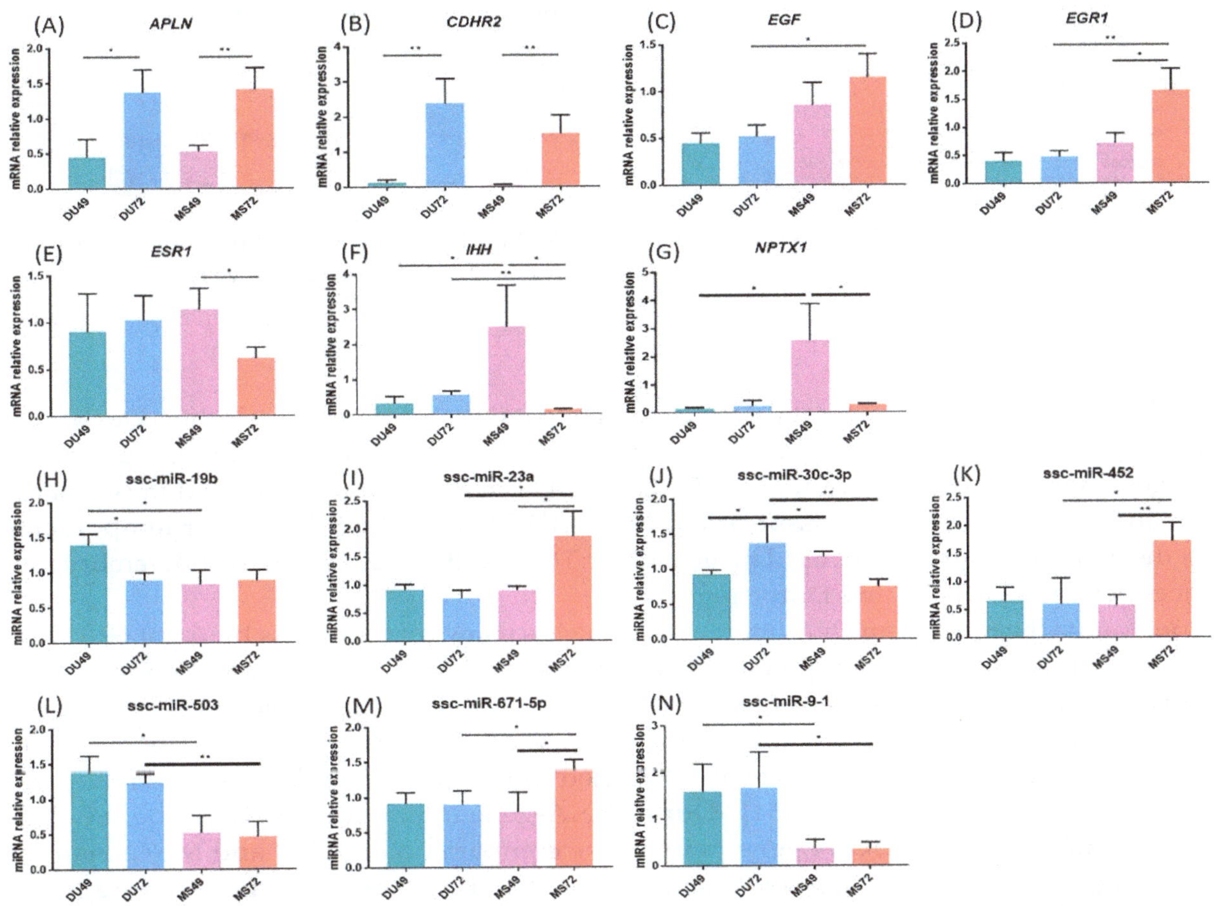

Figure 8. Quantitative PCR validation. The x-axis represents the different stages and breeds and the y-axis shows the fold changes in expression (* $p < 0.05$, ** $p < 0.01$). (**A**) *APLN*; (**B**) *CDHR2*; (**C**) *EGF*; (**D**) *EGR1*; (**E**) *ESR1*; (**F**) *IHH*; (**G**) *NPTX1*; (**H**) ssc-miR-19b; (**I**) ssc-miR-23a; (**J**) ssc-miR-30c-3p; (**K**) ssc-miR-452; (**L**) ssc-miR-503; (**M**) ssc-miR-671-5p; (**N**) ssc-miR-9-1. Relative expression values were normalized as descried in Materials and Methods.

3.6. Validation of miRNA-mRNA Interactions

The miRNA-mRNA interaction analyses suggest that the CDS region of *EGF* contains a potential target site for ssc-miR-503, and the 3′UTR region of *ESR1* contains a potential target site for ssc-miR-671-5p. We chose these two potential target relationships to test for binding between miRNA and mRNA targets. The nucleotide sequence at the potential target sites are identical in Meishan and Duroc pigs. We conducted dual luciferase assays to detect miRNA-mRNA interactions in HEK293T cells. As shown in Figure 9A, ssc-miR-503 mimics decreased luciferase activity compared with the negative control mimics when co-transfected with the *EGF* wild-type plasmid in HEK293T cells. In contrast, co-transfection with a plasmid containing a binding site mutation or the psiCHECK-2 plasmid had no effect on activity. The luciferase activity of co-transfected ssc-miR-671-5p mimics and *ESR1* plasmid yielded similar results (Figure 9B). These data support the conclusion that ssc-miR-503 directly targets the *EGF* gene CDS region and ssc-miR-671-5p directly targets the *ESR1* gene 3′UTR region.

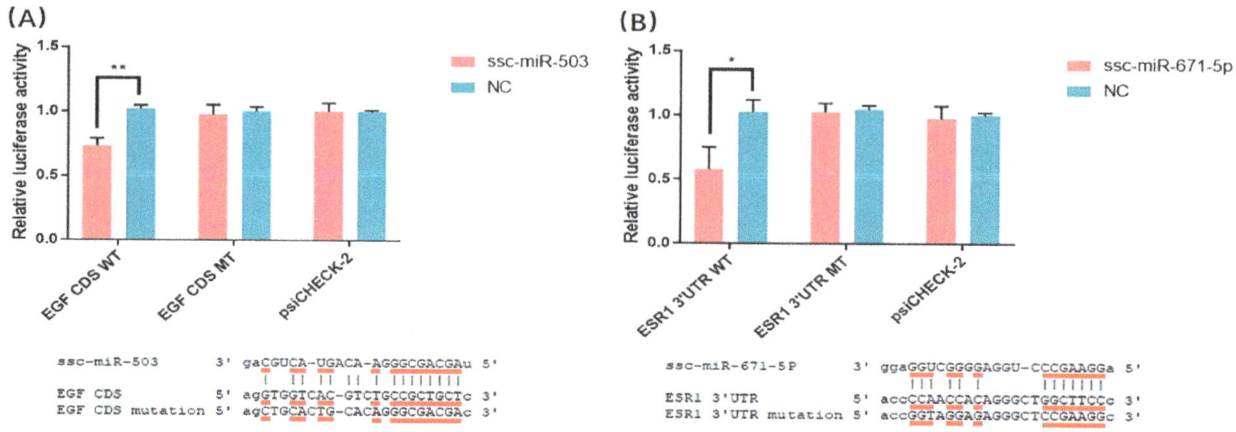

Figure 9. Validation of miRNA-gene interactions. (**A**) Validating *EGF* as a target of ssc-miR-503, (**B**) validating *ESR1* as a target of ssc-miR-671-5p. (* $p < 0.05$, ** $p < 0.01$). WT, wild type; MT, mutant type.

4. Discussion

Meishan pigs have lower prenatal mortality than commercial breeds, including Duroc [1,8]. The molecular mechanisms affecting embryonic mortality are still not well understood, especially during mid-late gestation. In this study, numerous DEGs and DE miRNAs were identified in comparisons of Meishan and Duroc endometrial tissue at GD49 and GD72, revealing that major temporal changes occur in the endometrium during mid-late gestation. These shifts may be involved in embryo loss rate differences between Meishan and Duroc pigs.

In our study, the miRNA-mRNA interaction analyses show that miRNAs bind not only the 3′UTR of target genes, but also the CDS region of the target genes, which is consistent with previous research results [54]. Some DE miRNAs identified in our study play a significant role in pregnancy loss. The miR-27a polymorphism, for example, is significantly associated with pregnancy loss in humans [52]. Consistent with our findings, Wessels et al. identified miR-323 and miR-27a as potentially involved in spontaneous embryo loss in pigs [51]. Our miRNA-gene interaction analyses identified a much large number of predicted miRNA-gene pairs in the comparison of MS49 and MS72 than in DU49 vs. DU72, suggesting that Meishan sows have a greater ability to regulate the uterine environment. In addition, unlike Duroc sows at same period or Meishan sows at GD49, DE genes in the Meishan endometrium at GD72 were associated with pregnancy hormones, hypoxia, and angiogenesis (Table 3). Genes associated with these terms may participate in blood vessel development and increase the density of blood vessels in the endometrium of Meishan pigs. For example, epidermal growth factor (*EGF*) can improve angiogenesis by directly stimulating endothelial cell proliferation [55], and *EGF* expression is higher at MS72 than at DU72. Studies have revealed that Meishan pigs have higher placental efficiency near the end of gestation because of dense vascularization, in comparison to commercial breeds [6]. Using the luciferase reporter system, we identified binding site for ssc-miR-503 in the *EGF* CDS region (Figure 9A). The relatively low expression of ssc-miR-503 in Meishan pigs at GD72 compared with Duroc pigs (Figure 8L) may enable expression of *EGF* (Figure 8C) and facilitate blood vessel development in the endometrium.

Angiogenesis occurs at the maternal-fetal interface and plays a crucial role in the development of growing conceptuses after embryo implantation [1]. In our study, GO analysis of DEGs between GD49 and GD72 in both Meishan and Duroc revealed significantly enriched GO terms involved in blood vessel development (Figure 2 and Table S10). The only gene found in common in the two comparisons was *APLN*, encoding apelin. Apelin is an endogenous peptide that functions as the ligand of the G protein-coupled receptor APJ, which may be involved in many physiological processes including angiogenesis [56]. Apelin plays a role in regulating the hemodynamics of pregnancy [57]. Increased expression of Apelin promotes the growth, migration, and angiogenesis of endothelial progenitor cells under conditions of hypoxia [58]. Using quantitative PCR, we found a higher level of *APLN* expression in GD72 than GD49 (Figure 8A). These results suggest that *APLN* plays a vital role in blood vessel development during mid-late gestation in the endometrium to meet increasing fetal demands for oxygen and nutrients. The top 20 most enriched GO terms for genes identified in the comparison between Meishan and Duroc at GD72 include several involved in blood vessel development. This is not the case for the corresponding comparison between the breed at GD49 (Figure 3 and Table S10). Some of these genes are expressed at higher levels in Meishan sows at GD72 such as *EGR1* [59], *EGF* [55], *ESM1* [60], and *LOXL2* [61], which may enhance endometrial angiogenesis in Meishan pigs. These results potentially explain the denser vascularization of the maternal-fetal interface in Meishan pigs during mid-late gestation (compared to commercial breeds). The same factors may also enhance the uterine receptivity of Meishan pigs.

Cytokine family members, such as interleukins and interferons, are known to function in porcine pregnancy [62,63]. In this study, we identified common DEGs between Meishan and Duroc at both GD49 and GD72 that are involved in the regulation of type I interferon signaling pathway (*IFI6, IFIT3, IFNA4, MX1, ISG15,* and *USP18*). *ISG15,* interferon (IFN)-stimulated gene 15, plays a vital role in promoting pregnancy maintenance and embryo development [64,65]. Embryo mortality increases in *ISG15-/-* mice [45], and *ISG15* may affect fetal growth, placenta development, and potential infection defense mechanisms in humans [64]. We found that *IGS15* has a higher expression level in Meishan than in Duroc pigs, which may increase the resistance of the Meishan breed to conditions such as maternal hypoxia or crowded uterus.

Estradiol-17β ($E_2\beta$), one of the most important steroid hormones in pregnancy, plays a key role in the high prolificacy of Meishan pigs. Lower levels of $E_2\beta$ secretion decrease the trophectoderm mitotic rate early in gestation in Meishan pigs, resulting in a smaller embryo and a higher survival rate [66]. Previous studies showed that $E_2\beta$ increases placental size and consequently decreases placental efficiency [67]. Reduced placental efficiency is thought to be the main cause of fetal loss during mid-late gestation in pigs [12]. In our study, enriched GO terms that were in common between DU72 vs. MS72 and MS49 vs. MS72 in the miRNA-mRNA interaction networks included genes and miRNAs involved in responding to estradiol (Figures 6 and 7 and Table 3). Down regulated genes in MS72 such as *IHH* and *ESR1* are known to affect placental development in mammals [68–70]. Estradiol regulates expression of *IHH* in the mammalian uterus and plays a role in development of the placenta [68,69]. The effects of estradiol in the uterus are mediated by *ESR1*, which is the predominant type of estrogen receptor in the mature uterus [70,71]. Based on the results of the dual luciferase assay, ssc-miR-671-5p appears to downregulate the expression of *ESR1* (Figure 9B). Lower expression of *IHH* and *ESR1* may be partly responsible for decreasing conceptus and placental size and promoting placental efficiency in Meishan pigs during late gestation. Other genes identified in our analyses (*DUSP1, EGFR, F7, PTCH1, SSTR1, WFDC1, BAD,* and *WNT7A*) may also contribute in important ways to the high placental efficiency of the Meishan breed.

5. Conclusions

In summary, RNA-sequencing analysis identified differences in gene and miRNA expression in endometrial tissue from Meishan and Duroc sows during mid-late gestation. miRNA-gene interactions were predicted and used to construct interaction networks. We identified several genes and miRNAs that affect conceptus and placental size by regulating estradiol and angiogenesis at the maternal-fetal interface. These effects may increase placental efficiency and decrease the fetal loss rate in Meishan sows during mid-late gestation. Our findings suggest different regulation patterns in the endometrium of Meishan and Duroc sows and reveal important molecular mechanisms affecting embryonic mortality in pigs during mid-late gestation.

Supplementary Materials:
Figure S1: Crucial pathways were clustered from DE miRNAs target genes, Table S1: PCR primers for qRT-PCR validation of differentially expressed genes and miRNAs between Meishan and Duroc sows at GD49 and GD72, Table S2: Sequences of primers used for amplification of wild type EGF CDS region and ESR1 3′UTR region, Table S3: The basic statistics for RNA-seq reads generated from endometrium tissues of 12 sows with Meishan and Duroc during the mid-late gestation, Table S4: Distribution of clean reads in the pig genome, Table S5: The basic statistics for RNA-seq reads generated from endometrium tissues of 12 sows with Meishan and Duroc during the mid-late gestation, Table S6: Annotated mature miRNAs and precursors, Table S7: Novel mature miRNA sequence and precursor sequence, Table S8: Differentially expressed genes in Meishan and Duroc sows' endometrium at GD49 and GD72, Table S9: Differentially expressed miRNAs in Meishan and Duroc sows' endometrium at GD49 and GD72, Table S10: List of DEGs enriched in GO categories in Meishan and Duroc sows' endometrium at GD49 and GD72, Table S11: List of DEGs enriched in KEGG pathways in Meishan and Duroc sows' endometrium at GD49 and GD72, Table S12: The predicted miRNA-gene pairs in Meishan and Duroc sows' endometrium at GD49 and GD72.

Author Contributions: Formal analysis, K.Y. and J.W.; investigation, K.Y., K.W., Q.T., and Y.L.; project administration, M.F.; resources, K.W., Q.T., and X.L.; writing—original draft, K.Y. and M.F. All authors have read and agreed to the published version of the manuscript.

Acknowledgments: The authors are grateful to Yunzhou Yang and Qiuyan Li for assistance with sample collection.

References

1. Kridli, R.T.; Khalaj, K.; Bidarimath, M.; Tayade, C. Placentation, maternal-fetal interface, and conceptus loss in swine. *Theriogenology* **2016**, *85*, 135–144. [CrossRef] [PubMed]
2. Freking, B.A.; Lents, C.A.; Vallet, J.L. Selection for uterine capacity improves lifetime productivity of sows. *Anim. Reprod. Sci.* **2016**, *167*, 16–21. [CrossRef]
3. Haley, C.S.; Lee, G.J.; Ritchie, M. Comparative reproductive performance in Meishan and Large White pigs and their crosses. *Anim. Sci.* **1995**, *60*, 259–267. [CrossRef]
4. Youngs, C.R.; Ford, S.P.; McGinnis, L.K.; Anderson, L.H. Investigations into the control of litter size in swine: I. Comparative studies on in vitro development of Meishan and Yorkshire preimplantation embryos. *J. Anim. Sci.* **1993**, *71*, 1561–1565. [CrossRef] [PubMed]
5. Anderson, L.H.; Christenson, L.K.; Christenson, R.K.; Ford, S.P. Investigations into the control of litter size in swine: II. Comparisons of morphological and functional embryonic diversity between Chinese and American breeds. *J. Anim. Sci.* **1993**, *71*, 1566–1571. [CrossRef] [PubMed]
6. Biensen, N.J.; Wilson, M.E.; Ford, S.P. The impact of either a Meishan or Yorkshire uterus on Meishan or Yorkshire fetal and placental development to days 70, 90, and 110 of gestation. *J. Anim. Sci.* **1998**, *76*, 2169–2176. [CrossRef] [PubMed]
7. Nina, J.; Wilson, M.E.; Ford, S. The impacts of uterine environment and fetal genotype on conceptus size and placental vascularity during late gestation in pigs. *J. Anim. Sci.* **1999**, *77*, 954–959. [CrossRef]
8. Bidarimath, M.; Tayade, C. Pregnancy and spontaneous fetal loss: A pig perspective. *Mol. Reprod. Dev.* **2017**, *84*, 856–869. [CrossRef]
9. Ford, S. Embryonic and fetal development in different genotypes in pigs. *J. Reprod. Fertil. Suppl.* **1997**, *52*, 165–176.
10. Leymaster, K.; Vallet, J.; Christenson, R. Number of fetuses and conceptus growth throughout gestation in lines of pigs selected for ovulation rate or uterine capacity. *J. Anim. Sci.* **2007**, *85*, 2093–2103. [CrossRef]
11. Waclawik, A.; Kaczmarek, M.M.; Blitek, A.; Kaczynski, P.; Ziecik, A.J. Embryo-maternal dialogue during pregnancy establishment and implantation in the pig. *Mol. Reprod. Dev.* **2017**, *84*, 842–855. [CrossRef] [PubMed]
12. Christenson, R.K.; Vallet, J.L.; Leymaster, K.; Young, L.D. Uterine function in Meishan pigs. *J. Reprod. Fertil. Suppl.* **1993**, *48*, 279–289. [PubMed]

13. Brennecke, J.; Hipfner, D.R.; Stark, A.; Russell, R.B.; Cohen, S.M. Bantam Encodes a Developmentally Regulated microRNA that Controls Cell Proliferation and Regulates the Proapoptotic Gene hid in Drosophila. *Cell* **2003**, *113*, 25–36. [CrossRef]
14. Dong, H.; Lei, J.; Ding, L.; Wen, Y.; Ju, H.; Zhang, X. MicroRNA: Function, detection, and bioanalysis. *Chem. Rev.* **2013**, *113*, 6207–6233. [CrossRef] [PubMed]
15. Krol, J.; Loedige, I.; Filipowicz, W. The widespread regulation of microRNA biogenesis, function and decay. *Nat. Rev. Genet.* **2010**, *11*, 597–610. [CrossRef] [PubMed]
16. Krawczynski, K.; Najmula, J.; Bauersachs, S.; Kaczmarek, M.M. MicroRNAome of porcine conceptuses and trophoblasts: Expression profile of micrornas and their potential to regulate genes crucial for establishment of pregnancy. *Biol. Reprod.* **2015**, *92*, 21. [CrossRef] [PubMed]
17. Su, L.; Liu, R.; Cheng, W.; Zhu, M.; Li, X.; Zhao, S.; Yu, M. Expression patterns of microRNAs in porcine endometrium and their potential roles in embryo implantation and placentation. *PLoS ONE* **2014**, *9*, e87867. [CrossRef]
18. Liu, R.; Wang, M.; Su, L.; Li, X.; Zhao, S.; Yu, M. The Expression Pattern of MicroRNAs and the Associated Pathways Involved in the Development of Porcine Placental Folds That Contribute to the Expansion of the Exchange Surface Area. *Biol. Reprod.* **2015**, *93*, 62. [CrossRef]
19. Wang, K.; Yang, K.; Xu, Q.; Liu, Y.; Li, W.; Bai, Y.; Wang, J.; Ding, C.; Liu, X.; Tang, Q.; et al. Protein expression profiles in Meishan and Duroc sows during mid-gestation reveal differences affecting uterine capacity, endometrial receptivity, and the maternal-fetal Interface. *BMC Genom.* **2019**, *20*, 991. [CrossRef]
20. Zhang, H.; Wang, S.; Liu, M.; Zhang, A.; Wu, Z.; Zhang, Z.; Li, J. Differential gene expression in the endometrium on gestation day 12 provides insight into sow prolificacy. *BMC Genom.* **2013**, *14*, 45. [CrossRef]
21. Gu, T.; Zhu, M.-J.; Schroyen, M.; Qu, L.; Nettleton, D.; Kuhar, D.; Lunney, J.K.; Ross, J.W.; Zhao, S.-H.; Tuggle, C.K. Endometrial gene expression profiling in pregnant Meishan and Yorkshire pigs on day 12 of gestation. *BMC Genom.* **2014**, *15*, 156. [CrossRef] [PubMed]
22. Huang, J.; Liu, R.; Su, L.; Xiao, Q.; Yu, M. Transcriptome Analysis Revealed the Embryo-Induced Gene Expression Patterns in the Endometrium from Meishan and Yorkshire Pigs. *Int. J. Mol. Sci.* **2015**, *16*, 22692–22710. [CrossRef] [PubMed]
23. Lin, H.; Wang, H.; Wang, Y.; Liu, C.; Wang, C.; Guo, J. Transcriptomic Analysis of the Porcine Endometrium during Embryo Implantation. *Genes* **2015**, *6*, 1330–1346. [CrossRef] [PubMed]
24. Kim, D.; Langmead, B.; Salzberg, S.L. HISAT: A fast spliced aligner with low memory requirements. *Nat. Methods* **2015**, *12*, 357–360. [CrossRef] [PubMed]
25. Liao, Y.; Smyth, G.K.; Shi, W. Featurecounts: An efficient general purpose program for assigning sequence reads to genomic features. *Bioinformatics* **2014**, *30*, 923–930. [CrossRef]
26. Love, M.I.; Huber, W.; Anders, S. Moderated estimation of fold change and dispersion for RNA-seq data with DESeq2. *Genome Biol.* **2014**, *15*, 550. [CrossRef]
27. Langmead, B.; Trapnell, C.; Pop, M.; Salzberg, S.L. Ultrafast and memory-efficient alignment of short DNA sequences to the human genome. *Genome Biol.* **2009**, *10*, R25. [CrossRef]
28. Friedlander, M.R.; Mackowiak, S.D.; Li, N.; Chen, W.; Rajewsky, N. miRDeep2 accurately identifies known and hundreds of novel microRNA genes in seven animal clades. *Nucleic Acids Res.* **2012**, *40*, 37–52. [CrossRef]
29. Wang, L.; Feng, Z.; Wang, X.; Wang, X.; Zhang, X. DEGseq: An R package for identifying differentially expressed genes from RNA-seq data. *Bioinformatics* **2010**, *26*, 136–138. [CrossRef]
30. Zhou, Y.; Zhou, B.; Pache, L.; Chang, M.; Khodabakhshi, A.H.; Tanaseichuk, O.; Benner, C.; Chanda, S.K. Metascape provides a biologist-oriented resource for the analysis of systems-level datasets. *Nat. Commun.* **2019**, *10*, 1523. [CrossRef]
31. Xie, C.; Mao, X.; Huang, J.; Ding, Y.; Wu, J.; Dong, S.; Kong, L.; Gao, G.; Li, C.Y.; Wei, L. KOBAS 2.0: A web server for annotation and identification of enriched pathways and diseases. *Nucleic Acids Res.* **2011**, *39*, W316–W322. [CrossRef] [PubMed]
32. Enright, A.J.; John, B.; Gaul, U.; Tuschl, T.; Sander, C.; Marks, D.S. MicroRNA targets in Drosophila. *Genome Biol.* **2003**, *5*, R1. [CrossRef] [PubMed]
33. Shannon, P.; Markiel, A.; Ozier, O.; Baliga, N.S.; Wang, J.T.; Ramage, D.; Amin, N.; Schwikowski, B.; Ideker, T. Cytoscape: A software environment for integrated models of biomolecular interaction networks. *Genome Res.* **2003**, *13*, 2498–2504. [CrossRef] [PubMed]

34. Onteru, S.K.; Fan, B.; Du, Z.Q.; Garrick, D.J.; Stalder, K.J.; Rothschild, M.F. A whole-genome association study for pig reproductive traits. *Anim. Genet.* **2012**, *43*, 18–26. [CrossRef] [PubMed]
35. Wang, L.; Zhang, L.; Li, Y.; Li, W.; Luo, W.; Cheng, D.; Yan, H.; Ma, X.; Liu, X.; Song, X.; et al. Data mining in networks of differentially expressed genes during sow pregnancy. *Int. J. Biol. Sci.* **2012**, *8*, 548–560. [CrossRef] [PubMed]
36. Lim, W.; Bae, H.; Bazer, F.W.; Song, G. Stimulatory effects of fibroblast growth factor 2 on proliferation and migration of uterine luminal epithelial cells during early pregnancy. *Biol. Reprod.* **2016**, *96*, 185–198. [CrossRef]
37. Hiramoto, K.; Yamate, Y.; Sato, E.F. Gp91phox NADPH oxidase modulates litter size by regulating mucin1 in the uterus of mice. *Syst. Biol. Reprod. Med.* **2017**, *63*, 130–139. [CrossRef]
38. Cordoba, S.; Balcells, I.; Castello, A.; Ovilo, C.; Noguera, J.L.; Timoneda, O.; Sanchez, A. Endometrial gene expression profile of pregnant sows with extreme phenotypes for reproductive efficiency. *Sci. Rep.* **2015**, *5*, 14416. [CrossRef]
39. Li, X.; Ye, J.; Han, X.; Qiao, R.; Li, X.; Lv, G.; Wang, K. Whole-genome sequencing identifies potential candidate genes for reproductive traits in pigs. *Genomics* **2020**, *112*, 199–206. [CrossRef]
40. Chen, Z.; Ye, S.; Teng, J.; Diao, S.; Yuan, X.; Chen, Z.; Zhang, H.; Li, J.; Zhang, Z. Genome-wide association studies for the number of animals born alive and dead in duroc pigs. *Theriogenology* **2019**, *139*, 36–42. [CrossRef]
41. Wang, Y.; Ding, X.; Tan, Z.; Xing, K.; Yang, T.; Wang, Y.; Sun, D.; Wang, C. Genome-wide association study for reproductive traits in a Large White pig population. *Anim. Genet.* **2018**, *49*, 127–131. [CrossRef] [PubMed]
42. Liu, L.Q.; Li, F.E.; Deng, C.Y. Short Communication: Molecular cloning and expression pattern of the porcine 5-aminolevulinate synthase 1 (ALAS1) gene and its association with reproductive traits. *Genet. Mol. Res.* **2016**, *15*. [CrossRef] [PubMed]
43. Wu, P.; Wang, K.; Yang, Q.; Zhou, J.; Chen, D.; Ma, J.; Tang, Q.; Jin, L.; Xiao, W.; Jiang, A.; et al. Identifying SNPs and candidate genes for three litter traits using single-step GWAS across six parities in Landrace and Large White pigs. *Physiol. Genom.* **2018**, *50*, 1026–1035. [CrossRef] [PubMed]
44. Chen, X.; Fu, J.; Wang, A. Expression of genes involved in progesterone receptor paracrine signaling and their effect on litter size in pigs. *J. Anim. Sci. Biotechnol.* **2016**, *7*, 31. [CrossRef]
45. Henkes, L.E.; Pru, J.K.; Ashley, R.L.; Anthony, R.V.; Veeramachaneni, D.N.; Gates, K.C.; Hansen, T.R. Embryo mortality in Isg15-/- mice is exacerbated by environmental stress. *Biol. Reprod.* **2015**, *92*, 36. [CrossRef]
46. Mucha, A.; Ropka-Molik, K.; Piorkowska, K.; Tyra, M.; Oczkowicz, M. Effect of EGF, AREG and LIF genes polymorphisms on reproductive traits in pigs. *Anim. Reprod. Sci.* **2013**, *137*, 88–92. [CrossRef]
47. Sharma, S.; Sharma, P.M.; Mistry, D.S.; Chang, R.J.; Olefsky, J.M.; Mellon, P.L.; Webster, N.J. PPARG regulates gonadotropin-releasing hormone signaling in LbetaT2 cells in vitro and pituitary gonadotroph function in vivo in mice. *Biol. Reprod.* **2011**, *84*, 466–475. [CrossRef]
48. Metodiev, S.; Thekkoot, D.M.; Young, J.M.; Onteru, S.; Rothschild, M.F.; Dekkers, J.C.M. A whole-genome association study for litter size and litter weight traits in pigs. *Livest. Sci.* **2018**, *211*, 87–97. [CrossRef]
49. Rothschild, M.; Jacobson, C.; Vaske, D.; Tuggle, C.; Wang, L.; Short, T.; Eckardt, G.; Sasaki, S.; Vincent, A.; McLaren, D.; et al. The estrogen receptor locus is associated with a major gene influencing litter size in pigs. *Proc. Natl. Acad. Sci. USA* **1996**, *93*, 201. [CrossRef]
50. Ashworth, C.J.; Haley, C.S.; Aitken, R.P.; Wilmut, I. Embryo survival and conceptus growth after reciprocal embryo transfer between Chinese Meishan and Landrace x Large White gilts. *J. Reprod. Fertil.* **1990**, *90*, 595–603. [CrossRef]
51. Wessels, J.M.; Edwards, A.K.; Khalaj, K.; Kridli, R.T.; Bidarimath, M.; Tayade, C. The microRNAome of pregnancy: Deciphering miRNA networks at the maternal-fetal interface. *PLoS ONE* **2013**, *8*, e72264. [CrossRef] [PubMed]
52. Rah, H.; Chung, K.W.; Ko, K.H.; Kim, E.S.; Kim, J.O.; Sakong, J.H.; Kim, J.H.; Lee, W.S.; Kim, N.K. miR-27a and miR-449b polymorphisms associated with a risk of idiopathic recurrent pregnancy loss. *PLoS ONE* **2017**, *12*, e0177160. [CrossRef] [PubMed]
53. Huang, X.; Han, X.; Huang, Z.; Yu, M.; Zhang, Y.; Fan, Y.; Xu, B.; Zhou, K.; Song, L.; Wang, X.; et al. Maternal pentachlorophenol exposure induces developmental toxicity mediated by autophagy on pregnancy mice. *Ecotoxicol. Environ. Saf.* **2019**, *169*, 829–836. [CrossRef] [PubMed]

54. Zhang, K.; Zhang, X.; Cai, Z.; Zhou, J.; Cao, R.; Zhao, Y.; Chen, Z.; Wang, D.; Ruan, W.; Zhao, Q.; et al. A novel class of microRNA-recognition elements that function only within open reading frames. *Nat. Struct. Mol. Biol.* **2018**, *25*, 1019–1027. [CrossRef] [PubMed]
55. Dunn, I.F.; Heese, O.; Black, P.M. Growth Factors in Glioma Angiogenesis: FGFs, PDGF, EGF, and TGFs. *J. Neuro-Oncol.* **2000**, *50*, 121–137. [CrossRef]
56. Zhang, L.; Takara, K.; Yamakawa, D.; Kidoya, H.; Takakura, N. Apelin as a marker for monitoring the tumor vessel normalization window during antiangiogenic therapy. *Cancer Sci.* **2016**, *107*, 36–44. [CrossRef]
57. Van Mieghem, T.; Doherty, A.; Baczyk, D.; Drewlo, S.; Baud, D.; Carvalho, J.; Kingdom, J. Apelin in Normal Pregnancy and Pregnancies Complicated by Placental Insufficiency. *Reprod. Sci.* **2016**, *23*, 1037–1043. [CrossRef]
58. Wen, Y.; Chen, R.; Zhu, C.; Qiao, H.; Liu, Y.; Ji, H.; Miao, J.; Chen, L.; Liu, X.; Yang, Y. MiR-503 suppresses hypoxia-induced proliferation, migration and angiogenesis of endothelial progenitor cells by targeting Apelin. *Peptides* **2018**, *105*, 58–65. [CrossRef]
59. Yang, W.; Lu, Z.; Zhi, Z.; Liu, L.; Deng, L.; Jiang, X.; Pang, L. Increased miRNA-518b inhibits trophoblast migration and angiogenesis by targeting EGR1 in early embryonic arrest. *Biol. Reprod.* **2019**, *101*, 664–674. [CrossRef]
60. Rocha, S.F.; Schiller, M.; Jing, D.; Li, H.; Butz, S.; Vestweber, D.; Biljes, D.; Drexler, H.C.; Nieminen-Kelha, M.; Vajkoczy, P.; et al. Esm1 modulates endothelial tip cell behavior and vascular permeability by enhancing VEGF bioavailability. *Circ. Res.* **2014**, *115*, 581–590. [CrossRef]
61. De Jong, O.G.; van der Waals, L.M.; Kools, F.R.W.; Verhaar, M.C.; van Balkom, B.W.M. Lysyl oxidase-like 2 is a regulator of angiogenesis through modulation of endothelial-to-mesenchymal transition. *J. Cell Physiol.* **2019**, *234*, 10260–10269. [CrossRef]
62. Geisert, R.D.; Lucy, M.C.; Whyte, J.J.; Ross, J.W.; Mathew, D.J. Cytokines from the pig conceptus: Roles in conceptus development in pigs. *J. Anim. Sci. Biotechnol.* **2014**, *5*, 51. [CrossRef]
63. Mathew, D.J.; Lucy, M.C.; Geisert, R.D. Interleukins, interferons, and establishment of pregnancy in pigs. *Reproduction* **2016**, *151*, R111–R122. [CrossRef]
64. Schanz, A.; Baston-Büst, D.M.; Heiss, C.; Beyer, I.M.; Krüssel, J.S.; Hess, A.P. Interferon stimulated gene 15 expression at the human embryo–maternal interface. *Arch. Gynecol. Obstet.* **2014**, *290*, 783–789. [CrossRef]
65. Zhao, S.; Wu, Y.; Gao, H.; Evans, A.; Zeng, S.-M. Roles of interferon-stimulated gene 15 protein in bovine embryo development. *Reprod. Fertil. Dev.* **2017**, *29*, 1209–1216. [CrossRef]
66. Wilson, M.E.; Ford, S.P. Differences in trophectoderm mitotic rate and P450 17alpha-hydroxylase expression between late preimplantation Meishan and Yorkshire conceptuses. *Biol. Reprod.* **1997**, *56*, 380–385. [CrossRef]
67. Wilson, M.E.; Ford, S.P. Effect of estradiol-17β administration during the time of conceptus elongation on placental size at term in Meishan pigs. *J. Anim. Sci.* **2000**, *78*, 1047–1052. [CrossRef]
68. Jiang, F.; Herman, G.E. Analysis of Nsdhl-deficient embryos reveals a role for Hedgehog signaling in early placental development. *Hum. Mol. Genet.* **2006**, *15*, 3293–3305. [CrossRef]
69. Katayama, S.; Ashizawa, K.; Gohma, H.; Fukuhara, T.; Narumi, K.; Tsuzuki, Y.; Tatemoto, H.; Nakada, T.; Nagai, K. The expression of Hedgehog genes (Ihh, Dhh) and Hedgehog target genes (Ptc1, Gli1, Coup-TfII) is affected by estrogenic stimuli in the uterus of immature female rats. *Toxicol. Appl. Pharmacol.* **2006**, *217*, 375–383. [CrossRef]
70. Geisert, R.D.; Whyte, J.J.; Meyer, A.E.; Mathew, D.J.; Juarez, M.R.; Lucy, M.C.; Prather, R.S.; Spencer, T.E. Rapid conceptus elongation in the pig: An interleukin 1 beta 2 and estrogen-regulated phenomenon. *Mol. Reprod. Dev.* **2017**, *84*, 760–774. [CrossRef]
71. Knapczyk-Stwora, K.; Durlej, M.; Duda, M.; Czernichowska-Ferreira, K.; Tabecka-Lonczynska, A.; Slomczynska, M. Expression of oestrogen receptor alpha and oestrogen receptor beta in the uterus of the pregnant swine. *Reprod. Domest. Anim.* **2011**, *46*, 1–7. [CrossRef] [PubMed]

2

Reconstruction and Analysis of Cattle Metabolic Networks in Normal and Acidosis Rumen Tissue

Maryam Gholizadeh [1], Jamal Fayazi [1,*], Yazdan Asgari [2], Hakimeh Zali [3] and Lars Kaderali [4]

1 Department of Animal Science, Faculty of Animal Science and Food Technology, Agricultural Sciences and Natural Resources University of Khuzestan, Mollasani, Ahvaz 6341773637, Iran; ma.gholizade@gmail.com
2 Department of Medical Biotechnology, School of Advanced Technologies in Medicine, Tehran University of Medical Sciences, Tehran 1416753955, Iran; Yasgari@tums.ac.ir
3 School of Advanced Technologies in Medicine, Shahid Beheshti University of Medical Sciences, Tehran 1416753955, Iran; Hakimehzali@gmail.com
4 Institute of Bioinformatics, University Medicine Greifswald, Felix-Hausdorff-Str. 8, 17475 Greifswald, Germany; lars.kaderali@uni-greifswald.de
* Correspondence: j_fayazi@asnrukh.ac.ir

Simple Summary: Economics of feedlot beef production dictate that beef cattle must gain weight at their maximum potential rate; this involves getting them quickly onto a full feed of high fermentable diet which can induce the ruminal acidosis disease. The molecular host mechanisms that occur as a response to the acidosis, are mostly unknown. For answering this question, the rumen epithelial transcriptome in acidosis and control fattening steers were obtained. By RNA sequencing we found the different expression profiles of genes in normal and acidosis induced steers. Then we constructed two metabolic networks for normal and acidosis tissue based on gene expression profile. Our results suggest that rapid shifts to diets rich in fermentable carbohydrates cause an increased concentration of ruminal volatile fatty acids (VFA) and toxins and significant changes in transcriptome profiles and metabolites of rumen epithelial tissue, with negative effects on economic consequences of poor performance and animal health.

Abstract: The objective of this study was to develop a system-level understanding of acidosis biology. Therefore, the genes expression differences between the normal and acidosis rumen epithelial tissues were first examined using the RNA-seq data in order to understand the molecular mechanisms involved in the disease and then their corresponding metabolic networks constructed. A number of 1074 genes, 978 isoforms, 1049 transcription start sites (TSS), 998 coding DNA sequence (CDS) and 2 promoters were identified being differentially expressed in the rumen tissue between the normal and acidosis samples ($p < 0.05$). The functional analysis of 627 up-regulated genes revealed their involvement in ion transmembrane transport, filament organization, regulation of cell adhesion, regulation of the actin cytoskeleton, ATP binding, glucose transmembrane transporter activity, carbohydrate binding, growth factor binding and cAMP metabolic process. Additionally, 111 differentially expressed enzymes were identified between the rumen epithelial tissue of the normal and acidosis steers with 46 up-regulated and 65 down-regulated ones in the acidosis group. The pathways and reactions analyses associated with the up-regulated enzymes indicate that most of these enzymes are involved in the fatty acid metabolism, biosynthesis of amino acids, pyruvate and carbon metabolism while most of the down-regulated ones are involved in purine and pyrimidine, vitamin B6 and antibiotics metabolisms. The degree distribution of both metabolic networks follows a power-law one, hence displaying a scale-free property. The top 15 hub metabolites were determined in the acidosis metabolic network with most of them involved in the fatty acid oxidation, VFA biosynthesis, amino acid biogenesis and glutathione metabolism which plays an important role in the stress condition. The limitations of this study were low number of animals and using only epithelial tissue (ventral sac) for RNA-seq.

Keywords: bovine ruminal acidosis; transcriptome profile; metabolic network

1. Introduction

In ruminant animals, such as cattle, healthy rumen plays an important role in production and economic efficiency. To meet the global demand for meat consumption, improving the production efficiency of cattle, would result in economic benefit. Economics of the feedlot meat production, dictates that cattle must gain weight at their maximum potential rate. This requires using intensive feeding strategies such as high energy as well as fermentable diets [1,2]. Economics also favor the processing of grain to increase the digestibility of starch. All of these factors set the stage for grain overload in the feedlot cattle and may induce digestive disorders leading to ruminal acidosis. The most common and primary form of ruminal acidosis in beef cattle fed high fermentable diets are thought to be lactic acidosis or acute acidosis [3,4]. Acute ruminal acidosis is often associated with an increase in lactic acid amount which yields pH depression and causes damage to the rumen epithelium as a consequence. Acidosis has a wide variety of clinical manifestations including the depressed feed intake, liver abscesses, diarrhea, inflammation and increases the morbidity and mortality of the stock [5–7].

The negative economic consequences resulted from the poor performance and animal health caused by acidosis, have made it one of the most prevalent animal welfare issues in the beef cattle industry. System level understanding of the bovine metabolism and relationship with their products is essential in cow breeding. However, several in vivo studies have demonstrated changes in some genes expression within the rumen tissue in response to the acidosis, but the functional and molecular host mechanisms are mostly unknown [8–10].

Systems biology is a field of biological science that aims to systematically study all the molecules and their interactions within an organism, tissue or living cell as a network. The metabolic network is a central area in systems biology and belongs to one of the best-studied biochemical networks [11]. The metabolic network is composed of small biomolecules (metabolites) and proteins (enzymes) that interact with them in order to catalyze a biochemical reaction. In fact, the goal of metabolic network reconstruction (MNR) is to identify the series metabolites involved in the biochemical reactions that define a cell metabolism [9].

Previous studies assayed gene expression changes in rumen epithelium under high grain concentration diets using microarray data or by PCR [3,4]. Li and coauthors studied the transcriptome changes in rumen tissue of young calves with feed induced acidosis using RNA-seq. Results from this study indicated that a starch-rich, diet-induced ruminal acidosis is accompanied by significant rumen epithelial transcriptome changes [7]. In this study we induced ruminal acidosis by feeding a starch-rich diet to beef cattle to analyze rumen epithelial tissue transcriptome. Furthermore, reconstruction of the metabolic networks of rumen epithelial tissue combined with systems biological analysis by implementing the advanced bioinformatics tools were carried out. The ventral sac of the rumen was chosen to assay changes as rumen is the first organ facing acidosis and the ventral sac has the highest capillary blood flow per unit weight mucosa of any location within the rumen [12]. The primary metabolic network in the bovine rumen tissue was reconstructed using the available bovine genome annotation and expression information. Moreover, acidosis network (subnetwork) reconstructed by mapping of differentially expressed genes (DEGs) were identified by rumen papillae tissue RNA-seq analysis with the primary network. Here we present a study that assays global transcriptional changes using RNA-seq based analysis that may contribute to enhance the understanding of ruminal acidosis. We hypothesize that the rumen tissue in acidosis cases has a more significant expression of genes involved in rapid structural changes to maintain hemostasis, nutrient absorption and energetic metabolism.

2. Materials and Methods

2.1. Data Collection and Pretreatment

Available genome annotation and expression information of a bovine's tissue-specific file was downloaded from the FTP site of the UniGene database (ftp://ftp.ncbi.nih.gov/repository/UniGene/) [13]. A total of 4159 genes expressed in the rumen tissue were selected from the downloaded files (Table S1). Each gene was then queried in the Uniprot database (http://www.ebi.uniprot.org/index.shtml) to verify whether it encodes an enzyme. At this step 654 enzyme encoding genes were detected. Since the ordinary organism-specific data have many errors and does not have complete enzymes and reaction information, the KEGG bovine's pathway information files were not directly used for the metabolic network construction [14,15]. The enzyme commission (EC) number of each gene was obtained from Uniprot [16], and the corresponding reaction for each enzyme (all substrates and products) and other information were queried in the KEGG (http://www.genome.jp/kegg/) and BRENDA (http://www.brenda-enzymes.org) databases and checked in Uniprot as well. Some genes coding the enzyme, such as endopeptidase, tryptase or protein disulfide isomerase have no definite reactions. Therefore, these enzymes were removed from the data set.

Ethical Approval: All applicable international, national, and/or institutional guidelines for the care and use of animals were followed.

2.2. Reconstruction and Topological Analysis of the Metabolic Networks

In order to complete the data set for MNR, several features of metabolites were considered. These features include the component names (metabolic description, KEGG, ChEBI and PubChem identifier), chemical formula and cellular compartment [2,17]. Metabolites names usually differ between databases and there is no universal identifier between them. For this reason, they are referred by their KEGG, ChEBI and PubChem ID. However, the formula of a metabolite used for matching metabolites, is not usable as a unique identifier. Two different metabolites aspects of function can have the same formula because there are various structures of atoms. The common small molecules (known as the currency metabolites such as ATP, NADH, H_2O, CO_2, etc.) were not removed from the network representation. A number of 15 compartments were assayed accounting for 1771 reactions and 1429 metabolites. The final form of data was converted into a Systems Biology Markup Language (SBML) file format [18] in order to be visualized with Cytoscape software (vs. 3.6.1). For the acidosis model, we have used RNA-Seq data in order to perform a mapping into the normal model. Hence, the differentially expressed genes have been mapped into the normal network to build an acidosis state. Again, the final network has been created in the SBML file format. Some topological properties of the reconstructed networks have been also analyzed using Network Analyzer and cytoHubba plugins in Cytoscape [19,20]. The node degree distribution is one of the most critical topological parameters that gives the number of nodes with a degree (link) k (for k = 0, 1, . . .) that is defined as:

$$P(k) \sim k^{-\gamma} \tag{1}$$

where P(k) determines the variation in the nearest neighbors (degree distribution) and γ determines many properties of the system [13]. Real networks display a scale-free property therefore, the degree of a real network follows a power-law distribution. In the scale-free network most nodes have only a few links, but a few nodes (called hubs) have a very large number of links. P(k) for a scale-free network has no well-defined peak, and for large k it decays as a power-law $P(k) \sim k^{-\gamma}$ appearing as a straight line with slope $-\gamma$ on a log-log plot [21,22]. The shortest path length for nodes n and m define as L (n, m), which determines the number of edges that have to be crossed between n and m. For the whole network, the mean path length represents the average of the shortest paths among all pairs of nodes that is also known as the characteristic path length and offers the expected distance between n and m [20]. On the other hand, in this investigation, we first identified DEGs between control and acidosis tissue by rumen papillae tissue RNA-seq analysis and then these DEGs mapped to the primary network to construct the acidosis metabolic network (subnetwork) using bottleneck algorithm and double screening scheme.

2.3. Animal Management and Rumen Tissue Collection

Six Holstein steers (230 days old) born in spring 2015, were randomly divided into two groups (control and acidosis) and managed in the Agricultural resources farm of the University of Tehran. The high fermentable diet to induce acidosis was fed for 130 days being composed of 10.5% alfalfa hay, 14.5% corn silage, 34% barley grain, 23.3% corn grain ground, 9% soybean meal, 3% wheat bran, 2.5% rice bran and 3.2% supplement. The control diet was composed of 10.5% alfalfa hay, 14.5% corn silage, 34% barley grain, 5.3% corn grain ground, 9% soybean meal, 10.5% wheat bran, 6% rice bran, 5% sucrose, 2.5% fat powder and 3.2% supplement [3,8]. The final average body weight before slaughter for the acidosis and control groups was 529 ± 20 and 570 ± 10 kg, respectively. Clinical signs of acidosis were observed in all three animals. The ruminal pH was measured in the 1st, 30th and 129th days by manually inserting a pH probe into the rumen through a cannula and average data over 6 time-points in a single day (−8, −4, 0, 2, 4 and 8 h relative to grain feeding) (Table 1). After slaughter, for each sample, a 4 cm^2 piece of rumen tissue was obtained from the central region of the ventral sac and rinsed with sterilized PBS buffer (pH = 6.8) before being placed in a 50 mL tube containing RNA later solution (Invitrogen, Carlsbad, CA, USA). The samples were then stored at −80 °C until further processing. The ventral sac of the rumen was chosen as sampling site because it has the highest capillary blood flow per unit weight mucosa of any location within the rumen [12].

Table 1. Ruminal pH of normal and acidosis groups.

Group	The 1st Day ‡	The 30th Day ‡	The 129th Day ‡
control	5.71; 5.89; 5.88	5.57; 5.97; 5.87	6.03; 6.17; 6.0
acidosis	5.95; 5.23; 5.64	5.97; 5.17; 5.57	5.19; 5.10; 5.24

‡ Each value represents the average pH of one animal.

2.4. RNA Extraction, Quantification and Sequencing

Total RNA was extracted from the tissue samples using Trizol Reagent kit (Invitrogen, Carlsbad, CA, USA) according to the manufacturer's protocol. The RNA integrity of the total RNA samples was assessed using the Agilent 2100 Bioanalyzer (Agilent Technologies, Santa Clara, CA, USA). Only samples with RNA integrity greater than 7 were subjected to RNA-seq library construction. Nanodrop 2000c spectrophotometer (Thermo Scientific, Wilmington, DE, USA) was used to measure the concentration and purity. RNA samples were stored at −80 °C until RNA-seq library preparation. RNA-sequencing library preparation was done using Illumina TruSeq ribo-zero gold kit following the manufacturer's instructions. For each sample, a cDNA library was generated from 100 ng of total RNA. mRNA was enriched by removing rRNA from the total RNA with Ribo-Zero TM Magnetic Kit. mRNA was fragmented into short fragments (about 200–700 bp), then positive-strand cDNA was synthesized by random hexamer primers using the fragments as templates. Buffer, dNTPS, RNase H and DNA polymerase *I* was added to synthesize the negative-strand cDNA. The double stranded cDNA was purified using the QiaQuick PCR extraction kit and then used for end-polishing. Sequencing adapters were ligated to fragments, then the second strand was degraded using UNG (Uracil-N-Glyosylase). The fragments were purified by agarose gel electrophoresis and enriched by PCR amplification. Sequencing was performed on Illumina HiSeqTM 2000 to obtain high quality, 100 bp paired-end reads.

2.5. Sequence Data Processing and Differential Gene Expression Analysis

The Tuxedo pipeline in the statistical environment R version 3.4.4, has been used for the RNA-seq data analysis. The pipeline consists of TopHat, [23] Cufflinks (v. 2.2.1) and cummeRbund (v. 3.4) software. Sequencing quality was assessed using FastQC (v. 0.11.2). The data were preprocessed using Trimmomatic (v.0.32) to filter out low-quality reads, remove adaptors and trim low-quality regions of reads with default parameters [24]. The transcriptome mapping and alignment of the reads to the bovine reference genome (ARS-UCD1.2) were performed using Bowtie2 (v. 2.0.1) and TopHat2 (v. 2.0.9) with default parameters, resulting into BAM format files [25]. Cufflinks was also used to assemble transcripts and estimate their abundance as Fragments Per Kilobase of transcript per Million mapped reads (FPKM, cut-off ≥ 1), which normalizes transcript expression for transcript length and the total number of reads per sample. All assemblies were merged into a single transcriptome annotation using Cuffmerge (v. 2.2.1). Cuffdiff (v. 2.2.1.5), as a part of the Cufflinks package, was used to identify the differentially expressed genes ($p < 0.05$). The list of significant genes was further screened using PANTHER for further functional annotation. Results were visualized using the cummeRbund R library. Then the differentially expressed genes of this step of data analysis used for mapping and reconstruction of an acidosis metabolic network (subnetwork) using a bottleneck algorithm and double screening scheme. Bottleneck index calculates the number of shortest paths going through a certain node. Therefore, nodes with the highest bottleneck index, represent the critical points of the network. In addition, they could also control most of the information flow in the network [26].

The ssizeRNA package of R [27] was used with 71 million clean reads and 2 fold changes in average, the dispersion parameter for each gene equals 0.1 and 10 simulations. The result was 0.86 for the power of test and 0.04 for the false discovery rate.

2.6. *Quantitative Real-Time PCR (qRT-PCR) Validation of Selected Differentially Expressed Genes*

The expression of four randomly selected DEGs related to butyrate metabolism (OXCT1), the CoA biosynthesis pathway (DPYD), fatty acid oxidation (CPT1A), and the pantothenate biosynthesis pathway (FOXF1) was detected using RT-qPCR to confirm their differential expression between acidosis and control groups. Primers targeting four selected DEGs were designed using Primer-BLAST (NCBI, Bethesda, MD), Primer Express software (Applied Biosystems, Foster City, CA). The specificity of primers was checked with BLAST (NCBI) and the UCSC In-Silico PCR program (Table S2). The RT-qPCR reactions were performed with SYBR Green (Fast SYBR®Green Master Mix; Applied Biosystems) using StepOnePlus™ Real-Time PCR System (Applied Biosystems, Foster City, CA, USA) with the fast cycle and the following program: 20 s pre-denaturalization at 95 °C, followed by 40 cycles of 3 s denaturation at 95 °C and 30 s annealing and extension at 60 °C. Gene expression values were normalized to the reference gene of β-actin in the same sample. The relative changes in each gene expression were calculated using the $2^{-\Delta\Delta CT}$ (cycle threshold, CT) method.

3. Results

3.1. *The Primary Metabolic Network*

The primary network was a directed metabolic-centric network that reconstructed for 15 compartments including the cytoplasm, nucleus, cytosol, Golgi apparatus, mitochondrion, membrane protein, endoplasmic reticulum, peroxisome, plasma membrane, lysosome, extracellular exosome, cytoskeleton, cell membrane, extracellular exosome and mitochondrial membrane. The network includes 1429 metabolites as the nodes and 535 genes catalyzing 1771 relevant reactions as the edges. The topological network parameters of the primary metabolic network are listed in Table 2. The network data file is attached to Table S3.

Node degree distribution of the primary network is plotted in Figure 1. The degree distribution of a real network follows a power-law distribution. In the present study, the correlation coefficient (R) is 0.91 ($p < 0.0001$), and γ is taken as 0.89. The small value of γ implies that the role of the hub nodes is more important in the network. In the networks with a power-law degree distribution, the majority of the nodes have only one or two neighbors while coexisting with many nodes with hundreds and some even with thousands of neighbors. For these networks there exists no typical node, and they are therefore often referred to as scale-free. Unlike exponential networks, scale-free networks are extremely heterogeneous, with their topology being dominated by a few highly connected nodes (hubs) which link the rest of the less connected nodes to the system. In this respect the emergence of power-law distribution is intimately linked to the growth of the network in which new nodes are preferentially attached to already established nodes, a property that is also thought to characterize the evolution of biological systems. Our results convincingly indicate the probability that a given substrate participates in k reactions follows a power-law distribution; in other words, metabolic networks belong to the class of scale-free networks. The shortest path length is one of the most critical topological parameters, also called the distance between two nodes [20]. The average path length in the present reconstruction is equal to 13.142. Radrich et al. (2010) demonstrated the value of the average path length becomes significantly higher when common small molecules are removed from the network [28]. The network diameter is the largest distance between the two nodes whose value in the present network is as long as 38. The results of the centrality analysis using cytoHubba plugin, have been provided in Ma and Zeng (2003) [17]. As it was expected, the top 20 nodes are the currency metabolites usually found in all metabolic networks. Removing the currency metabolites, the top 15 nodes with the highest degree centrality indexes are presented in Table 3.

Since the existing network is scale-free and hubs play a significant role in maintaining the network and system biology of the rumen epithelium tissue, assays were conducted to identify their functions [13]. The centralization for the present network is estimated as 0.119 which is given by $n/(n-2)(max(k)/(n-1)- Density$. Centralization is a measure of how a network is focused around central nodes or assesses if a network has star-like topology. This value is close to 0 for a decentralized network and approaches 1 for the network resembling a star. The value of heterogeneity (defined as the coefficient of variance of the connectivity) in our network is close to 0.862 which indicates that some metabolites are highly connected and most of them have a lesser number of connections.

Table 2. General properties of the constructed network.

Parameters	Values
Number of nodes	1429
Number of edges	1771
Network density	0.002
Network heterogeneity	0.862
Number of self-loops	0
Network diameter	38
Network centralization	0.019
Average path length	13.142
Average number of neighbors	2.476

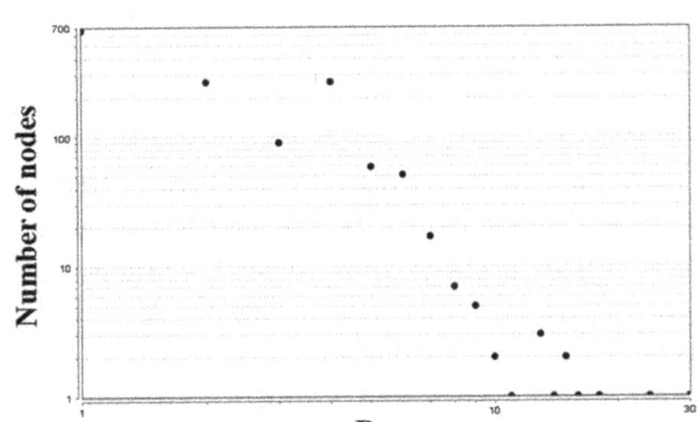

Figure 1. Topological properties of the reconstructed network for healthy ruminal epithelium. Node degree distribution.

Table 3. Top 15 hub metabolites of the metabolic network of bovine rumen tissue.

Metabolites	Neutral Formula	KeggID	PubChemID	Compartment
CoA	C21H36N7O16P3S	C00010	49600643	Mitochondrion
glutathione	C10H17N3O6S	C00051	3353	Cytoplasm
S-adenosyl-L-homocysteine	C14H20N6O5S	C00021	439155	Nucleus
S-adenosyl-L-methionine	C15H22N6O5S	C00019	34755	Nucleus
2-oxoglutarate	C5H6O5	C00026	3328	Mitochondrion
Acetyl-CoA	C23H38N7O17P3S	C00024	3326	Cytoplasm
D-fructose_6-phosphate	C6H13O9P	C00085	3385	Cytoplasm
ubiquinone	C19H26O4	C00399	5280346	Mitochondrion
formate	CH2O2	C00058	3358	Mitochondrion
electron-transferring flavoprotein	C27H30N9O15P2	C04253	6918	Mitochondrion
L-tryptophan	C11H12N2O2	C00078	3378	Cytoplasm
pyruvate	C3H4O3	C00022	3324	Mitochondrion
L-tyrosine	C9H11NO3	C00082	3382	Cytoplasm
tetrahydrofolate	C19H23N7O6	C00101	3401	Mitochondrion
L-serine	C3H7NO3	C00065	3365	Nucleus

3.2. Transcriptome Profiling of the Rumen Epithelium

The average of clean reads (after removing adaptor sequence and low-quality reads) was 71 million reads with an average of 70,975,460 ± 272,875 reads in normal samples and 71,142,189 ± 474,023 reads in the acidosis samples. The average of Q20 and GC percent were 97.96% and 50.87%, respectively (Table S4). The total number of expressed genes (genes with at least one read in every sample) for the rumen tissue was 24,494, and the overall read alignment rate to the bovine reference genome (ARS-UCD1.2) was evaluated as 87% ± 3.43%. A number of 1074 DEGs were identified in the rumen tissue between the normal and acidosis samples using the FDR-corrected p-value with a significance level of α =0.05 (Table S5). Then, DEGs mapped to the primary network to reconstruct the acidosis metabolic network. Further validation of the differential expression using reverse transcription quantitative real-time PCR (RT-qPCR) showed that the expression of OXCT1, DPYD, CPT1A and FOXF1 was consistent as detected with the RNA-seq data (Table S2). Additionally, we identified 978 isoforms (up-regulate: 584, down-regulate: 394), 1049 transcription start sites (TSS) (635 up-regulated and 414 down-regulated), coding DNA sequence (CDS) (607 up-regulated and 391 down-regulated) and 2 promoters (up-regulated) have differentially expressed in the rumen tissue between the normal and acidosis samples. Of the differentially expressed genes (DE), 627 were up-regulated, and 447 were down-regulated in acidosis. The functional analysis using PANTHER [29], indicated that the down-regulated genes control the biological processes such as; Biological regulation (GO: 0065007), Cellular process (GO: 0009987), Localization (GO: 0051179), Locomotion (GO: 0040011), Metabolic process (GO: 0008152), Multicellular organismal process (GO: 0032501), Response to stimulus (GO: 0050896). Conversely, PANTHER analysis for the up-regulated genes, illustrated that these genes involved in the biological adhesion (GO: 0022610), Biological regulation (GO: 065007), Cellular component organization (GO: 0071840), Developmental process (GO: 0032502), Immune system process (GO: 0002376), Multicellular organismal process (GO: 0032501), and Reproduction (GO: 0000003) (Table S6). There were 111 enzymes identified which presented significant changes in the expression under the acidosis condition among which, 46 enzymes were up-regulated and the remaining 65 down-regulated. The most significant up-regulated and down-regulated enzymes are presented in Tables 4 and 5. The PCA and volcano plots depict the global gene expression dynamics between the rumen epithelia of normal and acidosis steers. Figure 2A shows a volcano plot of the present data, depicting -log10 p-values over log2fold changes. The significantly differentially expressed genes are illustrated as red dots. The PCA plot (Figure 2C) shows that the gene expression profiles of the normal and acidosis tissues fall into distinct clusters. The result of cufflinks is converted to the InteractiVenn plot in order to identify the similarities and different genes expression between each sample group individually (Figure 2B).

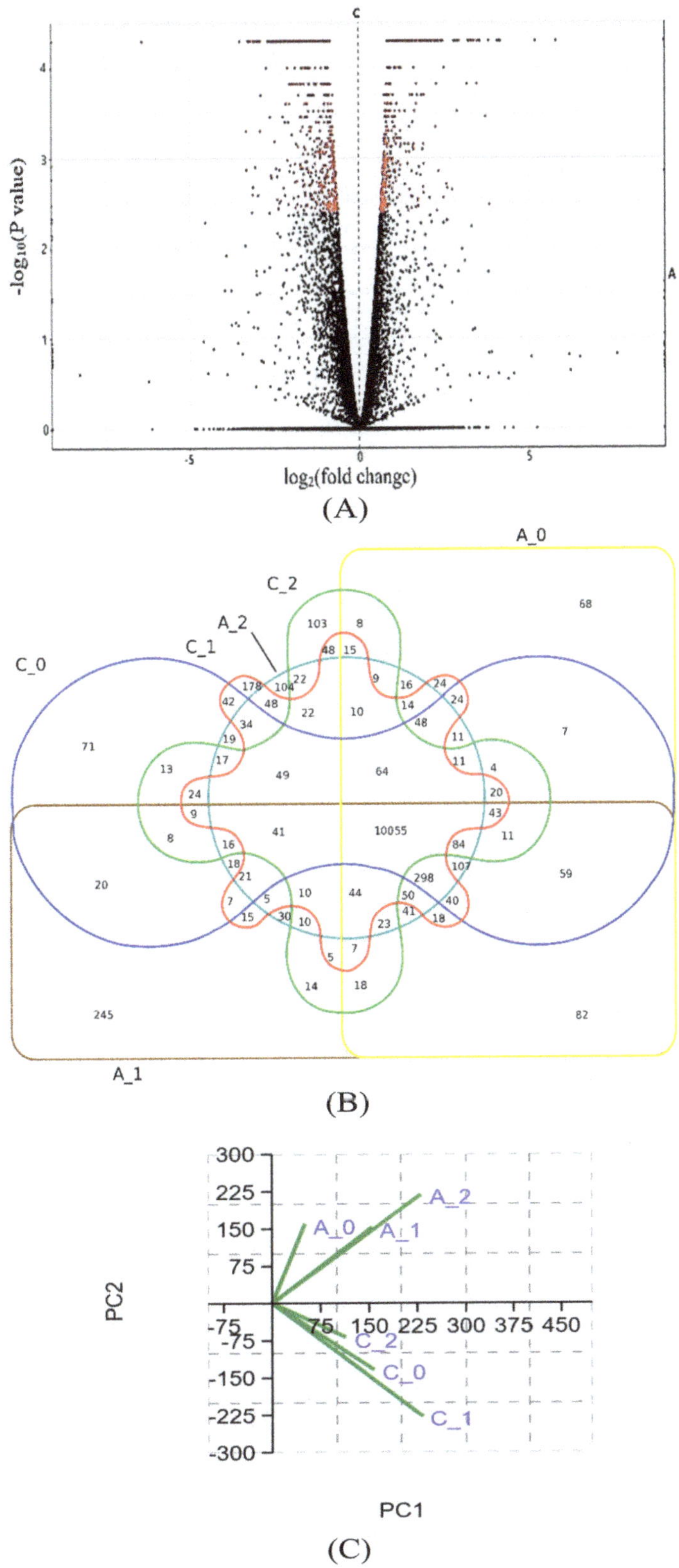

Figure 2. (A) Volcano plot is representing the relationship between fold-change and significance. Red points identified differentially expressed genes between control and acidosis samples. We identified 1074 differentially expressed genes ($p < 0.05$), of which 627 were up-regulated and 447 genes were down-regulated in the acidosis group. **(B)** InteractiVenn plot to identify similarities and different gene expressions between each sample (control samples: C_0, C_1, C_2 and acidosis samples: A_0, A_1, A_2). **(C)** PCA plot shows that gene expression profiles of normal and acidosis tissue fall into distinct clusters.

Table 4. The top up-regulated enzyme in the bovine rumen tissue under acidosis condition.

Gene	EC Number	Enzyme Name	Reaction	*P*-Value
HSPA6	4.2.1.134	very-long-chain (3R)-3-hydroxyacyl-CoA dehydratase	R10827	5.00E-05
GUCY1B1	4.6.1.2	guanylate cyclase	R00434	0.0001
PDK2	5.3.99.4	prostacyclin synthase	R02267	0.0019
PTGDS	5.3.99.2	prostaglandin-D synthase	R02266	0.0021
OXCT1	2.8.3.5	succinyl-CoA transferase	R01780	0.0002
SLC24A3	3.1.1.34	lipoprotein lipase	R01369	0.0001
PLA2G4A	3.1.1.4	phospholipase A2	R01313	0.004
PDE5A	3.1.4.35	cGMP phosphodiesterase	R01234	5.00E-05
CYP2W1	3.1.4.17	cyclic 3′,5′-phosphodiesterase	R03259	0.00085
PLCD3	3.1.4.11	Phosphorinositidase C	R03435	0.00355
EPHX1	3.3.2.9	microsomal epoxide hydrolase	R07627	0.00085
DYSF	3.6.1.5	ATP-diphosphatase	R10443	5.00E-05
GSTM3	2.5.1.18	glutathione transferase	R03522	0.0003
FOXF1	2.6.1.42	branched-chain-amino-acid transaminase	R01090	0.0002
PFKM	2.7.1.11	6-phosphofructokinase	R00756	0.0015
CHPT1	2.7.8.2	diacylglycerol cholinephosphotransferase	R01321	0.0012
CDS2	2.7.7.41	phosphatidate cytidylyl transferase	R01799	0.00215
MMP16	2.7.7.40	CDP ribitol pyro phosphorylase	R02921	0.0027
CEP112	2.7.12.2	MAP kinase kinase	R00162	0.0014
CHRNA3	1.14.13.39	nitric-oxide synthase (NADPH)	R00557	5.00E-05
TNS2	1.11.1.9	glutathione peroxidase	R00274	0.0018
TSPAN12	1.14.14.1	flavoprotein monooxygenase	R04122	5.00E-05
DPYD	1.3.1.2	dihydrouracil dehydrogenase (NADP+)	R00978	5.00E-05
ALDH9A1	1.2.1.3	aldehyde dehydrogenase (NAD+)	R00538	0.0001
AOC3	1.4.3.21	primary-amine oxidase	R01853	0.0001
SOX5	1.4.3.1	D-aspartate oxidase	R00359	5.00E-05
MAOB	1.4.3.4	monoamine oxidase	R11354	5.00E-05
CILP	1.7.1.7	GMP reductase	R01134	5.00E-05
AASS	1.5.1.8	lysine-2-oxoglutarate reductase	R00716	5.00E-05
MTHFR	1.5.1.20	methylenetetrahydrofolate	R01224	5.00E-05
GATM	2.1.4.1	glycine amidino transferase	R00565	0.00015
SCN7A	2.1.1.43	N-methyltransferase	R03938	5.00E-05
CPT1A	2.3.1.21	Palmitoyl carnitine transferase	R01923	5.00E-05

Table 5. The top down-regulated enzyme in the bovine rumen tissue under acidosis condition.

Gene	EC Number	Enzyme Name	Reaction	*P*-Value
ASNS	6.3.5.4	asparagine synthase	R00578	5.00E-05
CTPS1	6.3.4.2	CTP synthase	R00573	5.00E-05
TARS	6.1.1.3	threonyl-tRNA synthetase	R03663	0.00095
WARS	6.1.1.2	tryptophanyl-tRNA synthetase	R03664	0.0013
YARS	6.1.1.1	tyrosine-tRNA ligase	R02918	0.0003
ISYNA1	5.5.1.4	inositol-3-phosphate synthase	R07324	0.0005
CTH	4.4.1.1	cystathionine gamma-lyase	R01001	5.00E-05
SDS	4.3.1.17	L-serine ammonia-lyase	R00220	5.00E-05
CA3	4.2.1.1	carbonate dehydratase	R00132	0.00075
ODC1	4.1.1.17	ornithine decarboxylase	R00670	5.00E-05
ENPP4	3.6.1.29	bis(5′-adenosyl)-triphosphatase	R00187	0.0019
PPA1	3.6.1.1	inorganic diphosphatase	R00004	0.0001
ADA	3.5.4.4	adenosine deaminase	R01560	5.00E-05
ARG1	3.5.3.1	arginase	R00551	5.00E-05
GGH	3.4.19.9	folate gamma-glutamyl hydrolase	R04242	0.0033
PSPH	3.1.3.3	phosphoserine phosphatase	R00582	0.0003
PRDX6	3.1.1.4	phospholipase A2	R01313	0.00115
ABHD12	3.1.1.23	acyl glycerol lipase	R01351	0.0003
PDXK	2.7.1.35	pyridoxal kinase	R00174	0.0034
GK	2.7.1.30	glycerol kinase	R00847	5.00E-05
PSAT1	2.6.1.52	phosphoserine transaminase	R04173	5.00E-05

Table 5. *Cont.*

Gene	EC Number	Enzyme Name	Reaction	*P*-Value
GOT2	2.6.1.1	aspartate transaminase	R00355	0.0009
UPP1	2.4.2.3	uridine phosphorylase	R01876	5.00E-05
PNP	2.4.2.1	purine-nucleoside phosphorylase	R08368	0.00295
C1GALT1	2.4.1.122	D-galactosyltransferase	R05908	5.00E-05
DGAT2	2.3.1.20	diglyceride acyltransferase	R02251	0.0002
TALDO1	2.2.1.2	formaldehyde transketolase	R08575	0.0015
PYCR1	1.5.1.2	proline oxidase	R01248	5.00E-05
MTHFD2	1.5.1.15	methylenetetrahydrofolate dehydrogenase (NAD+)	R01218	5.00E-05
DHCR7	1.3.1.21	7-dehydrocholesterol reductase	R01456	0.0012
SCD	1.14.19.1	stearoyl-CoA 9-desaturase	R02222	5.00E-05
LPO	1.11.1.7	peroxidase	R03344	5.00E-05
PGD	1.1.1.44	phosphogluconate dehydrogenase	R03532	0.0022
DCXR	1.1.1.10	L-xylulose reductase	R01904	0.00155

3.3. *Acidosis Metabolic Network*

The acidosis metabolic network (as a subnetwork) based on the bottleneck and double screening scheme methods reconstruction which included 832 metabolites and 1021 reactions (Table S7). Node degree distribution of the network is plotted in Figure 3 and the topological network parameters are given in Table 6. The top 15 hubs identified by the bottleneck (BN) algorithm and double screening scheme (DSS) methods are shown in Table 7. Figure 4 depicts the graphical acidosis metabolic network representation. The search results in the KEGG pathway clarified that stearoyl-CoA, 2-oxoglutarate and lactate are intermediates in the fatty acid metabolism pathway (map01212), citrate cycle (map00020) and glycolysis (map00010), respectively and acetyl-CoA is a metabolite linking between the citrate cycle, fatty acid pathway and glycolysis.

Average path length 6.913

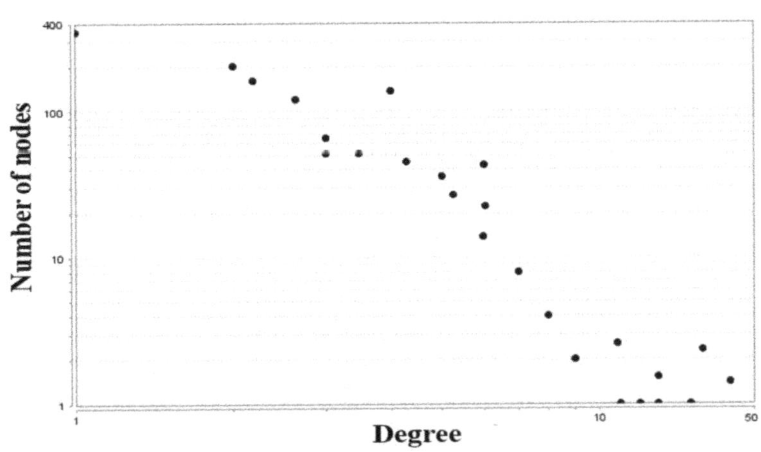

Figure 3. Node degree distribution of reconstructed metabolic network for acidosis ruminal epithelium.

Table 6. General properties of the constructed sub-network.

Parameters	Values
Number of nodes	832
Number of edges	1021
Network density	0.003
Network heterogeneity	0.733
Number of self-loops	0
Network diameter	28
Network centralisation	0.15
Average path length	6.913

Table 7. First 15 hub metabolites of the acidosis metabolic network.

Metabolites	Neutral Formula	KeggID	PubChemID	Compartment
Stearoyl-CoA	C39H70N7O17P3S	C00412	3702	Endoplasmic Reticulum
Ferrocytochrome b5	C36H61O11	C00996	4242	Membrane
glutathione	C10H17N3O6S	C00051	3353	Cytoplasm
2-oxoglutarate	C5H6O5	C00026	3328	Mitochondrion
protein-L-arginine	C7H13N5O2R2	C00613	3887	Nucleus
electron-transferring flavoprotein	C27H30N9O15P2	C04253	6918	Mitochondrion
ubiquinone	C19H26O4	C00399	5280346	Mitochondrion
S-adenosyl-L-methionine	C15H22N6O5S	C00019	34755	Cytoplasm
S-adenosyl-L-homocysteine	C14H20N6O5S	C00021	439155	Nucleus
Acetyl-CoA	C23H38N7O17P3S	C00024	3326	Cytoplasm
L-tryptophan	C11H12N2O2	C00078	3378	Cytoplasm
L-arginine	C7H13N5O2R2	C00613	3887	Mitochondrion
L-serine	C3H7NO3	C00065	3365	Cytoplasm
L-Lactate	C3H6O3	C00186	3486	Cytoplasm
Xanthine	C10H13N4O9P	C00655	73323	Nucleus

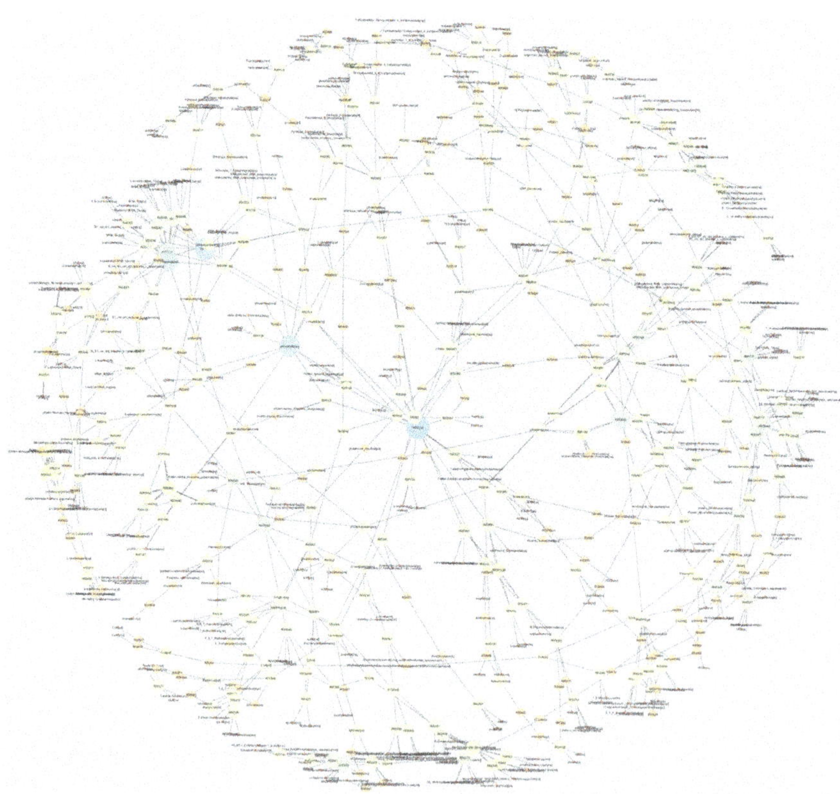

Figure 4. The reconstructed acidosis metabolic network for bovine.

To reconstruct the metabolic networks, we used a graph theory- based method that metabolites correspond to the nodes and reactions correspond to the edges as a metabolite-centric network. The existing networks were undirected because all the reactions were reversible and reconstructed for 15 compartments that most of the hub metabolites were in mitochondrion, cytoplasm, and nucleus. The small-world property, is one of the most critical real network properties. The networks with this property have a characteristic feature of regional specialization and relatively efficient information transfer [30]. This property characterized by a diameter network, which is defined as the average minimum path length between all pairs of nodes. The value of the diameter in the primary and subnetwork were 38 and 28, respectively. Embar et al (2016) reported the path length is of biological interest because it is assumed that a functionally related gene would be more closely connected in the metabolome in a gene network [31]. The average path length presents a measure of the network's overall navigability and it is assumed that functionally related genes would be more closely connected in the metabolome that the value in the primary and subnetwork which were equal to 13.142 and 6.913, respectively. Zhang et al. (2005) demonstrated that the diameter in all 13 real networks examined (biological, social, technological and linguistic), is higher than the random expectation [32]. Ma and Zeng (2003) revealed that the average path length and diameter have greater value in eukaryotes (9.75 and 33.1) than bacteria (7.23 and 20.6) [17]. Our results of topological analysis networks showed these two crucial parameters were nearer to the values of eukaryotes and far off bacteria.

4. Discussion

The profound understanding of the bovine rumen epithelium metabolism and changes during the ruminal acidosis have a tangible effect on cow breeding and provide crucial biological information. Ruminal acidosis is one of the most important disorders and production problems in the beef industry occurred by low ruminal pH in the range of 5.2 and 6 for a prolonged period [5]. Ruminal pH depends on the Henderson-Hasselbach balance that is defined as [9]:

$$pH = pKa + \log\,[(acid^{-})\,(H\,acid)^{-1}] \tag{2}$$

During the microbial ruminal fermentation, the population of microorganisms (bacteria, ciliate protozoa, anaerobic fungi and bacteriophages) ferments the carbohydrates and protein to the short-term intermediates such as sugars and amino acids [9]. The products of this initial degradation are metabolized to produce energy, methane, carbon dioxide, heat and volatile fatty acids (VFA). At the physiological pH of rumen fluid, 90–99% of the acid is in the dissociated form as a base conjugate (VFA-) and H+ that are absorbed through the rumen wall [33].

When cattle are fed with a high level of rapidly digestible carbohydrate to gain weight at their maximum level, VFA production exceeds the rumen acids absorption capacity which causes accumulated VFA and dropping ruminal pH. Lactic acid has a lower pKa (3.80) than other VFA, and its concentration is low during the ruminal acidosis. However, as the pH drops below 5.7, the rate of lactic acid formation exceeds the absorption capacity which causes a drastic decrease in pH value [34]. If this condition continues, the lactic acid entry into the blood system causes a big challenge in metabolism, changes the population of microorganisms, rumen motility, and the systemic body fluid balance. Recent management practices strive to minimize the ruminal acidosis occurrence, but the need for a better understanding of the disease's etiology, prevention and treatment are still felt [5,6].

The main goal of this study was the development of a system-level understanding of biological changes in rumen tissue under acidosis. For this purpose, the focus has been on the reconstruction of a bovine rumen tissue-specific network using the transcriptome data of bovines with and without acidosis in order to compare the acidosis-related metabolites and reactions based on the network topology. Studies conducted on the acidosis using the metabolic network method can provide new information about it through the identification of the novel metabolic function and metabolite biomarkers. Because a change in the concentration of one metabolite, leads to the change of several cascade reaction pathways that may yield the diverse phenotypes and disorders [35].

Of a total of 1074 DEGs, 111 genes encoded an enzyme. The analysis of pathway and reaction associated with up-regulated enzymes, indicated that most of these enzymes are involved in 3 central metabolic pathways including the fatty acid metabolism, biosynthesis of amino acids, pyruvate and carbon metabolism (Table 8). The map profiling results showed the significant up expression of the OXCT1 gene encoding succinyl-CoA transferase (2.8.3.5) under the acidosis. This enzyme takes part in butyrate metabolism within the mitochondrion and also the nucleus in order to produce butyrate as the microbial ferment product of a high starch diet and as one of the important substrates in the ruminant energy metabolism. Hernández et al. (2014) reported a significant increase in the rumen VFA concentration (150–225 mM) of the caws under ruminal acidosis [6]. Previous studies have reported that in acidosis lipopolysaccharide (LPS) derived from the rumen down-regulates stearoyl-CoA desaturase 1 expression (SCD1) and demonstrated up-regulation of the genes involved fatty acid β-oxidation [10]. The present results of the gene expression profile assay indicated the obvious pattern as up-regulation of the genes CPT1A, ALDH9A1and TSPAN12, associated with the fatty acid oxidation and down-regulation of SCD (stearoyl-CoA 9-desaturase) in the acidosis group.

Table 8. Most of the pathways enriched by up-regulated and down-regulated enzymes.

Significant Up-expression enzymes	Fatty acid metabolism	Butyrate metabolism (ec00650) Arachidonic acid metabolism (ec00590) Linoleic acid metabolism (ec00591) Fatty acid degradation (ec00071) Pantothenate and CoA biosynthesis (ec00770)
	Biosynthesis of amino acids	Valine, leucine and isoleucine metabolism (ec00280) Glycine, serine and threonine metabolism (ec00260)
	Pyruvate and carbon metabolism	Glycolysis/Gluconeogenesis (ec00010) Pentose phosphate (ec00030) Pyruvate metabolism (ec00620) Glutathione metabolism (ec00480) Fructose and mannose metabolism (ec00051) Galactose metabolism (ec00052) Synthesis and degradation of ketone bodies (ec00072) Pentose and glucuronate interconversions (ec00040)
Significant Down-expression enzymes	Nucleotide metabolism	Purine metabolism (ec00230) Pyrimidine metabolism (ec00240)
	Other metabolisms	Vitamin B6 metabolism (ec00750) Biosynthesis of antibiotics (ec01130)

The SCD enzyme is responsible for forming a double bond in Stearoyl-CoA and plays an important role in fatty acid biosynthesis, regulating the genes expression which is involved in the lipogenesis and also regulating the mitochondrial fatty acid oxidation [36]. Stearoyl-CoA is a long-chain acyl CoA ester that acts as an intermediate metabolite in the biosynthesis of the monounsaturated fatty acids. This oxidative reaction is catalyzed by the iron-containing, microsomal enzyme and SCD. NADH supplies the reducing equivalents for the reaction, the flavoprotein is cytochrome b5-reductase and the electron carrier is the heme protein cytochrome b5. The result of analysis using cytoHubba plugin show Stearoyl-CoA (C00412) and ferrocytochrome b5 (C00996) are hub metabolites in the subnetwork. Hubs in scale-free networks have a key biological role and significant effect in maintaining the network [13]. For this reason, improving the bovine acidosis researches seem to be useful. PDK2 (5.3.99.4), PTGDS (5.3.99.2), and TNS2 (1.11.1.9) take part in the arachidonic acid metabolism, whilst PLA2G4A (3.1.1.4) and TSPAN12 (1.14.14.1) participate in both arachidonic (ec00590) and linoleic acid (ec00591) metabolisms. Linoleic (18:2 Δ9, 12) and arachidonic acids (20:4 Δ5, 8, 11, 14) are essential long-chain polyunsaturated fatty acids (LC-PUFAs) and a type of linoleic acid (ω-6) can be converted to the arachidonic one [37]. In addition, both of them have important biological functions as a component of phospholipid membranes and arachidonic acid is also present in cell membranes of erythrocytes, neutrophils and monocytes.

The analysis results of the DEGs illustrated significant high expression of FOXF1 (2.6.1.42) and DPYD (1.3.1.2) involved in the Pantothenate and CoA biosynthesis pathway (ec00770). Pantothenate (vitamin B5) is the key precursor for the biosynthesis of coenzyme A [36]. This vitamin is a carrier protein, having a phosphopantetheine group that shuttles intermediates between the active sites of enzymes involved in the fatty acid and non-ribosomal peptide synthesis. Additionally, the results of the sub-network analysis showed that acetyl-CoA (C00024) and S-adenosyl-L-methionine (C00019) cytoplasmic are hub metabolites. The conversion of pyruvate into acetyl-CoA is a crucial step in the carbon energy metabolism that occurs by pyruvate formate-lyase (PFL; 2.3.1.54) or pyruvate ferredoxin oxidoreductase (PFO;1.2.7.1). Pyruvate formate-lyase comprises two cofactors of S-adenosyl methionine and thiamine diphosphate. The pathways results indicated that the biosynthesis of amino acids in the ruminal acidosis increases which might lead to the increased rumen pyruvate and VFA content in the acidosis condition [38]. In the rumen metabolism pyruvate and isobutyrate for valine, acetate for serine, propionate for arginine and histidine, butanoic acid for leucine and 2-methyl butyrate for isoleucine are the main substrates and precursors for the biogenesis of these amino acids. The result of GC–MS analysis by Xue et al. (2018) showed that the SARA-inducing diet significantly increases the propionate, pyruvate, lactate, glycine and valine contents in the rumen of dairy cows compare to the control one [10]. The results of pathways analysis indicated that TNS2 (glutathione peroxidase; 1.11.1.9) and GSTM3 (glutathione transferase; 2.5.1.18) participate in the glutathione metabolism (ec00480) and glutathione (C00051) was identified here as a hub metabolite in the subnetwork. Glutathione is a non-protein thiol for protecting cells against the oxidative stress condition via detoxification of the oxygen-derived free radicals using glutathione peroxidase (GSH-Px) and transferase enzymes [39]. These enzymes are antioxidants and play an important role in stress conditions such as immune and inflammatory response. Xue et al. (2018) detected significant differences in the glutathione metabolism of bovine rumen tissue between the control and SARA groups [10]. Also, Jiang et al. (2014) reported that glutathione transferase is significantly up-regulated in the goat liver tissue of the SARA group compared to the control one [8].

The pathway analysis of the down-regulated enzymes showed that most of these enzymes are involved in the purine and pyrimidine, vitamin B6 and biosynthesis of the antibiotic's metabolisms. Purine and pyrimidine are basic components of DNA and RNA in all living cells that participate in the wide range of biological processes. Purine and pyrimidine metabolism include the synthesis and degradation of these components with deferent enzymes and metabolites [40]. ADA (3.5.4.4: adenosine deaminase), PNP (2.4.2.1: purine-nucleoside phosphorylase) and ENPP4 (3.6.1.29: bis(5′-adenosyl)-triphosphatase) are involved in the purine metabolism pathway (ec00230). Also, UPP1 (2.4.2.3: uridine phosphorylase) and CTPS1 (6.3.4.2: CTP synthase) are involved in the pyrimidine metabolism pathway (ec00240). With the reduced expression of these enzymes, the appropriate substrates and other alternative products are accumulated (dATP and dGTP) and this leads to the toxic effects on the cells of the immune system. Cytidine triphosphate synthase (CTPS) catalyzes the last step in the pyrimidine biosynthesis which is the irreversible reaction of the converted uridine triphosphate (UTP) into CTP with the glutamine and ATP consumption. The results of the enzymes pathway analysis showed that the enzymes involved in the biosynthesis of antibiotics (ec01130) including ISYNA1 (5.5.1.4), SDS (4.3.1.17), ODC1 (4.1.1.17), ARG1 (3.5.3.1), PSPH (3.1.3.3), PSAT1 (2.6.1.52), TALDO1 (2.2.1.2), MTHFD2 (1.5.1.15) and PGD (1.1.1.44), have been down-regulated significantly. Antibiotics, one of the secondary metabolites, encompass a broad group of chemical compounds produced by fungi, bacteria and other organisms with a different molecular target. They have a wide range of functions that can selectively destroy or inhibit the growth of other microorganisms [41]. Sodium monensin is a polyether antibiotic produced by *Streptomyces cinnamonensis* which acts by inhibiting the growth of *Streptococcus bovis* (and other major lactic acid-producing strains except for Selenomonas) and leads to the lactate production decrement and rumen pH increment.

The pathways analysis results further indicated the down-regulation of PDXK (2.7.1.35: pyridoxal kinase) and PSAT1 (2.6.1.52: aspartate transaminase) was involved in vitamin B6 metabolism. Beef meat is a major source of B6, B12 Thiamin, Riboflavin and Niacin vitamins. Vitamin B6 is part of the vitamin B group and pyridoxal phosphate is the coenzyme form of this vitamin participating in the reactions related to the amino acid, glucose and lipid metabolism. Pyridoxal kinase (pdxK) is an essential enzyme in the terminal point of the vitamin B6 pathway which causes phosphorylation of pyridoxal to produce pyridoxal phosphate; therefore, a decreased expression of pdxK leads to vitamin B6 deficiency. Vitamin B6 deficiency causes impair weight gain, anorexia, and diarrhea which are some outward signs of an acidosis disorder.

5. Conclusions

In summary, this study gave a new comprehensive insight into the ruminal acidosis metabolic network. Transcriptome profiling and multivariate analysis were used to provide the metabolic network data of 6 rumen tissue Holstein cattle fed with normal and induced acidosis diets. As a first step, the normal metabolic network in the bovine rumen tissue was reconstructed using the available bovine genome annotation and expression information and then the differentially expressed genes of RNA-seq data analysis were used for mapping and reconstruction of an acidosis network. We identified the change in genes, enzymes and hub metabolites associated with acidosis in the rumen. With rapid shifts in the use of rich diets in fermentable carbohydrates, significant changes were found in the transcriptome profiles of the rumen epithelial tissue. The PCA of rumen data showed that acidosis samples are clustered closer to each other and separated from the control samples. Analysis of the pathways and reactions associated with up-regulated enzymes, showed that most of these enzymes are involved in the fatty acid metabolism, biosynthesis of amino acids, pyruvate and carbon metabolism. Additionally, most of the down-regulated enzymes are involved in the purine and pyrimidine, vitamin B6 and antibiotics metabolisms. The top 15 hub metabolites were determined in the acidosis network from which stearoyl-CoA and ferrocytochrome b5 are involved in the fatty acid oxidation, acetyl-CoA and S-adenosyl-L-methionine cytoplasmic in VFA biosynthesis, L-serine and L-arginine in amino acids biogenesis and glutathione in glutathione metabolism that play an important role in the stress condition such as the immune and inflammatory response. The network's distribution degree follows a power-law distribution hence displays a scale-free property. The values of diameter and average path length in this network were 28 and 6.913, respectively, indicating that the present network has small world property.

The achievements of the present study indicate that the use of a high grain diet could elicit a substantial change in the gene expression and transcriptome response in the ruminal tissue. Furthermore, the present results provide a fundamental understanding of which genes are associated with acidosis and may lead to the identification of a biomarker for selecting the best cows for animal breeding. Although significant genes were analyzed and known metabolic data collected from the different databases reconstructed the metabolic network, many factors such as CDS, promoter, TSS and isoform might affect the better interpretation and results of the system biology investigation. The limitations of this study were a low number of animals and using only epithelial tissue (ventral sac) for RNA-seq. Assaying liver and lung tissues are also suggested in order to provide more detailed information on the effects of acidosis on fattening steers.

Supplementary Materials:
file, Table S1: Total of genes expressed in the rumen tissue (UniGene database). Table S2: Fold change, Gene name, primer sequences and product length of qRT-PCR validated genes. Table S3: Primary metabolic network data. Table S4: Statical result of RNA-Seq data analysis. Table S5: The result of detected differentially expressed genes (DEGs) with the RNA-seq data analysis using Tuxedo package. Table S6: Enrichment analysis using KEGG, SP_PIR and GO. Reported p-values was corrected for multiple testing using the Benjamini-Hochberg procedure, enrichment analysis was performed using DAVID with default parameters. Table S7: Acidosis metabolic network data.

Author Contributions: Conceptualization, J.F. and M.G.; Methodology, M.G., J.F. and Y.A.; Software, M.G. and L.K.; Validation, L.K.; Formal Analysis, M.G. and L.K.; Investigation, M.G. and H.Z.; Resources, H.Z., M.G., J.F.; Writing-Original Draft Preparation, M.G. and J.F.; Writing-Review & Editing, M.G. and J.F.; Supervision, J.F.; Project Administration, J.F., M.G.; Funding Acquisition, J.F., H.Z., Y.A. All authors have read and agreed to the published version of the manuscript.

Acknowledgments: We thank Mohsen Sari, Mehdi Dehghan and Javad Khalifeh for their help and advice in animal raring period and sample preparation. We acknowledge Research deputy of Agricultural Sciences and Natural Resources University of Khuzestan.

References

1. Cottle, D.; Kahn, L. *Beef Cattle Production and Trade*; CSIRO Publishing: Clayton, Australia, 2014; p. 221.
2. Degtyarenko, K.; de Matos, P.; Ennis, M.; Hastings, J.; Zbinden, M.; McNaught, A.; Alcántara, R.; Darsow, M.; Guedj, M.; Ashburner, M. ChEBI: A database and ontology for chemical entities of biological interest. *Nucleic Acids Res.* **2008**, *36*, D344–D350. [CrossRef] [PubMed]
3. Steele, M.A.; Croom, J.; Kahler, M.; AlZahal, O.; Hook, S.E.; Plaizier, K.; McBride, B.W. Bovine rumen epithelium undergoes rapid structural adaptations during grain-induced subacute ruminal acidosis. *Am. J. Physiol. Regul. Integr. Comp. Physiol.* **2011**, *300*, R1515–R1523. [CrossRef] [PubMed]
4. Steele, M.A.; Vandervoort, G.; AlZahal, O.; Hook, S.E.; Matthews, J.C.; McBride, B.W. Rumen epithelial adaptation to high-grain diets involves the coordinated regulation of genes involved in cholesterol homeostasis. *Physiol. Genomics* **2011**, *43*, 308–316. [CrossRef] [PubMed]
5. Enemark, J.M. The monitoring, prevention and treatment of sub-acute ruminal acidosis (SARA): A review. *Vet. J.* **2008**, *176*, 32–43. [CrossRef]
6. Hernández, J.; Benedito, J.L.; Abuelo, A.; Castillo, C. Ruminal acidosis in feedlot: From aetiology to prevention. *Sci. World J.* **2014**, 702572. [CrossRef]
7. Li, W.; Gelsinger, S.; Edwards, A.; Riehle, C.; Koch, D. Transcriptome analysis of rumen epithelium and meta-transcriptome analysis of rumen epimural microbial community in young calves with feed induced acidosis. *Sci. Rep.* **2019**, *9*, 4744. [CrossRef]
8. Jiang, X.Y.; Ni, Y.D.; Zhang, S.K.; Zhang, Y.S.; Shen, X.Z. Identification of Differentially Expressed Proteins in Liver in Response to Subacute Ruminal Acidosis (SARA) Induced by High-concentrate Diet. *Asian Australas J. Anim. Sci.* **2014**, *27*, 1181–1188. [CrossRef]
9. Kleen, J.L.; Hooijer, G.A.; Rehage, J.; Noordhuizen, J.P. Subacute ruminal acidosis (SARA): A review. *J. Vet. Med. A Physiol. Pathol. Clin. Med.* **2003**, *50*, 406–414. [CrossRef]
10. Xue, F.; Pan, X.; Jiang, L.; Guo, Y.; Xiong, B. GC-MS analysis of the ruminal metabolome response to thiamine supplementation during high grain feeding in dairy cows. *Metabolomics* **2018**, *14*, 67. [CrossRef]
11. Bruggeman, F.J.; Westerhoff, H.V. The nature of systems biology. *Trends Microbiol.* **2007**, *15*, 45–50. [CrossRef]
12. O'Shea, E.; Waters, S.M.; Keogh, K.; Kelly, A.K.; Kenny, D.A. Examination of the molecular control of ruminal epithelial function in response to dietary restriction and subsequent compensatory growth in cattle. *J. Anim. Sci. Biotechnol.* **2016**, *7*, 53. [CrossRef] [PubMed]
13. Wang, C.; Wang, J.; Ju, Z.; Zhai, R.; Zhou, L.; Li, Q.; Li, J.; Li, R.; Huang, J.; Zhong, J. Reconstruction of metabolic network in the bovine mammary gland tissue. *Mol. Biol. Rep.* **2012**, *39*, 7311–7318. [CrossRef] [PubMed]
14. Kanehisa, M.; Goto, S. KEGG: Kyoto encyclopedia of genes and genomes. *Nucleic Acids Res.* **2000**, *28*, 27–30. [CrossRef] [PubMed]
15. Poolman, M.G.; Bonde, B.K.; Gevorgyan, A.; Patel, H.H.; Fell, D.A. Challenges to be faced in the reconstruction of metabolic networks from public databases. *Syst. Biol.* **2006**, *153*, 379–384. [CrossRef]
16. Stein, S.E.; Heller, S.R.; Tchekhovskoi, D. An open standard for chemical structure representation: The IUPAC chemical identifier. In Proceedings of the 2003 International Chemical Information Conference, Nimes, France, 1 October 2003; pp. 131–143.
17. Ma, H.; Zeng, A. Reconstruction of metabolic networks from genome data and analysis of their global structure for various organisms. *Bioinformatics* **2003**, *19*, 270–277. [CrossRef]
18. Bornstein, B.J.; Keating, S.M.; Jouraku, A.; Hucka, M. LibSBML: An API library for SBML. *Bioinformatics* **2008**, *24*, 880–881. [CrossRef]
19. Cline, M.S.; Smoot, M.; Cerami, E.; Kuchinsky, A.; Landys, N.; Workman, C.; Christmas, R.; Avila-Campilo, I.; Creech, M.; Gross, B.; et al. Integration of biological networks and gene expression data using Cytoscape. *Nat. Protoc.* **2007**, *2*, 2366–2382. [CrossRef]
20. Managbanag, J.R.; Witten, T.M.; Bonchev, D.; Fox, L.A.; Tsuchiya, M.; Kennedy, B.K.; Kaeberlein, M. Shortest-path network analysis is a useful approach toward identifying genetic determinants of longevity. *PLoS ONE* **2008**, *3*, e3802. [CrossRef]
21. Erdös, P.; Rényi, A. On the evolution of random graphs. *Publ. Math.* **1959**, *6*, 290–297.
22. Jeong, H.; Tombor, B.; Albert, R.; Oltvai, Z.N.; Barabási, A.L. The large-scale organization of metabolic networks. *Nature* **2000**, *407*, 651–654. [CrossRef]

23. Kim, D.; Pertea, G.; Trapnell, C.; Pimentel, H.; Kelley, R.; Salzberg, S.L. TopHat2: Accurate alignment of transcriptomes in the presence of insertions, deletions and gene fusions. *Genome Biol.* **2013**, *14*, R36. [CrossRef] [PubMed]
24. Bolger, A.M.; Lohse, M.; Usadel, B. Trimmomatic: A flexible trimmer for Illumina sequence data. *Bioinformatics* **2014**, *30*, 2114–2120. [CrossRef] [PubMed]
25. Langmead, B.; Salzberg, S.L. Fast gapped-read alignment with Bowtie 2. *Nat. Methods* **2012**, *9*, 357–359. [CrossRef] [PubMed]
26. Yu, H.; Kim, P.M.; Sprecher, E.; Trifonov, V.; Gerstein, M. The importance of bottlenecks in protein networks: Correlation with gene essentiality and expression dynamics. *PLoS Comput. Biol.* **2007**, *3*, e59. [CrossRef] [PubMed]
27. Bi, R.; Liu, P. Sample size calculation while controlling false discovery rate for differential expression analysis with RNA-sequencing experiments. *BMC Bioinform.* **2016**, *17*, 146. [CrossRef] [PubMed]
28. Radrich, K.; Tsuruoka, Y.; Dobson, P.; Gevorgyan, A.; Swainston, N.; Baart, G.; Schwartz, J.M. Integration of metabolic databases for the reconstruction of genome-scale metabolic networks. *BMC Syst. Biol.* **2010**, *4*, 114. [CrossRef]
29. Kong, R.S.; Liang, G.; Chen, Y.; Stothard, P.; Guan le, L. Transcriptome profiling of the rumen epithelium of beef cattle differing in residual feed intake. *BMC Genom.* **2016**, *17*, 592. [CrossRef]
30. Telesford, Q.K.; Joyce, K.E.; Hayasaka, S.; Burdette, J.H.; Laurienti, P.J. The ubiquity of small-world networks. *Brain Connect.* **2011**, *1*, 367–375. [CrossRef]
31. Embar, V.; Handen, A.; Ganapathiraju, M.K. Is the average shortest path length of gene set a reflection of their biological relatedness? *J. Bioinform. Comput. Biol.* **2016**, *14*, 1660002. [CrossRef]
32. Zhang, P.; Foerster, H.; Tissier, C.P.; Mueller, L.; Paley, S.; Karp, P.D.; Rhee, S.Y. MetaCyc and AraCyc. Metabolic pathway databases for plant research. *Plant Physiol.* **2005**, *138*, 27–37. [CrossRef]
33. Bergman, E.N. Energy contributions of volatile fatty acids from the gastrointestinal tract in various species. *Physiol. Rev.* **1990**, *70*, 567–590. [CrossRef] [PubMed]
34. Ash, R.; Baird, G.D. Activation of volatile fatty acids in bovine liver and rumen epithelium. *Biochem. J.* **1973**, *136*, 311–319. [CrossRef] [PubMed]
35. Braun, P.; Rietman, E.; Vidal, M. Networking metabolites and diseases. *Proc. Natl. Acad. Sci. USA* **2008**, *105*, 9849–9850. [CrossRef]
36. Leonardi, R.; Jackowski, S. Biosynthesis of Pantothenic Acid and Coenzyme A. *EcoSal Plus* **2007**, *2*. [CrossRef] [PubMed]
37. Artegoitia, V.M.; Foote, A.P.; Lewis, R.M.; Freetly, H.C. Rumen Fluid Metabolomics Analysis Associated with Feed Efficiency on Crossbred Steers. *Sci. Rep.* **2017**, *6*, 2864. [CrossRef]
38. Atteia, A.; van Lis, R.; Gelius-Dietrich, G.; Adrait, A.; Garin, J.; Joyard, J.; Rolland, N.; Martin, W. Pyruvate formate-lyase and a novel route of eukaryotic ATP synthesis in Chlamydomonas mitochondria. *J. Biol. Chem.* **2006**, *281*, 9909–9918. [CrossRef]
39. Abd Ellah, M.R.; Okada, K.; Yasuda, J. Oxidative stress and bovine liver diseases: Role of glutathione peroxidase and glucose 6-phosphate dehydrogenase. *Jpn. J. Vet. Res.* **2007**, *54*, 163–173.
40. Moffatt, B.A.; Ashihara, H. Purine and pyrimidine nucleotide synthesis and metabolism. *Arab. Book* **2002**, *1*, e0018. [CrossRef]
41. Clardy, J.; Fischbach, M.A.; Currie, C.R. The natural history of antibiotics. *Curr. Biol.* **2009**, *19*, R437–R441. [CrossRef]

3

Use of Large and Diverse Datasets for ^{1}H NMR Serum Metabolic Profiling of Early Lactation Dairy Cows

Timothy D. W. Luke [1,2], Jennie E. Pryce [1,2], Aaron C. Elkins [1], William J. Wales [3] and Simone J. Rochfort [1,2,*]

1 Agriculture Victoria Research, AgriBio, Centre for AgriBioscience, Bundoora 3083, Australia; tim.luke@agriculture.vic.gov.au (T.D.W.L.); jennie.pryce@agriculture.vic.gov.au (J.E.P.); aaron.elkins@agriculture.vic.gov.au (A.C.E.)
2 School of Applied Systems Biology, La Trobe University, Bundoora 3083, Australia
3 Agriculture Victoria Research, Ellinbank Centre, Ellinbank 3821, Australia; bill.wales@agriculture.vic.gov.au
* Correspondence: simone.rochfort@agriculture.vic.gov.au

Abstract: Most livestock metabolomic studies involve relatively small, homogenous populations of animals. However, livestock farming systems are non-homogenous, and large and more diverse datasets are required to ensure that biomarkers are robust. The aims of this study were therefore to (1) investigate the feasibility of using a large and diverse dataset for untargeted proton nuclear magnetic resonance (^{1}H NMR) serum metabolomic profiling, and (2) investigate the impact of fixed effects (farm of origin, parity and stage of lactation) on the serum metabolome of early-lactation dairy cows. First, we used multiple linear regression to correct a large spectral dataset (707 cows from 13 farms) for fixed effects prior to multivariate statistical analysis with principal component analysis (PCA). Results showed that farm of origin accounted for up to 57% of overall spectral variation, and nearly 80% of variation for some individual metabolite concentrations. Parity and week of lactation had much smaller effects on both the spectra as a whole and individual metabolites (<3% and <20%, respectively). In order to assess the effect of fixed effects on prediction accuracy and biomarker discovery, we used orthogonal partial least squares (OPLS) regression to quantify the relationship between NMR spectra and concentrations of the current gold standard serum biomarker of energy balance, β-hydroxybutyrate (BHBA). Models constructed using data from multiple farms provided reasonably robust predictions of serum BHBA concentration ($0.05 \leq \text{RMSE} \leq 0.18$). Fixed effects influenced the results biomarker discovery; however, these impacts could be controlled using the proposed method of linear regression spectral correction.

Keywords: NMR; metabotype; metabolomics; transition; ketosis; cattle; chemometrics; spectral correction

---

1. Introduction

Modern metabolomic techniques such as proton nuclear magnetic resonance (^{1}H NMR) spectroscopy allow high-throughput, synchronous characterization of the small metabolites present in biological matrices [1]. In dairy cows, the metabolome gives a snapshot of the complex interactions between host genetics, the rumen microbiome, and the environment at a given time point. ^{1}H NMR-based metabolomics therefore offers exciting opportunities to better understand and characterize the complex physiological and biochemical challenges facing cows in the transition period (defined as the three weeks before and after calving [2,3]) which is the period of greatest disease risk [4]. This in turn can facilitate identification of new molecular phenotypes (metabotypes) for genetic selection for improved animal health. These "intermediate phenotypes," so-termed because they sit between the genome and external phenotype [5], can then be integrated with genomic data to improve genomic prediction accuracies of complex traits [6,7]. The aim of metabotype identification is therefore to identify biomarkers that represent inter-animal variation free of confounding environmental factors.

Another aim of dairy cattle metabolomic studies is to identify biomarkers which enable early identification of health disorders in the transition period such as ketosis [8,9], hypocalcemia [10] and displaced abomasa [11]. Of particular interest are studies that have identified biomarkers that are predictive of transition period disorders, such as that by Hailemariam et al. [12], who identified a panel of three metabolites that could predict the occurrence of peri-parturient disease up to four weeks before calving. If robust, such predictive biomarkers would enable producers and veterinarians to implement preventive nutritional, management and/or veterinary interventions before the onset of disease.Unlike metabotype biomarkers used for genetic selection, the aim of biomarkers used for management purposes is to predict the external phenotype, and these must therefore capture all sources of phenotypic variation (i.e., host genetics, rumen microbiome, and the environment).

To date, most serum ^{1}H NMR-based metabolomic studies of livestock have involved relatively small numbers of animals, often of a single breed, and often located on a single farm. In their review, Goldansaz et al. [13] identified limited sample size and diversity as limitations of many livestock metabolomics studies and highlighted the need for larger and more diverse datasets to ensure models and biomarkers are robust. However this needs to be balanced against the need for careful experimental design to account for potential confounding from systematic environmental effects such as diet/nutritional management, parity and stage of lactation, which are known to affect the metabolic status of cows [13]. However, in order to achieve large datasets, it may be necessary to obtain samples from multiple different farms, especially when the prevalence of the condition being investigated is low (e.g., displaced abomasa). Previous studies have reported differences in the milk metabolome of animals from different geographical regions [14], farms [15], and of different breeds [16]. However, given that there is not a strong relationship between blood and milk metabolomes [17,18], these findings cannot be extrapolated to the blood serum/plasma metabolome. More information is therefore needed on the impact of systematic environmental effects on the serum metabolome of livestock.

Linear models are routinely used by quantitative geneticists to account for the influence of systematic environmental effects (also known as fixed effects) known to have significant effects on phenotypic variation [19], and thus disentangle genetic from non-genetic effects. Frequently used fixed effects include stage of lactation, parity, and herd-year-season. Similar approaches have recently been applied to metabolomic data, for example Wanichthanarak et al. [20], who used linear mixed-effects models and patient metadata to account for biological variation in metabolomics data, and Laine et al. [21], who used linear models to study the effect of pregnancy on mid-infrared spectral data derived from cows' milk.

The aim of this study was therefore to investigate the feasibility of using of large and diverse datasets in livestock metabolomics studies by examining the effects of fixed environmental and physiological effects on the ^{1}H NMR serum metabolome of clinically healthy dairy cows in early lactation. We propose a method that uses linear models to correct spectra for fixed effects and demonstrate its potential utility by quantifying the relationship between ^{1}H NMR spectra and the current gold-standard serum biomarker of energy balance, β-hydroxybutyrate (BHBA) [22,23].

2. Results

2.1. Dataset

Serum samples were collected from 707 early lactation cows (<30 d in milk) from 13 farms located in southeastern Australia. Descriptive statistics of the animals included in the experiment, including herd of origin, stage of lactation (reported as days in milk, or the number of days post-calving), parity and serum BHBA concentrations and are summarized in Table 1. Of particular interest were the BHBA results obtained from Farm 1, which had a greater mean and standard deviation than observed in other farms.

Table 1. Descriptive statistics of dataset used in this experiment, including farm details, number of cows (N), mean and standard deviation (shown in parentheses) of parity, days in milk (DIM), and serum β hydroxybutyrate (BHBA) concentrations obtained from dairy cows in the first 30 days of lactation from 13 farms in south eastern Australia.

Farm	N	Location	Parity	DIM	BHBA
1	129	Sth Gipp [1]	2.9 (1.1)	19.4 (7.2)	1.25 (0.69)
2	11	Sth Gipp	2.6 (1.2)	20.4 (8.1)	0.34 (0.12)
3	12	W Gipp [2]	2.6 (1.4)	22.8 (5.7)	0.33 (0.10)
4	11	W Gipp	3.1 (1.2)	17.9 (10.2)	0.54 (0.15)
5	18	MID [3]	2.9 (1.1)	22.6 (5.1)	0.61 (0.25)
6	248	W Gipp	2.1 (1.0)	16.7 (6.0)	0.55 (0.21)
7	9	GV [4]	2.6 (1.0)	13.9 (6.7)	0.53 (0.27)
8	24	MID	2.4 (1.2)	17.7 (8.2)	0.38 (0.09)
9	33	Sth Gipp	2.5 (1.2)	18.3 (7.2)	0.55 (0.33)
10	27	Sth Gipp	1.8 (1.1)	13.1 (7.7)	0.50 (0.14)
11	50	Tas [5]	2.6 (1.3)	18.6 (7.3)	0.42 (0.17)
12	123	MID	2.8 (1.2)	15.8 (8.6)	0.38 (0.15)
13	12	Tas	2.7 (0.8)	16.0 (7.6)	0.58 (0.22)
ALL	707	-	2.5 (1.2)	17.4 (7.3)	0.63 (0.46)

[1] South Gippsland Region, [2] West Gippsland Region, [3] Macalister Irrigation District, [4] Goulburn Valley Region, [5] Tasmania.

2.2. ^{1}H NMR Spectroscopy of Serum Samples

^{1}H NMR spectra were complex; however, more than 20 metabolites could be identified. Spectra were dominated by organic acids, amino acids, glucose and phospholipid intermediates (Figure S1 and Table S1).

2.3. Preliminary Data Analysis Using Principal Component Analysis

Preliminary data analysis and outlier identification was performed using principal component analysis (PCA). Plots of the first 2 principal components (PCs) identified several samples located outside the 95% confidence level. These spectra were manually inspected, and a single outlier with erroneous phasing was identified and removed from subsequent analyses.

PCA was repeated after outlier removal. The first 13 PCs explained greater than 90% of the variation in the spectra. Scores plots of the first three PCs, which explain 47.64%, 15.59%, and 7.45% of variation, respectively, are shown in Figure 1a–c. There was obvious clustering of samples by herd of origin. Samples from Farm 1 showed greater variation than those from the other farms. The separation between farm clusters was most obvious along PC1 and PC2. Visual comparisons based on stage of lactation (defined as weeks in milk (WIM)) and parity were also performed, but no obvious clustering or separation was observed. Loadings plots of the first three PCs show that energy metabolites BHBA, lactate, acetate and glucose, have the largest influences on spectral differences (Figure 1d–f), with smaller influences from the branched chain amino acids, lipoproteins, glycine, creatine, and betaine.

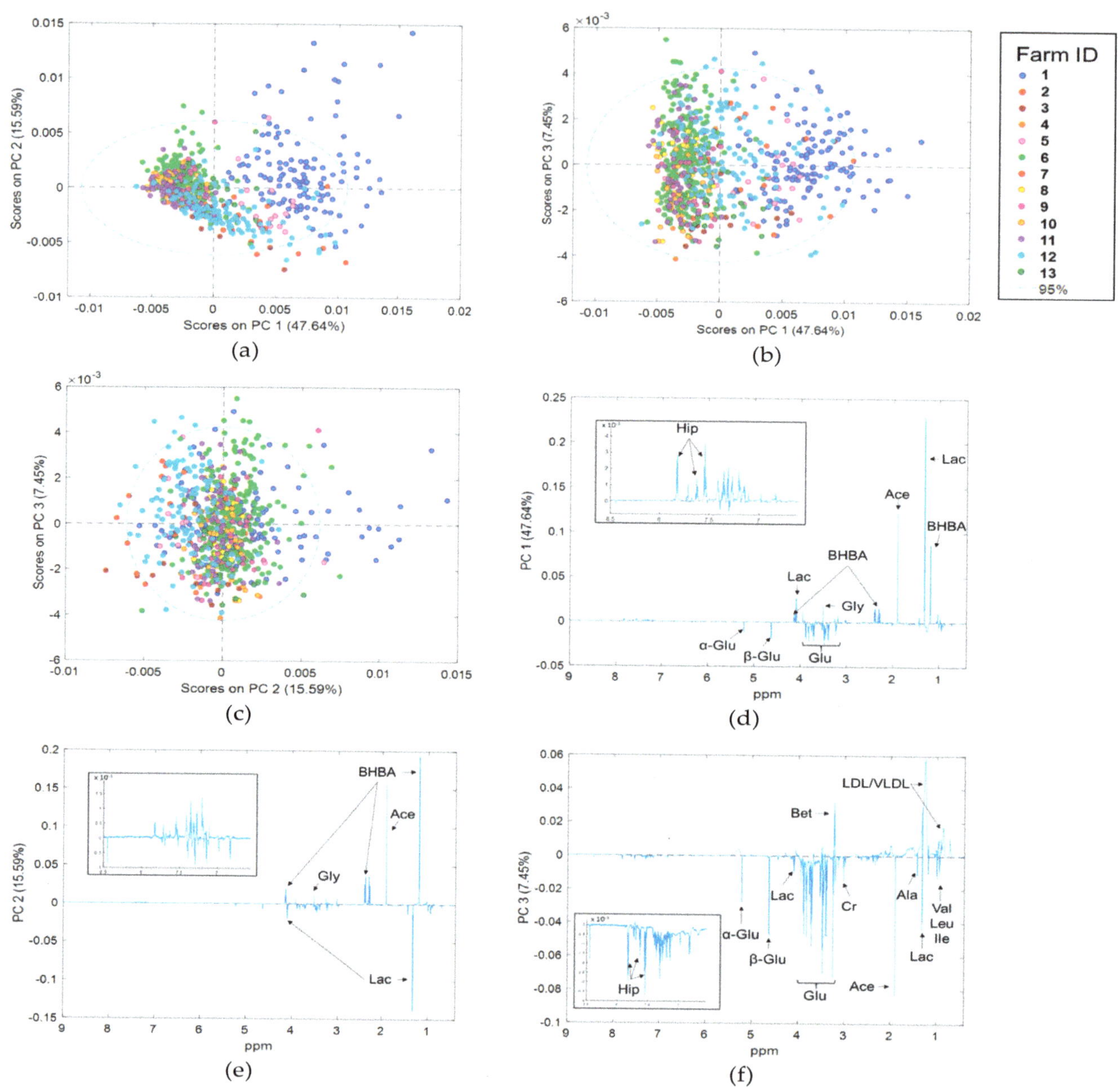

Figure 1. Results of principal component analysis (PCA) of 707 proton nuclear magnetic resonance (^{1}H NMR) spectra of serum obtained from dairy cows in early lactation; (**a**) principal component (PC) 1 vs. PC 2 scores, (**b**) PC 1 vs. PC 3 scores, (**c**) PC 2 vs. PC 3 scores, (**d**) PC 1 loadings, (**e**) PC 2 loadings, and (**f**) PC 3 loadings plots. Scores plots are colored by farm of origin. The δ 6.5 to 8.5 region of loadings plots have been magnified for clarity purposes. α-Glu = α glucose, Ace = acetate, Ala = alanine, β-Glu = β glucose, Bet = betaine, BHBA = β hydroxybutyrate, Cr = creatine, Glu = glucose, Gly = glycine, Hip = hippurate, Ile = isoleucine, Lac = lactate, Leu = leucine, Val = valine, VLDL/LDL = Very low density lipoprotein and low density lipoprotein.

2.4. *Principal Component Analysis of Spectra Corrected for Fixed Effects*

Principal component analysis (PCA) was repeated on spectra that had been corrected for (1) WIM, (2) Parity, (3) Herd, and (4) WIM, Parity and Herd simultaneously (hereafter referred to as all fixed effects) (Models 1 to 4). Results derived from spectra corrected separately for WIM and Parity are nearly identical to uncorrected spectra (Figures S2 and S3). By contrast, scores plots derived from PCA of spectra corrected for Herd (Figure S4), and spectra corrected for all fixed effects (Figure 2a–c), show no obvious clustering of samples by Herd, WIM or Parity. There is, however, still considerable separation of samples along all three PC axes, suggesting that significant inter-animal variation in the serum metabolome exists after accounting for fixed effects. Compared to the uncorrected data; (1) more PCs were required to explain >90% of spectral variation (24 vs. 13), (2) the percentage of variation captured by PC1 was lower (25.70% vs. 47.64%), and (3) the percentage of variation captured by PC2 and PC3 was higher (16.92% vs. 15.59% and 11.79% vs. 7.45%, respectively). Loadings plots are shown in Figure 2d–f. Interestingly, separation of samples along PC1 (25.70%) is due almost entirely to lactate. Loadings on PC2 and PC3 are similar to uncorrected data.

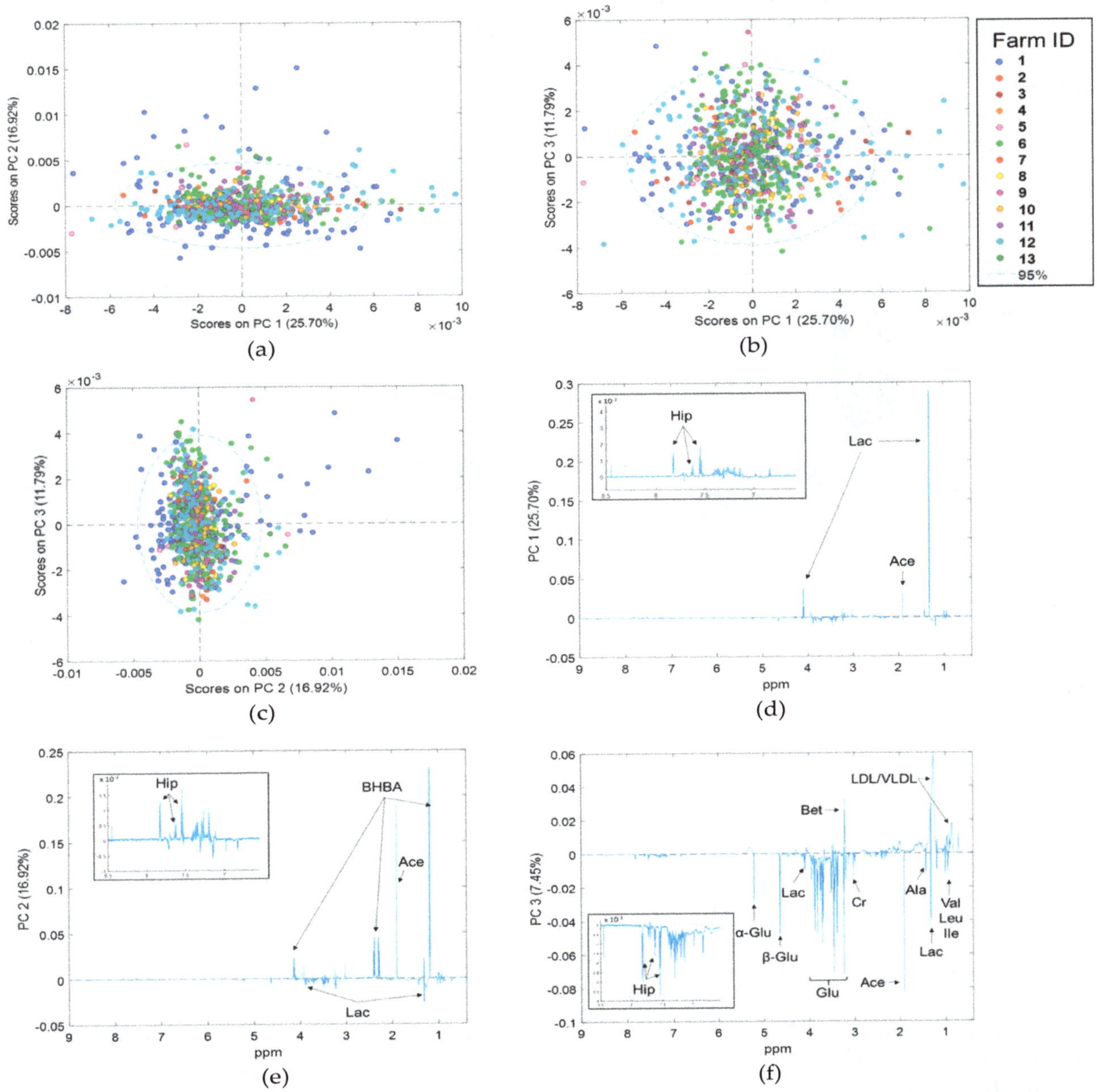

Figure 2. Results of PCA of 707 ^{1}H NMR spectra of serum, corrected for herd of origin, week of lactation, and parity obtained from dairy cows in early lactation; (**a**) PC 1 vs. PC 2 scores, (**b**) PC 1 vs. PC 3 scores, (**c**) PC 2 vs. PC 3 scores, (**d**) PC 1 loadings, (**e**) PC 2 loadings, and (**f**) PC 3 loadings plots. Scores plots are colored by farm of origin. The δ 6.5 to 8.5 region of loadings plots have been magnified for clarity purposes. α-Glu = α glucose, Ace = acetate, Ala = alanine, β-Glu = β glucose, Bet = betaine, BHBA = β hydroxybutyrate, Cr = creatine, Glu = glucose, Gly = glycine, Ile = isoleucine, Lac = lactate, Leu = leucine, Val = valine, VLDL/LDL = Very low density lipoprotein and low density lipoprotein.

2.5. *Effect of Stage of Lactation, Parity, and Herd Effects on* [1]*H NMR Spectra*

In order to quantify the effect of each fixed effect on NMR spectra, we calculated Pearson's correlations between scores for the first three PCs from the previously described PCAs (Figure 3). The largest differences (i.e., lowest correlations) were seen between uncorrected spectra, and spectra corrected for Herd (r = 0.43). This suggests that there are significant differences between those 2 spectral datasets, and that Herd, therefore, has a significant effect on the serum NMR metabolome. This is consistent with the clustering of samples by farm in the original PCA (Figure 1a–c). By comparison, the correlations between PC scores derived from uncorrected spectra, and spectra corrected for WIM and Parity, were high (0.99 and 0.97, respectively). This suggests that these spectra are nearly identical, and that these fixed effects have minimal influence on the serum metabolome. Correlations between PC2 scores were consistent with those observed between PC1 scores, and correlations between PC3 scores were all high (≥0.89).

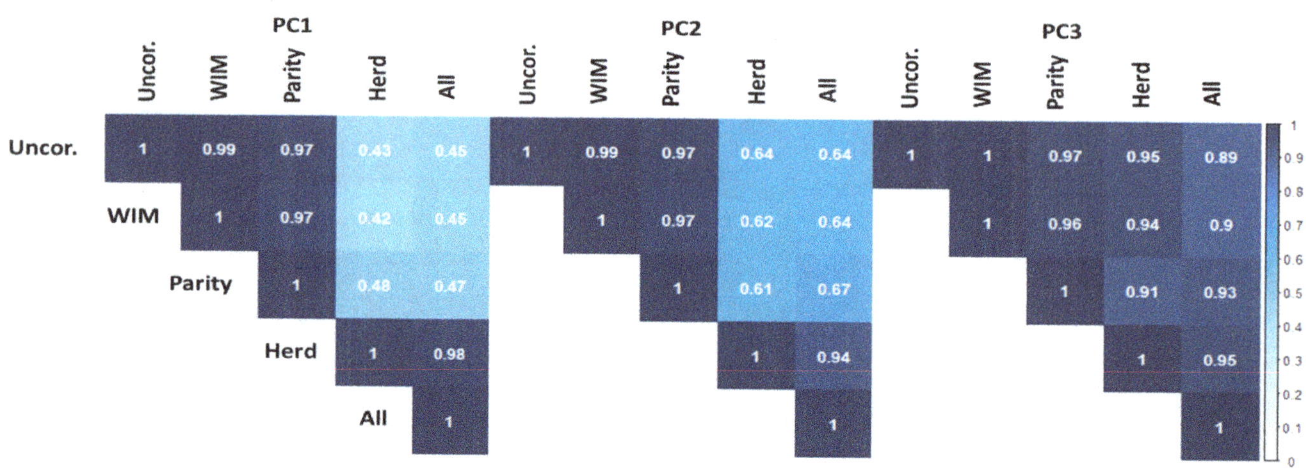

Figure 3. Pearson's correlations between scores derived from PCA of uncorrected ^{1}H NMR spectra of bovine serum, and the same spectra corrected using linear regression for week of lactation (WIM), parity, herd of origin, and WIM, parity, and herd simultaneously (All). Color map shows strength of Pearson's correlation.

To test the statistical significance of fixed effects on ^{1}H NMR spectra, we used conditional Wald F statistics derived from multiple linear regression models on the first three PC scores (Model 5). The higher the F statistic, the greater the effect of that variable on the PC score, and the lower the P value, the greater the statistical significance. Results derived from these models are summarized in Table 2. PC1 results were consistent with the results of Pearson's correlations, showing that Herd had the greatest effect. Interestingly, results for PC2 and PC3 differed slightly from Pearson's correlations. While Herd had a relatively large and significant ($P < 0.001$) impact on both PC2 and PC3, the effect of Parity was nearly as great on PC2 scores and greater on PC3 scores.

Table 2. Results of multiple linear regression models of principal component (PC) scores derived from PCA of ^{1}H NMR spectra, against weeks in milk (WIM), parity, and herd of origin. Conditional Wald F statistics (F-con) and corresponding *P* values describe the magnitude and statistical significance of each fixed effect, respectively.

	PC1 (47.64%)		PC2 (15.59%)		PC3 (7.45%)	
Fixed Effect	**F-con**	***P* Value**	**F-con**	***P* Value**	**F-con**	***P* Value**
WIM	2.66	0.047	5.42	0.001	2.14	0.094
Parity	2.78	0.041	20.39	<0.001	15.19	<0.001
Herd	158.29	<0.001	26.78	<0.001	6.66	<0.001

R^2 values obtained from models 1–4 were used to investigate which regions of the NMR spectra were most strongly influenced by the fixed effects. As the signal intensity at each chemical shift was treated as a separate response variable, the R^2 values from Models 1, 2, and 3 describe the effect of WIM, parity, and herd on each of the 24,349 chemical shifts, respectively. These R^2 values were color-coded, and overlaid on an average NMR spectrum. Plots showing the effects of WIM and Parity were unremarkable (all $R^2 < 0.2$, Figure S5), however R^2 values obtained from Model 2 showed that approximately 10–20% of the variation in glucose and acetate concentration could be explained by parity. The plot showing the effect of herd is shown in Figure 4. The strongest effect was seen in the downfield region of the spectrum, with close to 80% of variation in the concentration of some phenolic compounds being explained by Herd. Of these, hippurate could be clearly identified. Peaks at δ 7.31 and 7.39 were tentatively assigned to 3-phenyllactate, but the peak at δ 7.22 could not be identified. Lactate, acetate, BHBA, betaine, pyruvate, glycine, and glucose concentrations were also strongly influenced by herd effect.

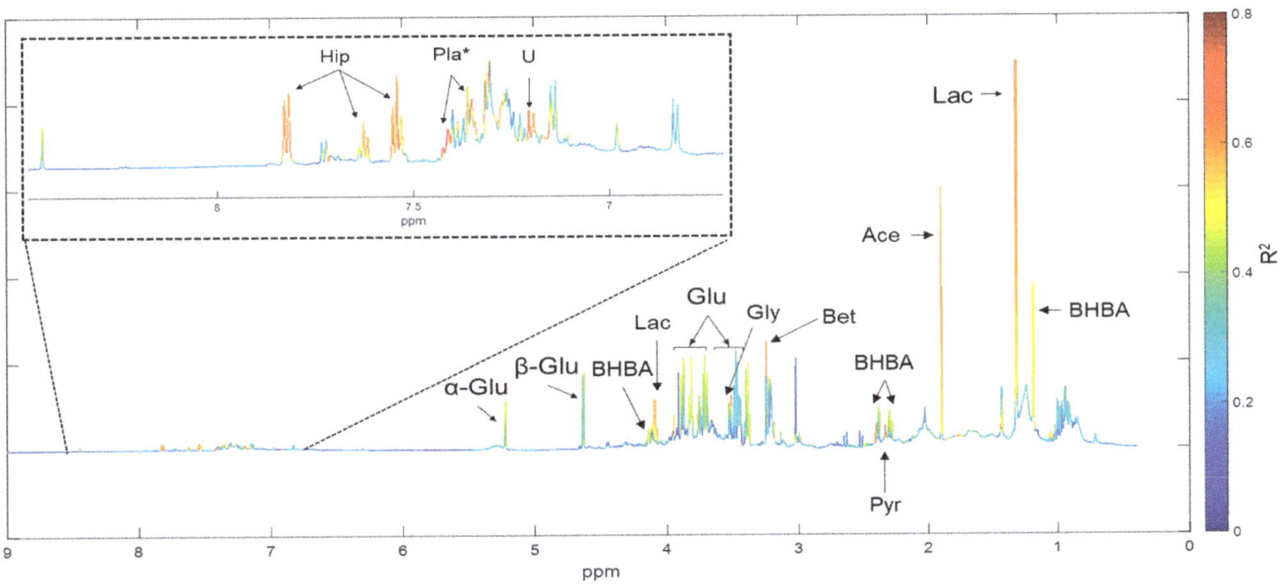

Figure 4. Average ^{1}H NMR spectrum of bovine serum. Color coding represents the percentage of variation in the signal at each chemical shift intensity that can be explained by herd of origin. The δ 6.5 to 8.5 region has been magnified for clarity purposes. Ace = acetate, Bet = betaine, BHBA = β hydroxybutyrate, Gly = glycine, Hip = hippurate, Lac = lactate, Pla = 3-phenyllactate, Pyr = pyruvate, U = unidentified peak. * indicates tentative identification.

The results of ANOVA-simultaneous component analysis (ASCA) were consistent with results of linear regression spectral correction and are shown in Table S2. Herd had the greatest effect (43.99, P = 0.02), followed by parity (4.10, P = 0.02) and WIM (1.37, P = 0.02). When ASCA was performed on corrected spectra, the effect of the fixed effect(s) was reduced to zero. For example, when ASCA was performed on spectra corrected for Herd, the effect of herd was zero (P = 1.00), but the effects of WIM (1.68, P = 0.02) and parity (3.30, P = 0.02) were retained.

2.6. *Robustness of ^{1}H NMR Predictions of Serum BHBA*

Our results show that ^{1}H NMR spectra can be used to predict serum BHBA concentration with good accuracy. This result is expected, as BHBA is directly quantifiable from NMR spectra. The overall robustness of our approach was assessed using a "leave-one-farm-out" external validation of OPLS models built using uncorrected data. This involved iteratively setting aside data from one farm, training models using data from the remaining 12 farms, then using the withheld data for external validation. R^2 results were variable ($0.30 \le R^2 \le 0.99$), however RMSE values remained relatively low (≤ 0.18) (Table 3). Interestingly, RMSE values were considerably higher when Farm 1 data were withheld for validation.

Table 3. Results of leave-one-farm out external validation of orthogonal partial least squares (OPLS) regression models predicting serum BHBA concentration from uncorrected ^{1}H NMR spectra. Validation farm specifies the identity of the data used for validation, N the number of animals used in calibration and validation datasets. The coefficient of determination (R^2) and root mean square error (RMSE) are reported for each calibration/validation subset.

			Calibration			Cross Validation		External Validation		
Validation Farm	P	LV	N	R^2	RMSE	R^2	RMSE	N	R^2	RMSE
-	<0.05	3	707	0.95	0.10	0.95	0.10	-	-	-
1	<0.05	5	578	0.87	0.08	0.85	0.08	129	0.96	0.18
2	<0.05	3	696	0.95	0.10	0.95	0.10	11	0.59	0.10
3	<0.05	4	695	0.96	0.09	0.96	0.10	12	0.78	0.06
4	<0.05	3	696	0.95	0.10	0.95	0.10	11	0.93	0.09
5	<0.05	3	689	0.96	0.10	0.95	0.10	18	0.99	0.09
6	<0.05	3	459	0.96	0.11	0.96	0.11	248	0.87	0.10
7	<0.05	3	698	0.95	0.10	0.95	0.10	9	0.98	0.05
8	<0.05	3	683	0.95	0.10	0.95	0.10	24	0.30	0.07
9	<0.05	3	674	0.95	0.10	0.95	0.10	33	0.95	0.11
10	<0.05	3	680	0.95	0.10	0.95	0.10	27	0.85	0.09
11	<0.05	3	657	0.95	0.10	0.95	0.10	50	0.82	0.08
12	<0.05	3	584	0.97	0.09	0.96	0.09	123	0.52	0.12
13	<0.05	3	695	0.95	0.10	0.95	0.10	12	0.98	0.05

2.7. *Influence of Fixed Effects on Interpretation of ^{1}H NMR Metabolomic Data*

The impact of fixed effects on the interpretation of ^{1}H NMR metabolomic data was determined by comparing the results of OPLS models built using (1) data from Farm 1 only (used as a control), (2) uncorrected data from all farms, and (3) data from all farms corrected for all fixed effects. Fixed effects appeared to have minimal effect on the predictive ability of models. We observed similar 10-fold cross validation prediction accuracies for all 3 datasets (Table 4). Interestingly, RMSE results were quite close to the results of the leave-one-farm out external validation (0.05 ≤ RMSE ≤ 0.18).

Table 4. Results obtained from 10-fold cross validation of OPLS regression models predicting serum BHBA concentration from ^{1}H NMR spectra using data from Farm 1 only, uncorrected data from all farms, and data from all farms corrected for the effect of herd. Number of cows (N), number of latent variables included in each mode (LV), coefficient of determination (R^2) and root mean square error (RMSE) of calibration (C) and 10-fold cross validation (CV) are shown.

Dataset	N	LVs	P Value [1]	R^2_C	$RMSE_C$	R^2_{CV}	$RMSE_{CV}$
Farm 1 Uncorrected	129	4	<0.001	0.98	0.10	0.97	0.12
All Data Uncorrected	707	4	<0.001	0.96	0.09	0.96	0.10
All Data Corrected for Herd	707	4	<0.001	0.93	0.09	0.93	0.09

[1] P-value derived from permutation testing (50 iterations) and pairwise Wilcoxon signed rank test.

The influences of fixed effects on biomarker discovery were investigated by comparing loadings on LV1. Results obtained using only Farm 1 data were used as a reference and show a strong positive correlation between BHBA concentration and acetate, and strong negative correlations with lactate and glucose (Figure 5a,b). Loadings from the complete dataset corrected for all fixed effects were very similar (Figure 5e,f). Results from uncorrected data, however, were quite different (Figure 5c,d), with BHBA being positively correlated with lactate and glycine. Examination of scores plots shows obvious clustering and separation by herd (especially Farm 1) when uncorrected data are used (Figure S6a), but not when corrected data are used (Figure S6b). Results from the original PCA showed that samples from Farm 1 clustered at the positive end of PC1, and that lactate and glycine both had strong positive influences on PC1 loadings. Therefore, it is possible that OPLS results are confounded by a strong herd effect when uncorrected data are used.

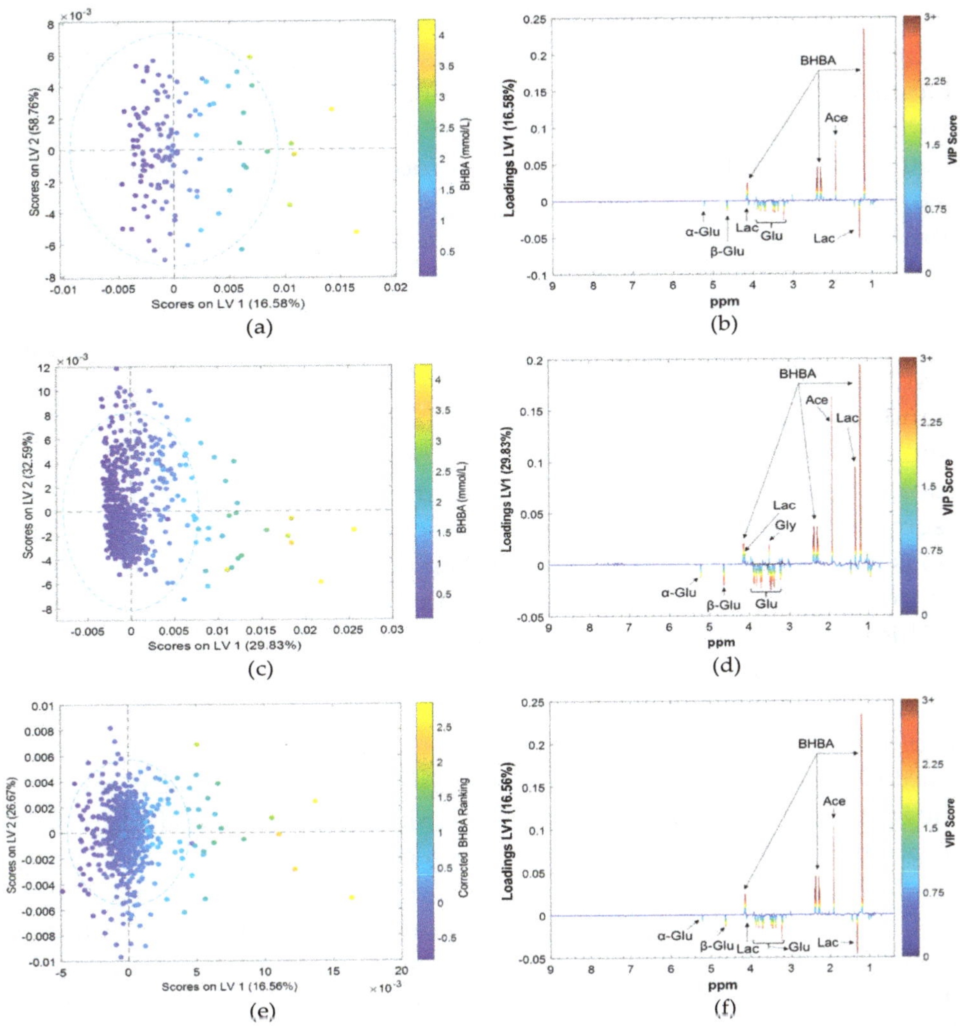

Figure 5. Results of OPLS regressions of serum BHBA concentration against ^{1}H NMR spectrum of bovine serum: (**a**) Farm 1 (N = 179) LV1 vs. LV2 scores and (**b**) LV1 loadings, (**c**) all farms (N = 707) uncorrected data LV1 vs. LV2 scores and (**d**) LV1 loadings, and (**e**) all farms data LV1 vs. LV2 scores and (**d**) LV1 loadings.

3. Discussion

To the best of the authors' knowledge, this is the first large-scale serum metabolomics study to investigate the impact of systematic environmental and physiological fixed effects on the ^{1}H NMR serum metabolome of clinically healthy dairy cattle. Our results indicate that herd-specific environmental factors have much greater effects on the serum metabolome of early lactation dairy cows than physiological factors such as WIM and parity. We demonstrate that, while confounding from herd effects can significantly influence the results of biomarker discovery, models built using data collected from multiple farms can give robust predictions of external phenotypes such as BHBA. In order to overcome the potential confounding of fixed effects on biomarker discovery, we propose a method to correct ^{1}H NMR spectra prior to multivariate analysis using multiple linear regression.

3.1. *Differences in ^{1}H NMR Spectra Between Herds*

Our results clearly demonstrate that there are significant differences in the serum metabolomes of animals from different herds. The fact that energy metabolites BHBA, lactate, acetate and glucose dominated PCA loadings (Figure 2d–f), and that herd effect accounted for a large percentage of the variation seen in lactate, acetate, pyruvate, glucose, and BHBA concentrations, (Figure 4) suggests that metabolic state, in particular energy balance, varied significantly between farms.

The importance of lactate was particularly interesting. Lactate was one of the most abundant metabolites identified in this experiment. This is very different to the findings of Sun et al. [8], who reported that lactate was one of the weakest signals in serum ^{1}H NMR spectra obtained from early-lactation cows fed a total mixed ration. One possible explanation for the very high concentrations of lactate seen in our dataset could be ruminal lactate production. During spring, dairy cows in pastoral farming systems of southeastern Australia are typically fed rations high in fermentable carbohydrate, and low in neutral detergent fiber. As a consequence, ruminal acidosis is common [24]. Serum concentrations of lactate, and in particular D-lactate from microbial fermentation, have been shown to increase following experimental induction of ruminal acidosis [25,26]. Without the use of a shift reagent and specialized experiments it is not possible to differentiate between the different lactate isomers by ^{1}H NMR [27]. We therefore plan to quantify the relative contributions of L- and D- lactate to better understand the cause of high lactate concentrations in our dataset.

The strong influence of Herd on the concentration of phenolic compounds could also be consistent with ruminal acidosis. Signal intensities in the downfield region of 2D spectra were weak, meaning clear identification of some of the phenolic peaks in our dataset was not possible. Our tentative identification of 3-phenyllactate is consistent with the findings of Yang et al. [26], who demonstrated that beef steers fed high starch (corn) diets had higher plasma concentrations of phenyllactate compared to those fed low starch diets. This study also identified L-phenyllalanine biosynthesis and metabolism as important metabolic pathways in high starch feeding. We plan to (1) enrich samples and repeat 2D analyses and (2) perform LCMS-based metabolomics on a subset of samples to identify these compounds.

Nearly 80% of the variation seen in hippurate concentration could be explained by herd effect (Figure 4). Hippurate is formed by the conjugation of glycine and benzoic acid, and has been associated with microbial degradation of dietary compounds [28]. Concentrations of hippurate increase with increased consumption of phenolic compounds [13], which are present in relatively high concentrations in pasture species. Milk hippurate concentration has been proposed as a biomarker of pasture/forage intake in goats [29], and it is possible that our results represent differences in feeding regimens between farms. Hippurate has also been proposed as a biomarker for gut microbiome diversity in humans [30], and our results may indicate differences in the gastrointestinal health of animals from different farms (i.e., ruminal acidosis). Detailed information of ration formulations is very difficult to define in grazing systems as pasture quality and intake vary considerably within and between herds. This information was therefore not available for the herds in our dataset and more data are required to further investigate this finding.

Results of the initial PCA showed that data from Farm 1 were significantly different to, and showed more variation than, data from the other farms. The reasons for these differences are hard to determine from our dataset, as Farm 1 differed in environment/management, breed and reference BHBA concentrations (and therefore it is assumed animal metabolic status). Given that we also observed clustering and separation of the 12 Holstein-Friesian herds in the initial PCA (Figure 2), it appears that herd-specific environmental factors have a larger effect on the serum metabolome than breed. However, Liao et al. [31] recently reported clear differences in the serum metabolomes (GC-MS) of three different breeds of beef steers, all the same age, fed the same ration, and managed under the same conditions. Further data are therefore required to investigate if there are differences between the serum metabolomes of different dairy breeds.

Pre-analytical sample handling and processing have been shown to have significant effects on human metabolomic data [32], and considerable efforts are made to streamline and standardize sample collection and processing protocols [33–35]. Standardizing protocols in livestock studies provides its own challenges, when relatively large number of samples are being taken at once, often in diverse, challenging and remote locations. While all attempts were made to ensure consistency, there were some unavoidable differences in the way samples from different herds were handled (for example time, between blood sample collection and centrifugation varied from approximately 2–4 h). It is therefore possible that some of the variation between farms seen in our data could be due to pre-analytical sample handling. However, overall our results suggest that metabolomic differences between animals from different farms are due largely to differences in diet/nutritional management. We plan to collect more samples from animals receiving different diets to investigate this further.

3.2. *Effect of Lactation Stage and Parity on Serum Metabolome*

Our results suggest that stage of lactation appeared to have a minimal effect on the NMR spectra. This is consistent with the findings of Ilves et al. [17] who found that the mass spectrometry (MS) based plasma metabolome of dairy cows was more heavily influenced by animal individuality than by lactation stage. By contrast, several authors report that both the NMR and MS-derived milk metabolome changes across lactation [17,36]. This suggests that blood-based metabolomics may be more suitable for identification of individual animal-specific differences within a population, and therefore provide more robust metabotypes for genetic selection.

Parity appeared to have a small but significant ($P < 0.05$) effect on the overall ^{1}H NMR serum metabolome. We could find no other reports in the literature describing the effect of parity on the entire serum metabolome. However, our results are consistent with other studies that showed parity has a significant effect on the concentration of several metabolites in serum including glucose, creatinine, urea and BHBA [37–39]. This suggests that parity should be taken into consideration when undertaking metabolomic studies in dairy cows.

3.3. *Accuracy of OPLS Models for Predicting Serum BHBA Concentration*

Despite the significant influence of fixed effects on the serum metabolome, results obtained from the leave-one-farm out external validation suggest that prediction models constructed with data from multiple farms are quite robust. R^2 values varied significantly depending on which farm was used for validation ($0.30 \leq R^2 \leq 0.99$); however, the R^2 is known to be affected by the range of the dataset, and RMSE is often considered to be a better predictor of model performance [40]. Promisingly, external validation RMSE results ($0.05 \leq R^2 \leq 0.18$) were close to those obtained from 10-fold cross validation of models built using only Farm 1 data (RMSE = 0.12) and all data (RMSE = 0.10). The fact that prediction errors were highest when Farm 1 data were withheld for validation suggests that the increased variation observed in Farm 1 data represents valuable biological variation rather than confounding/noise.

Correcting data for fixed effects had very little impact on the predictive ability of OPLS models. Furthermore, when corrected spectra were used, y-values also had to be corrected, making interpretation of phenotypic values difficult. Interestingly, Wanichthanarak, et al. [20] found that "readjusting" mass spectroscopy metabolite signals using patient metadata and linear mixed models improved the sensitivity and specificity of classification of human tissue samples with and without colorectal cancer. Conversely, Posma et al. [41] found that adjusting NMR data for confounding factors lead to a loss of predictive power for cardiovascular risk in a large-scale human NMR metabolomic dataset. Whether using NMR spectra corrected using linear regression will improve the performance of classification models (as opposed to regression against a continuous variable as used in this study) requires further investigation. Overall, our results suggest that models constructed using uncorrected data collated from multiple farms may be appropriate for prediction of external phenotypes which are influenced by both genetic and environmental factors.

3.4. *Impact of Fixed Effects on the Interpretation of Metabolomic Data for Biomarker and Metabotype Discovery*

Loadings from OPLS models built using uncorrected spectra from Farm 1, and spectra from all farms corrected for fixed effects, were consistent with the literature. BHBA and glucose concentrations have been shown to be negatively correlated in the serum of cows in early lactation dairy cows [42]. L-lactate is an important gluconeogenic substrate in dairy cows [43,44], so it follows that lactate concentration is also negatively correlated with BHBA concentration. Our results are also consistent with the findings of Sun et al. [8] who showed that cows with subclinical (1.2 < BHBA < 2.9 mmol/L) and clinical ketosis (BHBA > 2.9 mmol/L) had lower lactate and glucose concentrations and higher BHBA and acetate concentrations than the healthy controls.

The fact that loadings were different when uncorrected spectra from all farms were used demonstrates that herd-specific environmental effects can influence the results of biomarker discovery. How significant this is ultimately depends on the research question being asked. If the study aim is to identify biomarkers of external phenotypes (i.e., biomarkers that represent both genetic and environmental factors which are used for management purposes such as disease prediction), then the impact of environmental effects is important and must be captured. However, if the aim is to identify biomarkers indicative of inter-animal differences free of environmental confounding, or to understand biological processes, our results suggest that the influence of environmental effects could lead to erroneous results. This is consistent with the findings of Posma et al. [41] who showed that differences in fixed effects between subjects from the north and south of China explained some metabolite associations, which had previously been attributed to cardiovascular disease risk. This study also reported that adjusting metabolomic data for confounding using an algorithm called Covariate-Adjusted Projection to Latent Structures (CA-PLS) improved model interpretability and led to the identification of more robust biomarkers. Our results are also consistent with other studies that have explored the impacts of data pretreatments on the interpretation of metabolomics data. For example, van den Berg et al. [45] showed that pretreatment methods such as scaling, centering and transformations can greatly affect the outcome of metabolomic analyses (including the biological ranking of important metabolites) and have the potential to enhance biological interpretability. Similarly, Emwas et al. [46] concluded that the choice of spectral processing and post-processing depended on many factors including the aim of the experiment and the quality of data.

We believe that our approach has particular application in animal breeding, where the aim is to understand the biological processes that underpin economically important traits [47] and to identify metabotypes that represent inter-animal variation independent of confounding from systematic environment effects. Even with the advent of genomic selection, livestock genetic studies require relatively large numbers of animals to ensure there is adequate genetic variation in the study population [48,49]. The same is likely to be true for metabotype discovery studies. Such large datasets can be hard to compile, especially when the trait of interest is difficult and/or expensive to measure. As well as collecting data from multiple farms, another potential solution is data sharing through international collaboration. This is routinely done by geneticists; for example, de Haas et al. [50] used data from Holstein cattle in Europe, North America and Australasia to improve genomic prediction accuracies for feed intake. The ability to correct metabolomic data for factors such as experimental batch, diet, herd, year and season should allow similar collaborations in metabotype studies.

4. Materials and Methods

All procedures undertaken in this study were conducted in accordance with the Australian Code of Practice for the Care and Use of Animals for Scientific Purposes (National Health and Medical Research Council, 2013). Approval to proceed was granted by the Agricultural Research and Extension Animal Ethics Committee of the Department of Jobs, Precincts and Resources Animal Ethics Committee (DJPR, 475 Mickleham Road, Attwood, Victoria 3049, Australia), and the Tasmanian Department of Primary Industries, Parks, Water and Environment (DPIPWE Animal Biosecurity and Welfare Branch, 13 St Johns Avenue, New Town, Tasmania 7008, Australia). AEC project approval codes 2017-05 and 2018-07.

4.1. Sample Collection

A single 10 mL blood sample was taken from 708 clinical healthy cows, located on 13 farms in south-eastern Australia between September 2017 and July 2019. All cows had been calved 30 days or less at the time of sampling. Cows on all farms except Farm 1 were Australian Holstein-Friesians, while cows on Farm 1 were crossbred animals (including Holstein–Friesian, Jersey, and Australian Red breeds). All farms operated a feeding system reliant on grazed pasture plus other forages, and concentrates fed in the bail at milking time.

Blood samples were collected from the coccygeal vein into 10 mL serum clot activator vacutainer tubes (Becton Dickinson, Franklin Lakes, NJ, USA). Samples were allowed to clot at room temperature, before being centrifuged at 1000 g for 20 min at 20 °C. Sera were divided into two aliquots. The first aliquot was refrigerated at 4 °C then shipped on ice to a commercial laboratory for BHBA analysis. The second aliquot was stored at –20 °C until processing for NMR spectroscopy.

4.2. Reference BHBA Measurements

Serum BHBA concentrations were determined using a colorimetric enzymatic kinetic assay [51]. All assays were performed by Regional Laboratory Services (Benalla, Victoria, Australia) using a Kone 20 XT clinical chemistry analyzer (Thermo Fisher Scientific, Waltham, MA, USA). The uncertainty of measurement (at a 95% confidence level) was ± 0.060 mmol/L at 0.85 mmol/L.

4.3. Chemicals

Methanol (>99.9% pure) and dipotassium hydrogen phosphate (anhydrous) were purchased from Fisher Chemical (Fair Lawn, NJ, USA). Sodium 2,2-dimethyl- 2-silapentane-5-sulfonate (DSS-d6, 98%) and deuterium oxide (D_2O, 98%) were purchased from Cambridge Isotope Laboratories, Inc. (Tewksbury, MA, USA).

4.4. Sample Preparation for NMR Spectroscopy

Serum samples were thawed at room temperature for one hour and were prepared for NMR spectroscopy using a methanol protein precipitation method described by Nagana Gowda and Raftery [52]. Briefly, 300 μL of serum was mixed with 600 μL of methanol, vortexed (Ratek multi tube vortex mixer, MTV1), incubated at –20 °C for 20 min, then centrifuged to pellet proteins (11,360 g, 21 °C, 30 min). A 600 μL aliquot of supernatant was then transferred to a clean 2 mL microcentrifuge tube and dried under vacuum at 21 °C overnight using a SpeedVac Savant SPD 2010 Concentrator (Thermo Fisher Scientific, Waltham, MA, USA). Dried extracts were then reconstituted in a D_2O phosphate buffer solution (100 mM K_2HPO_4) containing 0.25 mM DSS-d6 as an internal standard. A 550 μL aliquot was transferred to 5 mm NMR tube for analysis.

4.5. ^1H NMR Data Acquisition

Routine 1D proton spectra were obtained on a Bruker Ascend 700 MHz spectrometer equipped with cryoprobe and SampleJet automatic sample changer (Bruker Biospin, Rheinstetten, Germany). A Bruker noesypr1d pulse sequence was used over –0.76 ppm to 10.32 ppm spectral range with 256 scans collected after eight dummy scans at 298K, with a total acquisition time of 2.11 seconds per increment and a relaxation delay (D1) of 2.00 seconds. The overall number of data points was 32,768. A line broadening of 0.3 Hz was applied to all spectra prior to Fourier transformation. Spectra were manually phased then baseline corrected in Topspin v.3.6.1 (Bruker Biospin, Rheinstetten, Germany). Samples were referenced to the internal standard (DSS-d6) at δ 0.00.

4.6. ^1H NMR Spectral Processing & Multivariate Statistical Analysis

NMR spectra were imported into MatLab v.R2017b (Mathworks, Natick, WA, USA) using the ProMetab v.1.1 script [53]. Each raw spectrum consisted of 31,313 data points between −0.60 and 10.00 ppm.

Statistical analyses were performed in MatLab utilizing the PLS Toolbox v. 8.5.2 (Eigenvector Research Inc., Manson, WA, USA). The spectral region containing the residual water peak (δ 4.68–5.00) was removed. Spectra were aligned using the correlation optimized warping algorithm [54] to account for chemical shift drift, then normalized to total signal area to account for inherent concentration differences between samples. After normalization, spectral regions containing methanol (δ 3.32–3.36) and DSS-d6 (δ 0.4––0.60) peaks, and the non-informative region beyond 9.00 ppm were removed. Finally, spectra were baseline corrected using automatic weighted least squares, and scaled by mean centering. After editing, a total of 24,349 chemical shift datapoints were included in subsequent statistical analyses.

For multivariate analyses, unsupervised principal component analysis (PCA) was used. Peaks of interest were identified using the Chenomx NMR suite software v.8.4 (Chenomx Inc., Edmonton, AB, Canada), comparison to the literature, and 2D NMR analysis.

4.7. Correction of ^{1}H NMR Spectra for the Effects of Systematic Environemtal and Physiological Effects

In order to investigate the effects on spectra of systematic environmental effects (also known as fixed effects) spectra were "corrected" using linear regression models. When correcting for a single categorical fixed effect, this is equivalent to scaling data using the "class centering" pre-processing step. Rather than mean centering, which involves subtracting the global mean from each variable, class centering subtracts the mean of each class. This allows investigation of intra-class variation by removing the effects of inter-class variation [55]. The advantage of using linear models rather than class centering is that the effect of multiple fixed effects or classes can be modelled simultaneously.

The approach we took was based on the principals of quantitative genetic models, where

$$\text{Phenotypic observation} = \text{environmental effects} + \text{genetic effects} + \text{residual effects} \quad (1)$$

In this study, we only want to remove the effect of environmental factors (as it is the variation in NMR spectra under genetic influence that we are interested in), so the equation can be further simplified to

$$\text{Phenotypic observation} = \text{genetic effects} + \text{residual effects} \quad (2)$$

The "corrected phenotype" (i.e., the phenotypic observation with the effects of the environmental effects removed) is defined as the residuals from the above model. For the purposes of this study each chemical shift was treated as a separate phenotype, with the signal intensity at each chemical shift being an individual phenotypic observation. The "corrected spectra" was a matrix of the residuals of each model.

A 707 × 24,349 matrix of signal intensities of pre-processed spectra was imported into the R statistical software package v 3.6.2 [56]. Each row in the matrix represented a single sample, and each column represented 1 of the 24,349 chemical shifts between δ 0.40 and δ 8.99 that made up an individual spectrum. The following 4 linear models were applied to each of the 24,349 columns in the matrix (i.e., the signal intensity at each chemical shift was treated as the response variable in a separate regression model):

$$y_{il} = \mu + WIM_i + e_{il} \text{ (Model 1)} \quad (3)$$

$$y_{jl} = \mu + P_j + ej_{jl} \text{ (Model 2)} \quad (4)$$

$$y_{kl} = \mu + H_k + e_{kl} \text{ (Model 3)} \quad (5)$$

$$y_{ijkl} = \mu + WIM_i + P_j + H_k + e_{ijkl} \text{ (Model 4)} \quad (6)$$

where y is the signal intensity at a given chemical shift, μ is the mean, WIM is weeks in milk (4 levels, defined as 1, 2, 3, or 4), P is parity (4 levels, defined as 1, 2, 3, or ≥ 4), H is the effect of herd (13 levels, with a range of 9 to 248 cows per herd), and e is the random error term. This resulted in four separate 707 × 24,349 matrices containing spectra corrected for the effects of WIM, parity, herd, and all fixed effects, respectively.

The R^2 values from each regression model were stored in a separate vector. This resulted in four vectors each containing 24,349 R^2 values; each value representing the percentage of variation in signal intensity explained by the fixed effect(s) at a given chemical shift.

4.8. Quantifying the Effect of Stage of Lactation, Parity and Herd on ^{1}H NMR Spectra

A separate PCA was performed on each of the 4 corrected spectral datasets (as described in 4.7). Scores of the first three PCs were extracted for each model, and for the PCA model constructed using uncorrected data. We then calculated Pearson's correlations between scores derived from the 5 PCAs using the corrplot package [57] in R v 3.6.2 [56]. This resulted in three correlation matrices (one for each PC). The lower the Pearson's correlation coefficient, the greater the differences between PC scores, the greater the differences between the two spectral datasets and therefore the greater the significance of the fixed effect(s).

An alternative approach to investigating the influence of fixed effects is to use multiple linear regression on PC scores from uncorrected spectra. The advantage of this approach is that all fixed effects can be fitted simultaneously, and the statistical significance of each fixed effect can be calculated. The model used was

$$y_{ijkl} = \mu + WIM_i + P_j + H_k + e_{ijkl} \text{ (Model 5)} \quad (7)$$

where y is the PC score (on either PC1, PC2, or PC3) and μ, WIM, P, H, and e are the mean, fixed effect, and error terms described previously. The statistical significance of each fixed effect was determined using conditional Wald F statistics in ASReml v 4.2 (VSN International Ltd., Hemel Hempstead, UK). Conditional F statistics are used in multiple linear regression to infer the significance of a given fixed effect assuming that the effect of remaining predictor variables have been accounted for [58].

Finally, we validated our results using the analysis of variance (ANOVA) simultaneous component analysis (ASCA) method in the PLS Toolbox [55]. ASCA is a generalization of ANOVA used to quantify the variation induced by fixed experimental design factors on complex multivariate datasets [59]. ASCA was performed on all spectral datasets (corrected and uncorrected). Statistical significance was determined using permutation testing (50 iterations).

4.9. The Relationships between ^{1}H NMR Spectra and Existing Energy Balance Biomarker Concentrations

In order to assess the utility of large and diverse datasets in livestock metabolomics studies, we used orthogonal partial least squares (OPLS) regression to compare ^{1}H NMR spectra to serum BHBA concentrations determined by colorimetric assay. The aims of this analysis were (1) to assess the robustness of OPLS models built using uncorrected data and (2) investigate the influence of systematic environmental effects on the interpretation of ^{1}H NMR spectra when used for untargeted metabolomic analyses.

4.9.1. Robustness of OPLS Models to Predict External Phenotypes Using Uncorrected Data

The robustness of OPLS models constructed using large and diverse datasets was assessed using a leave-one-farm-out external validation. This involved setting aside data from one farm, training OPLS models using data from the remaining 12 farms, then using the withheld data for external validation. This process was repeated until data from each farm was used as an external validation set once. Model performance was assessed using the R^2 and RMSE of calibration, cross validation (venetian blind CV with 10 data splits, and one sample per split), and external validation. The statistical significance of OPLS models was determined using permutation testing (cross validated, Wilcoxon test). Only uncorrected data were used for this part of the analysis.

4.9.2. Influence of Fixed Effects on Interpretation of ^{1}H NMR Metabolomic Data

To assess the impact of fixed effects on the results of untargeted metabolomic analyses we compared the results of OPLS models constructed from (1) uncorrected data from Farm 1 only

(N = 129), (2) uncorrected data from all farms, and (3) data from all farms corrected for all fixed effects (Model 4). Farm 1 data was used to simulate a more "typical" metabolomics experiment in which confounding from environmental effects is controlled through experimental design.

When corrected spectra were used, reference BHBA concentrations were corrected for the same fixed effects (Model 4). The residuals of this model represent the "corrected BHBA" concentration, which is the expected BHBA concentration of an individual accounting for differences in WIM, Parity and Herd. This poses some challenges in terms of interpretation, as negative residual values (i.e., negative BHBA concentrations) are possible. However, for the purposes of genetic evaluations, the ranking of an animal, or the relative phenotypic value, is of more interest than an absolute value. The corrected value can therefore be considered a "corrected phenotypic ranking."

The impact of fixed effects on the ability of NMR spectra to predict external phenotypes (i.e., to classify animals or predict biomarker concentrations for management purposes) was assessed by comparing the predictive ability of OPLS models. The influences of fixed effects on biomarker discovery were investigated using scores and loadings on LV1 which show the magnitude and direction of relationships between BHBA concentration and ^1H NMR spectral features. Variable importance of projection (VIP) scores were used to identify the most statistically significant spectral features in each model. Variables with VIP scores greater than one were considered significant [60].

5. Conclusions

In this study we investigated the feasibility of using large and diverse datasets for untargeted ^1H NMR serum metabolomic profiling of clinically healthy dairy cows in early lactation. In particular, we investigated the effects of systematic environmental factors on the serum metabolome. We used linear regression to correct spectra for (1) herd of origin; (2) parity; (3) WIM; and (4) herd, parity, and WIM simultaneously. Corrected and uncorrected spectra were then analyzed using PCA. Comparison of PCA results showed that herd of origin had a much greater impact on the serum metabolome than either parity or WIM. In order to simulate the impact of these effects in untargeted metabolomics, we used OPLS regression to quantify the relationship between both corrected and uncorrected NMR spectra, and the current gold-standard biomarker of energy balance in dairy cows, BHBA. Our results showed that (1) models constructed using uncorrected data from multiple farms provided reasonably robust predictions of serum BHBA concentration, (2) environmental effects can alter the results of biomarker discovery, and (3) that correcting spectra for environmental effects using linear regression may be useful when the aim of analysis is to investigate phenotypic variation free of confounding from environmental effects (e.g., identification of metabotypes for genetic selection).

Supplementary Materials:
Table S1. ^1H NMR chemical shifts (δ) and multiplicity of metabolites in bovine serum run in deuterated water (D2O).; Table S2. Results of ANOVA-simultaneous component analysis (ASCA) of uncorrected ^1H NMR spectra of bovine serum.; Figure S1. Representative 700MHz ^1H NMR spectrum (δ 0.4 to 9.0) of serum obtained from a Holstein–Friesian cow in early lactation.; Figure S2. Results of PCA of 707 ^1H NMR spectra of serum obtained from dairy cows in early lactation, corrected for weeks in milk using linear regression; (a) PC 1 vs. PC 2 scores, (b) PC 1 vs. PC 3 scores, (c) PC 2 vs. PC 3 scores, (d) PC 1 loadings, (e) PC 2 loadings, and (f) PC 3 loadings plots.; Figure S3. Results of PCA of 707 ^1H NMR spectra of serum obtained from dairy cows in early lactation, corrected for Parity using linear regression; (a) PC 1 vs. PC 2 scores, (b) PC 1 vs. PC 3 scores, (c) PC 2 vs. PC 3 scores, (d) PC 1 loadings, (e) PC 2 loadings, and (f) PC 3 loadings plots.; Figure S4. Results of PCA of 707 ^1H NMR spectra of serum obtained from dairy cows in early lactation, corrected for Herd using linear regression; (a) PC 1 vs. PC 2 scores, (b) PC 1 vs. PC 3 scores, (c) PC 2 vs. PC 3 scores, (d) PC 1 loadings, (e) PC 2 loadings, and (f) PC 3 loadings plots.; Figure S5. Average ^1H NMR spectrum of bovine serum. Color-coding represents the percentage of variation in the signal at each chemical shift intensity that can be explained by (a) WIM and (b) Parity: Figure S6: Results of OPLS regressions of serum BHBA concentration against ^1H NMR spectrum of bovine serum (n = 707): (a) LV1 vs. LV2 scores for uncorrected data (b) CV predicted vs. measured BHBA (c) LV1 vs. LV2 scores for corrected data (d) CV predicted vs. measured corrected BHBA ranking.

Author Contributions: Conceptualization, T.D.W.L., J.E.P., W.J.W., and S.J.R.; Formal analysis, T.D.W.L. and S.J.R.; Funding acquisition, J.E.P.; Investigation, T.D.W.L., A.C.E., and S.J.R.; Methodology, T.D.W.L., A.C.E., and S.J.R.; Project administration, J.E.P., W.J.W., and S.J.R.; Resources, S.J.R.; Supervision, J.E.P., W.J.W., and S.J.R.; Visualization, T.D.W.L.; Writing—original draft, T.D.W.L.; Writing—review and editing, J.E.P., A.C.E., W.J.W., and S.J.R. All authors have read and agreed to the published version of the manuscript.

Acknowledgments: The authors thank Simone Vassiliadis for assistance with ^{1}H NMR protocol and metabolite identification, Di Mapleson, Brigid Ribaux and the staff at Ellinbank Dairy Research Centre (Ellinbank, Australia) for their technical expertise and assistance, Erika Oakes and the staff at Datagene (Bundoora, Australia) for their work coordinating this study, and the farmers who took part in this project.

References

1. Wishart, D.S. Metabolomics: Applications to food science and nutrition research. *Trends Food Sci. Technol.* **2008**, *19*, 482–493. [CrossRef]
2. Drackley, J.K. Biology of dairy cows during the transition period: The final frontier? *J. Dairy Sci.* **1999**, *82*, 2259–2273. [CrossRef]
3. Grummer, R.R. Impact of changes in organic nutrient metabolism on feeding the transition dairy cow. *J. Anim. Sci.* **1995**, *73*, 2820–2833. [CrossRef]
4. LeBlanc, S.J.; Lissemore, K.D.; Kelton, D.F.; Duffield, T.F.; Leslie, K.E. Major advances in disease prevention in dairy cattle. *J. Dairy Sci.* **2006**, *89*, 1267–1279. [CrossRef]
5. Houle, D.; Govindaraju, D.; Omholt, S. Phenomics: The next challenge. *Nat. Rev. Gen.* **2010**, *11*, 855–866. [CrossRef]
6. Daetwyler, H.D.; Xiang, R.; Yuan, Z.; Bolormaa, S.; Vander Jagt, C.J.; Hayes, B.J.; van der Werf, J.H.J.; Pryce, J.E.; Chamberlain, A.J.; Macleod, I.M.; et al. Integration of functional genomics and phenomics into genomic prediction raises its accuracy in sheep and dairy cattle. In Proceedings of the Association for the Advancement of Animal Breeding and Genetics, Armidale, NSW, Australia, 27 October–1 November 2019; pp. 11–14.
7. Xiang, R.; Berg, I.V.D.; Macleod, I.M.; Hayes, B.J.; Prowse-Wilkins, C.P.; Wang, M.; Bolormaa, S.; Liu, Z.; Rochfort, S.J.; Reich, C.M.; et al. Quantifying the contribution of sequence variants with regulatory and evolutionary significance to 34 bovine complex traits. *Proc. Natl. Acad. Sci. USA* **2019**, *116*, 19398–19408. [CrossRef]
8. Sun, L.W.; Zhang, H.Y.; Wu, L.; Shu, S.; Xia, C.; Xu, C.; Zheng, J.S. 1H-Nuclear magnetic resonance-based plasma metabolic profiling of dairy cows with clinical and subclinical ketosis. *J. Dairy Sci.* **2014**, *97*, 1552–1562. [CrossRef]
9. Zhang, G.; Dervishi, E.; Dunn, S.; Mandal, R.; Liu, P.; Han, B.; Wishart, D.; Ametaj, B. Metabotyping reveals distinct metabolic alterations in ketotic cows and identifies early predictive serum biomarkers for the risk of disease. *Metabolomics* **2017**, *13*, 1–15. [CrossRef]
10. Sun, Y.; Xu, C.; Li, C.; Cheng, X.; Xu, C.; Wu, L.; Zhang, H. Characterization of the serum metabolic profile of dairy cows with milk fever using (1)H-NMR spectroscopy. *Vet. Quart.* **2014**, *34*, 1–18. [CrossRef]
11. Basoglu, A.; Baspinar, N.; Tenori, L.; Licari, C.; Gulersoy, E. Nuclear magnetic resonance (NMR)-based metabolome profile evaluation in dairy cows with and without displaced abomasum. *Vet. Quart.* **2020**, *40*, 1–15. [CrossRef]
12. Hailemariam, D.; Mandal, R.; Saleem, F.; Dunn, S.M.; Wishart, D.S.; Ametaj, B.N. Identification of predictive biomarkers of disease state in transition dairy cows. *J. Dairy Sci.* **2014**, *97*, 2680–2693. [CrossRef] [PubMed]
13. Goldansaz, S.A.; Guo, A.C.; Sajed, T.; Steele, M.A.; Plastow, G.S.; Wishart, D.S. Livestock metabolomics and the livestock metabolome: A systematic review. *PLoS ONE* **2017**, *12*, e0177675. [CrossRef] [PubMed]
14. Sacco, D.; Brescia, M.A.; Sgaramella, A.; Casiello, G.; Buccolieri, A.; Ogrinc, N.; Sacco, A. Discrimination between southern Italy and foreign milk samples using spectroscopic and analytical data. *Food Chem.* **2009**, *114*, 1559–1563. [CrossRef]
15. Tenori, L.; Santucci, C.; Meoni, G.; Morrocchi, V.; Matteucci, G.; Luchinat, C. NMR metabolomic fingerprinting distinguishes milk from different farms. *Food Res. Int.* **2018**, *113*, 131–139. [CrossRef]
16. Sundekilde, U.K.; Frederiksen, P.D.; Clausen, M.R.; Larsen, L.B.; Bertram, H.C. Relationship between the metabolite profile and technological properties of bovine milk from two dairy breeds elucidated by NMR-based metabolomics. *J. Agric. Food Chem.* **2011**, *59*, 7360–7367. [CrossRef]

17. Ilves, A.; Harzia, H.; Ling, K.; Ots, M.; Soomets, U.; Kilk, K. Alterations in milk and blood metabolomes during the first months of lactation in dairy cows. *J. Dairy Sci.* **2012**, *95*, 5788–5797. [CrossRef]
18. Maher, A.D.; Hayes, B.; Cocks, B.; Marett, L.; Wales, W.J.; Rochfort, S.J. Latent biochemical relationships in the blood-milk metabolic axis of dairy cows revealed by statistical integration of 1H NMR spectroscopic data. *J. Proteome Res.* **2013**, *12*, 1428. [CrossRef]
19. Mrode, R.A. *Linear Models for the Prediction of Animal Breeding Values*, 3rd ed.; CABI: Wallingford, UK, 2014.
20. Wanichthanarak, K.; Jeamsripong, S.; Pornputtapong, N.; Khoomrung, S. Accounting for biological variation with linear mixed-effects modelling improves the quality of clinical metabolomics data. *Comput. Struct. Biotechnol. J.* **2019**, *17*, 611–618. [CrossRef]
21. Laine, A.; Bastin, C.; Grelet, C.; Hammami, H.; Colinet, F.G.; Dale, L.M.; Gillon, A.; Vandenplas, J.; Dehareng, F.; Gengler, N. Assessing the effect of pregnancy stage on milk composition of dairy cows using mid-infrared spectra. *J. Dairy Sci.* **2017**, *100*, 2863–2876. [CrossRef]
22. Oetzel, G.R. Monitoring and testing dairy herds for metabolic disease. *Vet. Clin. North Am. Food Anim. Pr.* **2004**, *20*, 651–674. [CrossRef]
23. Ospina, P.A.; Nydam, D.V.; Stokol, T.; Overton, T.R. Associations of elevated nonesterified fatty acids and β-hydroxybutyrate concentrations with early lactation reproductive performance and milk production in transition dairy cattle in the northeastern United States. *J. Dairy Sci.* **2010**, *93*, 1596–1603. [CrossRef] [PubMed]
24. Lean, I.; Westwood, C.T.; Playford, M.C. Livestock disease threats associated with intensification of pastoral dairy farming. *N. Z. Vet. J.* **2008**, *56*, 261–269. [CrossRef] [PubMed]
25. Schwaiger, T.; Beauchemin, K.A.; Penner, G.B. Duration of time that beef cattle are fed a high-grain diet affects the recovery from a bout of ruminal acidosis: Short-chain fatty acid and lactate absorption, saliva production, and blood metabolites. *J. Anim. Sci.* **2013**, *91*, 5743–5753. [CrossRef] [PubMed]
26. Yang, Y.; Dong, G.; Wang, Z.; Wang, J.; Zhang, Z.; Liu, J. Rumen and plasma metabolomics profiling by UHPLC-QTOF/MS revealed metabolic alterations associated with a high-corn diet in beef steers. *PLoS ONE* **2018**, *13*, e0208031. [CrossRef] [PubMed]
27. Zhang, L.; Martins, A.F.; Zhao, P.; Tieu, M.; Esteban-Gómez, D.; McCandless, G.T.; Platas-Iglesias, C.; Sherry, A.D. Enantiomeric recognition of d- and l-lactate by CEST with the aid of a paramagnetic shift reagent. *J. Am. Chem. Soc.* **2017**, *139*, 17431–17437. [CrossRef] [PubMed]
28. Lees, H.J.; Swann, J.R.; Wilson, I.D.; Nicholson, J.K.; Holmes, E. Hippurate: The natural history of a mammalian–microbial cometabolite. *J. Proteome Res.* **2013**, *12*, 1527–1546. [CrossRef] [PubMed]
29. Carpio, A.; Bonilla-Valverde, D.; Arce, C.; Rodríguez-Estévez, V.; Sánchez-Rodríguez, M.; Arce, L.; Valcárcel, M. Evaluation of hippuric acid content in goat milk as a marker of feeding regimen. *J. Dairy Sci.* **2013**, *96*, 5426–5434. [CrossRef]
30. Pallister, T.; Jackson, M.A.; Martin, T.C.; Zierer, J.; Jennings, A.; Mohney, R.P.; MacGregor, A.; Steves, C.J.; Cassidy, A.; Spector, T.D.; et al. Hippurate as a metabolomic marker of gut microbiome diversity: Modulation by diet and relationship to metabolic syndrome. *Sci. Rep.* **2017**, *7*, 13670. [CrossRef]
31. Liao, Y.; Hu, R.; Wang, Z.; Peng, Q.; Dong, X.; Zhang, X.; Zou, H.; Pu, Q.; Xue, B.; Wang, L. Metabolomics profiling of serum and urine in three beef cattle breeds revealed different levels of tolerance to heat stress. *J. Agric. Food Chem.* **2018**, *66*, 6926–6935. [CrossRef]
32. Yin, P.; Lehmann, R.; Xu, G. Effects of pre-analytical processes on blood samples used in metabolomics studies. *Anal. Bioanal. Chem.* **2015**, *407*, 4879–4892. [CrossRef]
33. Beckonert, O.; Keun, H.C.; Ebbels, T.M.; Bundy, J.; Holmes, E.; Lindon, J.C.; Nicholson, J.K. Metabolic profiling, metabolomic and metabonomic procedures for NMR spectroscopy of urine, plasma, serum and tissue extracts. *Nat. Protoc.* **2007**, *2*, 2692–2703. [CrossRef] [PubMed]
34. Emwas, A.-H.; Luchinat, C.; Turano, P.; Tenori, L.; Roy, R.; Salek, R.M.; Ryan, D.; Merzaban, J.S.; Kaddurah-Daouk, R.; Zeri, A.C.; et al. Standardizing the experimental conditions for using urine in NMR-based metabolomic studies with a particular focus on diagnostic studies: A review. *Metabolomics* **2015**, *11*, 872–894. [CrossRef] [PubMed]
35. Jobard, E.; Trédan, O.; Postoly, D.; André, F.; Martin, A.-L.; Elena-Herrmann, B.; Boyault, S. A systematic evaluation of blood serum and plasma pre-analytics for metabolomics cohort studies. *Int. J. Mol. Sci.* **2016**, *17*, 2035. [CrossRef]

36. Klein, M.S.; Almstetter, M.F.; Schlamberger, G.; Nürnberger, N.; Dettmer, K.; Oefner, P.J.; Meyer, H.H.D.; Wiedemann, S.; Gronwald, W. Nuclear magnetic resonance and mass spectrometry-based milk metabolomics in dairy cows during early and late lactation. *J. Dairy Sci.* **2010**, *93*, 1539–1550. [CrossRef] [PubMed]
37. Cozzi, G.; Ravarotto, L.; Gottardo, F.; Stefani, A.L.; Contiero, B.; Moro, L.; Brscic, M.; Dalvit, P. Short communication: Reference values for blood parameters in holstein dairy cows: Effects of parity, stage of lactation, and season of production. *J. Dairy Sci.* **2011**, *94*, 3895–3901. [CrossRef] [PubMed]
38. Luke, T.D.W.; Rochfort, S.; Wales, W.J.; Bonfatti, V.; Marett, L.; Pryce, J.E. Metabolic profiling of early-lactation dairy cows using milk mid-infrared spectra. *J. Dairy Sci.* **2019**, *102*, 1747–1760. [CrossRef]
39. Morales Pineyrua, J.T.; Farina, S.R.; Mendoza, A. Effects of parity on productive, reproductive, metabolic and hormonal responses of Holstein cows. *Anim. Reprod. Sci.* **2018**, *191*, 9–21. [CrossRef]
40. Davies, A.; Fearn, T. Back to basics: Calibration statistics. *Spectrosc. Eur.* **2006**, *18*, 31–32.
41. Posma, J.M.; Garcia-Perez, I.; Ebbels, T.M.D.; Lindon, J.C.; Stamler, J.; Elliott, P.; Holmes, E.; Nicholson, J.K. Optimized phenotypic biomarker discovery and confounder elimination via covariate-adjusted projection to latent structures from metabolic spectroscopy data. *J. Proteome Res.* **2018**, *17*, 1586–1595. [CrossRef]
42. Lean, I.J.; Farver, T.B.; Troutt, H.F.; Bruss, M.L.; Galland, J.C.; Baldwin, R.L.; Holmberg, C.A.; Weaver, L.D. Time series cross-correlation analysis of postparturient relationships among serum metabolites and yield variables in Holstein cows. *J. Dairy Sci.* **1992**, *75*, 1891–1900. [CrossRef]
43. Aschenbach, J.R.; Kristensen, N.B.; Donkin, S.S.; Hammon, H.M.; Penner, G.B. Gluconeogenesis in dairy cows: The secret of making sweet milk from sour dough. *IUBMB Life* **2010**, *62*, 869–877. [CrossRef] [PubMed]
44. Drackley, J.K.; Overton, T.R.; Douglas, G.N. Adaptations of glucose and long-chain fatty acid metabolism in liver of dairy cows during the periparturient period. *J. Dairy Sci.* **2001**, *84*, E100–E112. [CrossRef]
45. Van den Berg, R.A.; Hoefsloot, H.C.J.; Westerhuis, J.A.; Smilde, A.K.; van der Werf, M.J. Centering, scaling, and transformations: Improving the biological information content of metabolomics data. *BMC Genom.* **2006**, *7*, 142. [CrossRef] [PubMed]
46. Emwas, A.-H.; Saccenti, E.; Gao, X.; McKay, R.; Santos, V.; Roy, R.; Wishart, D. Recommended strategies for spectral processing and post-processing of 1D 1 H-NMR data of biofluids with a particular focus on urine. *Off. J. Metab. Soc.* **2018**, *14*, 1–23. [CrossRef]
47. Fontanesi, L. Metabolomics and livestock genomics: Insights into a phenotyping frontier and its applications in animal breeding. *Anim. Front.* **2016**, *6*, 73–79. [CrossRef]
48. Boichard, D.; Brochard, M. New phenotypes for new breeding goals in dairy cattle. *Anim. Int. J. Anim. Biosci.* **2012**, *6*, 544–550. [CrossRef]
49. Egger-Danner, C.; Cole, J.B.; Pryce, J.E.; Gengler, N.; Heringstad, B.; Bradley, A.; Stock, K.F. Invited review: Overview of new traits and phenotyping strategies in dairy cattle with a focus on functional traits. *Anim. Int. J. Anim. Biosci.* **2015**, *9*, 191–207. [CrossRef]
50. De Haas, Y.; Pryce, J.E.; Calus, M.P.; Wall, E.; Berry, D.P.; Lovendahl, P.; Krattenmacher, N.; Miglior, F.; Weigel, K.; Spurlock, D.; et al. Genomic prediction of dry matter intake in dairy cattle from an international data set consisting of research herds in Europe, North America, and Australasia. *J. Dairy Sci.* **2015**, *98*, 6522–6534. [CrossRef]
51. McMurray, C.H.; Blanchflower, W.J.; Rice, D.A. Automated kinetic method for D-3-hydroxybutyrate in plasma or serum. *Clin. Chem.* **1984**, *30*, 421–425. [CrossRef]
52. Gowda, N.G.A.; Raftery, D. Quantitating metabolites in protein precipitated serum using NMR spectroscopy. *Anal. Chem.* **2014**, *86*, 5433–5440. [CrossRef]
53. Viant, M.R. Improved methods for the acquisition and interpretation of NMR metabolomic data. *Biochem. Biophys. Res. Commun.* **2003**, *310*, 943–948. [CrossRef] [PubMed]
54. Nielsen, N.-P.V.; Carstensen, J.M.; Smedsgaard, J. Aligning of single and multiple wavelength chromatographic profiles for chemometric data analysis using correlation optimised warping. *J. Chromatogr.* **1998**, *805*, 17–35. [CrossRef]
55. Eigenvector. *PLS Toolbox, R2017b*; Eigenvector Research Inc.: Manson, WA, USA, 2017.
56. Team, R.C. *R: A Language and Environment for Statistical Computing*; R Foundation for Statistical Computing: Vienna, Austria, 2019.
57. Wei, T.; Simko, V. R Package "Corrplot": Visualization of a Correlation Matrix, (Version 0.84). Available online: https://github.com/taiyun/corrplot (accessed on 29 April 2020).
58. Gilmour, A.R.; Gogel, B.J.; Cullis, B.R.; Thompson, R. *ASReml User Guide Release 4.1 Functional Specification, 4.1*; HP1 1ES; VSN International Ltd.: Hemel Hempstead, UK, 2015; p. 67.

59. Smilde, A.; Jansen, J.; Hoefsloot, H.; Lamers, R.-J.; van Der Greef, J.; Timmerman, M. ANOVA-simultaneous component analysis (ASCA): A new tool for analyzing designed metabolomics data. *Bioinformatics* **2005**, *21*, 3043–3048. [CrossRef]
60. Chong, I.-G.; Jun, C.-H. Performance of some variable selection methods when multicollinearity is present. *Chemom. Intellig. Lab. Syst.* **2005**, *78*, 103–112. [CrossRef]

4

Metabolomics Approach Reveals the Effects of Breed and Feed on the Composition of Chicken Eggs

Tatsuhiko Goto [1,2,*], Hiroki Mori [2], Shunsuke Shiota [3] and Shozo Tomonaga [3]

[1] Research Center for Global Agromedicine, Obihiro University of Agriculture and Veterinary Medicine, Obihiro, Hokkaido 080-8555, Japan

[2] Department of Life and Food Sciences, Obihiro University of Agriculture and Veterinary Medicine, Obihiro, Hokkaido 080-8555, Japan; s27224@st.obihiro.ac.jp

[3] Graduate School of Agriculture, Kyoto University, Kyoto 606-8502, Japan; shiota.shunsuke.68v@st.kyoto-u.ac.jp (S.S.); shozo@kais.kyoto-u.ac.jp (S.T.)

* Correspondence: tats.goto@obihiro.ac.jp.

Abstract: Chicken eggs provide essential nutrients to consumers around the world. Although both genetic and environmental factors influence the quality of eggs, it is unclear how these factors affect the egg traits including egg metabolites. In this study, we investigated breed and feed effects on 10 egg traits, using two breeds (Rhode Island Red and Australorp) and two feed conditions (mixed feed and fermented feed). We also used gas chromatography–mass spectrometry (GC–MS/MS) to analyze 138 yolk and 132 albumen metabolites. Significant breed effects were found on yolk weight, eggshell weight, eggshell colors, and one albumen metabolite (ribitol). Three yolk metabolites (erythritol, threitol, and urea) and 12 albumen metabolites (erythritol, threitol, ribitol, linoleic acid, isoleucine, dihydrouracil, 4-hydroxyphenyllactic acid, alanine, glycine, N-butyrylglycine, pyruvic acid, and valine) were significantly altered by feed, and a significant interaction between breed and feed was discovered in one albumen metabolite (N-butyrylglycine). Yolk and albumin had higher levels of sugar alcohols when hens were fed a fermented diet, which indicates that sugar alcohol content can be transferred from diet into eggs. Linoleic acid was also enriched in albumen under fermented feed conditions. This study shows that yolk and albumen metabolites will be affected by breed and feed, which is the first step towards manipulating genetic and environmental factors to create "designer eggs."

Keywords: albumen; breed; chicken; feed; metabolome; yolk

1. Introduction

Approximately 80 million tons of chicken eggs are produced every year, and they are a crucial source of animal protein in many developing countries [1,2]. Globally, 821 million people are malnourished [3], so international efforts to increase egg production are of high priority. In more developed countries, consumers are increasingly concerned with the quality of eggs and the bioavailability of favorable functional ingredients [4,5]. Increasing the quantity and quality of eggs around the world is key not only to alleviating hunger, but also to providing important dietary nutrients and keeping up with a rapidly changing international livestock industry. Currently, huge scientific effort is devoted to modifying agricultural produce traits to create "designer foods" [6]; eggs are no exception, and considerable effort is currently underway to modify egg traits and produce "designer eggs" to meet consumer demand [7,8].

Both genetic and environmental factors influence the quantity and quality of eggs [9–11]. Heritability estimates of egg traits such as overall weight, albumen weight, and yolk weight, are around 0.30–0.70 [12–14], which indicates that genetic factors are crucial to the regulation of egg traits. These general egg traits, which are mass and quality of egg components (yolk, albumen, and eggshell), are important for quality of egg itself as the product of egg market. Environmental factors including age, nutrition, stress, disease, medication, and production system also have important roles in modifying egg traits [9,10]. Thus, studies seeking to enhance egg traits should be cognizant of both genetic and environmental factors.

Egg enrichment with omega-3 polyunsaturated fatty acids (n-3 PUFA), which provide various human health benefits, is of current interest [15]. Evidence suggests that fatty acids quantity can be altered through changes to hens' diet [16]. Other research suggests that responses to dietary enrichment and conversion into egg metabolites may be breed-specific [17]. Both breed and feed appear to regulate the abundance of metabolites in egg yolk and albumen. Recently, we also have reported that egg yolk amino acid was modified by both breed and feed [18]. This is very likely to be the tip of the iceberg regarding the effects of genetic and environmental factors on egg composition.

Metabolomics, the study of the set of metabolites present in various tissues, is used to identify novel metabolites or changes in metabolite ratios in tissues [19]. Metabolomic analyses are used in a variety of fields, including biomedicine [20,21], plant science [19], food science [22,23], and ecological/environmental science [24,25]. Typical metabolite analyses use gas chromatography– mass spectrometry (GC–MS), liquid chromatography–mass spectrometry (LC–MS), capillary electrophoresis–mass spectrometry (CE–MS), and nuclear magnetic resonance (NMR). Metabolome analyses have previously been used to identify changes caused by stress and/or feed in mice using CE–MS [26] and GC–MS [27], so we expect a metabolomics approach to provide useful insight into the role of breed and feed in determining egg composition.

As studies into the metabolomes of livestock animals accumulate, a central database—the Livestock Metabolome Database—has been developed to compile this information. Research into cattle metabolic constitutes a large proportion of this work (76 of 149 articles), as does research into animal health, nutrition, and production (97 of 149 papers; [28]). For instance, effects of long-distance transportation in serum metabolites have been studied in cattle [29] and heat stress-induced metabolomic changes have been investigated in chicks [30]. Relatively few reports target non-major livestock breeds, and there is a need to populate the Livestock Metabolome Database with basic metabolome data for all livestock breeds.

In this study, we analyzed the yolk and albumen metabolome of eggs and general egg traits from two different breeds of hens under two different feed conditions. We used the GC–MS/MS technique to measure metabolome content, and tested how egg metabolites are influenced by specific genetic and environmental factors, i.e., breed and feed, respectively.

2. Results

2.1. *Egg Traits*

To test the effects of breed and feed on egg traits, 10 traits were investigated (Table 1). Two-way mixed design analysis of variance (ANOVA) revealed a significant effect of breed on yolk weight ($F_{1,12}$ = 8.098, P = 0.0147), eggshell weight ($F_{1,12}$ = 7.287, P = 0.0193), lightness of eggshell color ($F_{1,12}$ = 6.022, P = 0.0304), redness of eggshell color ($F_{1,12}$ = 10.818, P = 0.0065), and yellowness of eggshell color ($F_{1,12}$ = 14.394, P = 0.0026). Rhode Island Red (RIR) eggs showed higher yolk weight and lower eggshell weight than Australorp (AUS) eggs. Eggshell color in RIR eggs had less lightness, and more redness and yellowness than that of AUS. There were no significant effects of feed or interaction terms between breed and feed on any of the egg traits.

Table 1. Egg traits at two stages in Rhode Island Red and Australorp.

Traits	Rhode Island Red (RIR)		Australorp (AUS)		*P*-Value from ANOVA Mixed Design		
	Mixed	Fermented	Mixed	Fermented	Main-Effect		Interaction-Effect
	33 Weeks	41 Weeks	33 Weeks	41 Weeks	Breed	Feed	Breed × Feed
Egg weight (g)	54.2 ± 2.0	55.5 ± 2.5	52.6 ± 4.7	56.7 ± 4.9	0.068	0.573	0.918
Length of long axis of the egg (mm)	56.2 ± 2.4	57.8 ± 1.0	54.8 ± 2.1	57.8 ± 1.6	0.067	0.421	0.563
Length of short axis of the egg (mm)	42.0 ± 0.9	41.8 ± 1.0	41.4 ± 1.4	41.2 ± 1.2	0.065	0.974	0.953
Yolk weight (g)	15.5 ± 0.5	17.0 ± 1.2	13.4 ± 1.1	15.5 ± 1.1	0.015	0.287	0.826
Eggshell weight (g)	6.0 ± 0.4	6.7 ± 0.5	6.8 ± 0.8	6.3 ± 1.4	0.019	0.657	0.186
Albumen weight (g)	29.7 ± 1.7	29.5 ± 2.0	29.5 ± 3.0	31.1 ± 2.7	0.360	0.404	0.971
Eggshell thickness (mm)	0.38 ± 0.01	0.40 ± 0.05	0.42 ± 0.03	0.38 ± 0.03	0.227	0.725	0.667
Eggshell color L *	62.5 ± 5.0	66.3 ± 5.4	68.5 ± 2.0	75.5 ± 1.5	0.030	0.705	0.950
Eggshell color a *	14.3 ± 2.9	12.2 ± 2.6	9.5 ± 1.3	6.3 ± 0.7	0.006	0.376	0.600
Eggshell color b *	22.3 ± 3.5	20.6 ± 3.0	15.2 ± 1.8	12.5 ± 2.1	0.003	0.453	0.886

2.2. Egg Metabolite Traits

Egg metabolite traits from yolk (138 metabolites) and albumen (132 metabolites) were semi-quantified using GC−MS/MS. Full results were shown for yolk and albumen in Supplementary Tables S1 and S2, respectively. Controlling for multiple comparisons in each sample, some metabolites were found to be metabolome-wide significantly altered by breed and feed ($Q < 0.1$).

Albumen ribitol was significantly affected by breed (Table 2), with RIR eggs containing significantly higher ribitol levels than AUS eggs. Three metabolites in yolk and 12 metabolites in albumen had significant effects of feed (Table 3). Erythritol and threitol were significantly altered by feed in both the yolk and the albumen. Urea was altered in the yolk samples only, whereas isoleucine, dihydrouracil, linoleic acid, 4-hydroxyphenyllactic acid, alanine, glycine, N-butyrylglycine, pyruvic acid, ribitol, and valine were altered in albumen samples only. For threitol, erythritol, dihydrouracil, linoleic acid, pyruvic acid, and ribitol, the fermented feed group in both chicken breeds showed significantly higher metabolite content than mixed feed. On the other hand, the fermented feed group had significantly lower contents of urea, isoleucine, 4-hydroxyphenyllactic acid, alanine, glycine, N-butyrylglycine, and valine than the mixed feed group. There was a significant interaction between breed and feed on N-butyrylglycine in the albumen; N-butyrylglycine content in RIR chickens was higher with mixed feed than with fermented feed, but this effect was reversed in the AUS samples (Table 4).

Table 2. Metabolite with breed-induced changes ($Q < 0.1$).

Metabolite	HMDB [1]	Relative Area (Mean ± Standard Deviation (SD))				Mixed-Design ANOVA		
		RIR	RIR	AUS	AUS	Breed		
		Mixed	Fermented	Mixed	Fermented	*P*-Value	*Q*-Value	
Albumen_Ribitol	HMDB0000508	−0.17 ± 0.65	0.59 ± 0.49	−0.84 ± 1.27	0.42 ± 0.59	0.0005	0.0628	*

[1] The Human Metabolome Database [31]. * $Q < 0.1$.

Table 3. Metabolites with feed-induced changes ($Q < 0.1$).

Metabolite	HMDB [1]	Relative Area (Mean ± SD)				Mixed Design ANOVA		
		RIR	RIR	AUS	AUS	Feed		
		Mixed	Fermented	Mixed	Fermented	*P*-Value	*Q*-Value	
Yolk_Urea	HMDB0000294	0.47 ± 0.22	−0.78 ± 0.56	0.81 ± 1.24	−0.50 ± 0.42	0.002	0.085	*
Yolk_Threitol	HMDB0004136	−0.84 ± 0.27	0.89 ± 0.33	−1.02 ± 0.13	0.97 ± 0.37	0.001	0.082	*
Yolk_Erythritol	HMDB0002994	−0.77 ± 0.34	0.93 ± 0.45	−1.05 ± 0.09	0.89 ± 0.39	0.001	0.075	*
Albumen_Isoleucine	HMDB0000172	0.26 ± 0.77	−0.89 ± 0.63	1.20 ± 0.41	−0.57 ± 0.39	0.001	0.020	*
Albumen_Dihydrouracil	HMDB0000076	−0.57 ± 0.93	0.69 ± 0.93	−0.77 ± 0.48	0.65 ± 0.43	0.000	0.001	*
Albumen_Erythritol	HMDB0002994	−0.67 ± 0.20	1.13 ± 0.72	−1.04 ± 0.12	0.58 ± 0.30	0.001	0.017	*
Albumen_Linoleic acid	HMDB0000673	−0.02 ± 0.88	0.81 ± 1.34	−0.75 ± 0.32	−0.04 ± 0.25	0.000	0.008	*
Albumen_4-Hydroxyphenyllactic acid	HMDB0000755	0.41 ± 0.79	−1.05 ± 0.52	1.05 ± 0.43	−0.40 ± 0.57	0.007	0.091	*
Albumen_Alanine	HMDB0000161	0.54 ± 1.29	−0.75 ± 0.39	0.85 ± 0.55	−0.64 ± 0.18	0.007	0.082	*
Albumen_Glycine	HMDB0000123	0.44 ± 1.28	−0.70 ± 0.73	0.83 ± 0.42	−0.57 ± 0.25	0.004	0.066	*
Albumen_N-Butyrylglycine	HMDB0000808	0.12 ± 0.78	−0.34 ± 0.72	−0.04 ± 1.02	0.26 ± 1.24	0.000	0.007	*
Albumen_Pyruvic acid	HMDB0000243	−0.06 ± 1.36	0.19 ± 0.99	−0.17 ± 1.07	0.04 ± 0.39	0.005	0.083	*
Albumen_Ribitol	HMDB0000508	−0.17 ± 0.65	0.59 ± 0.49	−0.84 ± 1.27	0.42 ± 0.59	0.007	0.083	*
Albumen_Threitol	HMDB0004136	−0.70 ± 0.17	1.11 ± 0.74	−1.01 ± 0.14	0.59 ± 0.34	0.001	0.020	*
Albumen_Valine	HMDB0000883	0.54 ± 1.27	−0.77 ± 0.36	0.89 ± 0.49	−0.67 ± 0.10	0.006	0.088	*

[1] The Human Metabolome Database [31]. * $Q < 0.1$.

Table 4. Metabolite with breed × feed-induced changes ($Q < 0.1$).

Metabolite	HMDB [1]	Relative Area (Mean ± SD)				Mixed Design ANOVA		
		RIR	RIR	AUS	AUS	Breed × Feed		
		Mixed	Fermented	Mixed	Fermented	*P*-Value	Q-Value	
Albumen_N-Butyrylglycine	HMDB0000808	0.12 ± 0.78	−0.34 ± 0.72	−0.04 ± 1.02	0.26 ± 1.24	0.0000	0.0048	*

[1] The Human Metabolome Database [31]. * $Q < 0.1$.

3. Discussion

This study was designed to test the effects of breed and feed on 10 egg traits, 138 yolk metabolite traits, and 132 albumen metabolite traits, in RIR/AUS hens fed with either mixed feed or fermented feed. There were significant breed effects on yolk weight, eggshell weight, eggshell colors, and one albumen metabolite. Three yolk metabolites and 12 albumen metabolites were significantly altered by feed, and a significant interaction between breed and feed affected levels of one albumen metabolite. Using the metabolome technique, this study has demonstrated that certain egg properties, including metabolites in the yolk and albumen, can be changed by both genetic and environmental factors.

Since general egg traits, which are mass and quality of egg components (yolk, albumen, and eggshell), are crucial to maintain egg quality, we investigated 10 egg traits in this study. Egg traits including weight, length, width, albumen weight, and eggshell thickness showed no significant differences between RIR and AUS hens in this study, although AUS eggshell weight was higher than RIR. Average body weight at 35 weeks of age in RIR hens was more than twice that of AUS hens (3.69 and 1.58 kg in RIR and AUS, respectively). Yolk weight in RIR was significantly greater than that in AUS, corresponding to differences in body weight. We previously reported that Oh-Shamo (2.91 kg), and classical type of White Leghorn hens (1.54 kg) with average body weight at 36 weeks of age produced 53.8 ± 4.2 g and 47.4 ± 2.3 g of egg weight at 300 days of age, respectively [32,33]. Egg weight in AUS hens was 56.7 ± 4.9 g at around 300 days (41 weeks). Thus, AUS hens have a potential to produce eggs that are relatively larger than expected from their body weight [18]. On the other hand, significant breed effects were found on eggshell colors (lightness, redness, and yellowness), with RIR hens laying darker brown eggs than AUS hens. White to brown variation in eggshell color is a heritable quantitative trait [9–11], and eggshell colors of Oh-Shamo, White Leghorn, Hy-line Brown, and Onagadori chickens were reported [33–35]. Comparing this study to previous work, eggs appear to decrease in color from brown to white in the order of Hy-Line Brown > RIR > Oh-Shamo > AUS > Onagadori > White Leghorn [33–35]. Further efforts must be needed to reveal the genetic basis of eggshell coloration in chickens using population genomics and genome-wide scan.

The present metabolome analyses revealed significant changes in the amount of sugar alcohols (polyols) present in eggs. Erythritol and threitol in the yolk and albumen appear to be affected by dietary changes, specifically moving to a fermented diet. Sugar alcohols are naturally found in small quantities in fruits, vegetables, mushrooms, and fermented foods such as wine, beer, sake, and soy sauce [36], and can be produced by several yeasts and fungi [37]. The fermented feed used in this study was made with wheat, pumpkin, yam, soybean, potato, rice bran, fish meal, beet lees, scallop shell, and other materials, and lactic acid bacteria were used to ferment the mixture [18]; the amount of sugar alcohols in fermented feed would be higher than in mixed feed. Eggs produced under the fermented feed treatment had high sugar alcohol content, which is likely due to the transfer of these nutrients from feed to eggs. The same trend was also apparent for ribitol, which was found at a higher concentration in eggs produced on the fermented diet. There was also a significant breed effect, and ribitol levels in RIR eggs were higher than those of AUS eggs. These results indicate that different genetic backgrounds can affect the transfer of nutrients into eggs.

After 7 weeks on the fermented feed treatment, erythritol contents in both the yolk and albumen were significantly higher than on mixed feed. Erythritol is a four-carbon sugar alcohol (polyol) that has a sweetness 60% to 80% that of sucrose, and is used as a low-calorie sweetener [38]. This may indicate that yolk and albumen taste are affected by feed type. In humans, erythritol is fast absorbed through the small intestine, and >90% is excreted intact in urine. The remaining 10% is fermented in the large intestine by colonic microorganisms [36]. Polyols act as probiotics like a fiber, and can help normalize intestine function [36]. There are massive evidences that the intake of erythritol would not cause adverse effects in humans [38]. Artificial sweeteners and/or polyols may be helpful in diabetes and weight control, and eggs enriched with erythritol may be useful as a functional food.

Linoleic acid levels in egg albumen were significantly increased by the fermented feed diet. Linoleic acid is a long-chain polyunsaturated fatty acid (PUFA), and an essential nutrient in wide range of animals including humans. PUFA is vital for body functions and plays an important role in the formation and functioning of cell membranes, cell physiology, signaling, immunity, and reproduction [39]. Many animals lack the ability to synthesize linoleic acid de novo, so dietary intake of linoleic acid is necessary [39]. Enrichment of eggs with PUFAs, including linoleic acid, is relatively common and increases the nutritional functionality of enriched eggs [15]. Linoleic acid is found at relatively high levels in seed oils [40]. Fermented feed may derive its high linoleic acid content from its constituent plant matter, which includes pumpkin seeds, or from its relatively high proportion of crude fat. However, although isoleucine, alanine, glycine, and valine levels in the fermented feed were also higher than those of mixed feed, albumin isoleucine, alanine, glycine, and valine contents of eggs under the fermented feed diet were significantly lower than those of the mixed feed diet. More works are required to understand the changes to these amino acids levels in albumen.

There was a significant interaction between breed and feed on albumin N-butyrylglycine levels. N-butyrylglycine is an acylglycine [41], which is formed by the conjugation of acyl-CoA esters with glycine [42]. Acylglycines are used as diagnostic markers of inborn errors of metabolism [41], and previous studies have shown that acylglycine levels are sensitive in the urine of spontaneously hypertensive rats on a high-fat diet [42]. In this study, we found that acylglycine content was higher under the mixed feed treatment than fermented feed in RIR hens, but the opposite was true for AUS hens. Since N-butyrylglycine is involved in fatty acid metabolism [42], this may indicate that the combination of breed and feed can affect fatty acid metabolism. Actually, linoleic acid in albumen was significantly altered in this study. Such interaction effects of breed and feed have been documented in the enriched fatty acid levels of yolks in other studies [17]. In the modern chicken industry, hybrid chickens rather than pure breeds are often used for egg and meat production. This study indicates that the combination of breed and feed should be considered to modulate metabolite levels in yolk and albumen.

In conclusion, we found significant breed and feed effects on yolk weight, eggshell weight, eggshell colors, and several metabolites in the yolk and albumen. Feed alone had a notable impact on sugar alcohol and fatty acid levels, which were enriched in both yolk and albumen under the fermented feed treatment. We illustrated that both genetic and environmental factors are critical to determining egg composition and should be considered in the efforts to meet consumer needs and develop nutritionally functional designer eggs.

4. Materials and Methods

4.1. Study Animals

Rhode Island Red (RIR; $n = 5$) and Australorp (AUS; $n = 5$) hens were procured from the Animal Research Center of the Hokkaido Research Organization, Japan. They were introduced to the experimental farm of the Obihiro University of Agriculture and Veterinary Medicine, Japan, at 22 weeks of age. After introduction, all hens were reared in individual cages under a photoperiod cycle of 16 h light and 8 h dark, with free access to diet and water. Body weight at 35 weeks of age was 3.69 ± 0.57 and 1.58 ± 0.09 kg (mean ± standard deviation) for RIR and AUS, respectively ($F_{1,8} = 67.324$, $P = 3.6 \times 10^{-5}$). Animal management was carried out following the Guide for the Use of Experimental Animals in Universities (The Ministry of Education, Science, Sports, and Culture, Tokyo, Japan 1987) and the Standards Related to the Care and Management of Experimental Animals (Prime Ministers' Office, Tokyo, Japan, 1980). This study (authorization number 18–15) was approved from the Experimental Animal Committee of the Obihiro University of Agriculture and Veterinary Medicine.

4.2. Experimental Conditions and Sampling

To test the effects of breed and feed on egg metabolome, RIR and AUS hens were reared under two different feed conditions. Mixed feed (Rankeeper; Marubeni Nisshin Feed Co., Ltd., Tokyo, Japan) was provided for 11 weeks, from introduction until 33 weeks of age. Fermented feed (Kusanagi Farm

Limited Company, Obihiro, Japan) was provided for 9 weeks, from 34 weeks of age until the end of the experiment. The fermented feed was made from many feed materials such as potato peel, cottons and seeds of pumpkin, sake lees, and wheat, especially using a silage preparation additive, WS360 (Protocol Japan Ltd., Obihiro, Japan), which contains lactic acid bacteria and cellulolytic enzyme [18]. The ingredients of both mixed and fermented feeds were analyzed at the Institute of Chemurgy, in the Tokachi Federation of Agricultural Cooperatives, Japan (Supplementary Table S3). Eggs from each breed (RIR and AUS) were collected at two stages: at the end of the mixed feed period (33 weeks of age) and near the end of the fermented feed period (41 weeks of age, see Figure 1). Five eggs were collected from each breed at each stage, totaling 20 eggs over the course of the study.

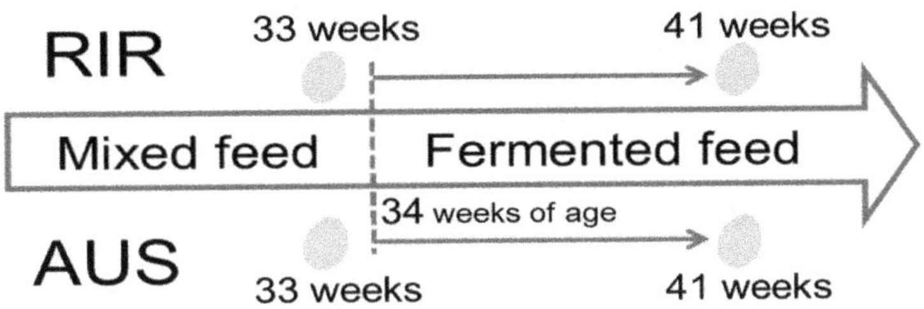

Figure 1. Experimental design. Eggs from Rhode Island Red (RIR) and Australorp (AUS) hens, which had been fed mixed feed, were collected at 33 weeks of age (mixed feed). Feed was switched at 34 weeks of age to fermented feed. Eggs were again collected at 41 weeks of age (fermented feed). Five eggs were collected at each stage from each breed; egg traits and yolk and albumen metabolite traits were measured in the 20 eggs. These data were analyzed using two-way mixed design analysis of variance (ANOVA) with breed group as the between-subjects factor and feed group as the within-subject factor.

4.3. Measuring Egg Properties

Ten egg traits were measured, including egg weight, length of the long axis, length of the short axis, eggshell weight, yolk weight, albumen weight, eggshell thickness, and eggshell lightness (L*), redness (a*), and yellowness (b*). Size was measured using a digital caliper (P01 110–120; ASONE, Japan). Eggshell colors were measured using a chromameter (CR-10 Plus Color Reader; Konica Minolta Japan, Inc., Tokyo, Japan), and eggshell thickness was measured with a Peacock dial pipe gauge P-1 (Ozaki MFG Co., Ltd., Tokyo, Japan). After weighing, yolk and albumen were kept separately at −30 °C pending further analysis.

4.4. Metabolomic Analysis of Egg Yolk and Albumen

Prior to metabolomic analysis, yolk and albumen samples were dried using a freeze dryer (FD-550P, Eyela; Tokyo Rikakikai Co. Ltd., Tokyo, Japan). Samples were crushed into a powder form, and 10 mg of each sample was used for metabolomic analysis. Powder samples were mixed with 250 μL of pretreatment liquid (methanol/H_2O/chloroform 5:2:2) and 5 μL of internal standard solution (2-isopropylmalic acid, 1 mg/mL). After vortex mixing, the sample tubes were shaken at 1200 rpm at 37 °C for 30 min under dark light conditions. After centrifugation at 16,000× *g*, 4 °C for 5 min, the supernatants (160 μL) were collected and placed in new tubes with 200 μL of H_2O and mixed using a vortex. The tubes were again centrifuged at 16,000× *g*, 4 °C for 5 min. The supernatants (200 μL) were collected into new tubes and centrifuged at room temperature in a vacuum (Centrifugal Evaporator RD400; Yamato Scientific Co. Ltd., Tokyo, Japan) for 20 min. Then, the tubes were placed in −80 °C until freezing. The tubes were centrifuged at room temperature in a vacuum for 7–8 h. Methoxyamine hydrochloride in pyridine (20 mg/mL, 40 μL) was added to each tube and mixed using the vortex. The tubes were shaken for 90 min at 1200 rpm, 30 °C, under dark conditions, for oximation. N-methyl-N-trimethylsilyltrifluoroacetamine (MSTFA; 20 μL) was added to each tube and vortex-mixed. The tubes were shaken at 1200 rpm at 37 °C for 45 min in the dark to prepare trimethylsilyl (TMS) derivatives. GC–MS/MS analyses were carried out using a GCMS-TQ8050 (Shimadzu, Kyoto, Japan) with BPX-5 column (Length 30 m; 0.25 mm I.D.; df = 0.25 μm) (SGE, Melbourne, Australia) according to the method in Smart Metabolites Database (Shimadzu, Kyoto, Japan).

Data processing was performed with Smart Metabolites Database (Shimadzu, Kyoto, Japan), MS-DIAL ver 3.08 [43] and MRMPROBS program ver. 2.42 [44]. Peaks of 40 samples (20 eggs × 2 samples) were recorded over the mass range 45–600 m/z. Peaks were automatically detected via MS-DIAL with peak detection options that minimum peak height is 2000. A data quality check was conducted using thresholds, which are $-10 < \mathrm{RI} < 10$, dot prod > 0.8, and presence > 0.6 and then manually checked. Finally, 138 and 132 metabolites for yolk and albumen were identified, respectively. Relative quantity of metabolites was calculated using the peak area of each metabolite relative to an internal standard (2-isopropylmalic acid).

4.5. Statistical Methods

To identify the effects of breed and feed on egg properties, data were analyzed using a two-way mixed-design analysis of variance (ANOVA) with breed group (RIR and AUS) as the between-subjects factor and feed group (mixed feed and fermented feed) as the within-subject (repeated) factor [18,45–47]. Main- and interaction-effects of breed and feed on egg properties were tested ($P < 0.05$). Data are presented as the mean and standard deviation. Statistical analyses were conducted using R [48]. In metabolites in particular, individual values were standardized with their mean and standard deviation.

To control the P value for multiple comparisons in metabolites, the false discovery rate was determined using the approach of Benjamini and Hochberg [49]. The P values were adjusted using the Benjamini–Hochberg correction, and post-adjustment are referred to as Q values. The metabolome-wide significance threshold was set at $Q < 0.1$, following previous studies [26,27,50,51].

Author Contributions: Conceptualization, T.G. and S.T.; methodology, T.G. and S.T.; software, S.S. and S.T.; validation, T.G., H.M., S.S. and S.T.; formal analysis, T.G. and H.M.; investigation, T.G., H.M., S.S. and S.T.; resources, T.G. and S.T.; data curation, T.G.; writing—original draft preparation, T.G.; writing—review and editing, T.G. and S.T.; visualization, T.G.; supervision, T.G.; project administration, T.G.; funding acquisition, T.G.

Acknowledgments: We thank Masaaki Hanada and Masafumi Tetsuka for the use of laboratory equipments, and all members of the Animal Breeding Research Group at the Obihiro University of Agriculture and Veterinary Medicine for their continuous support.

References

1. FAOSTAT. 2019. Available online: http://www.fao.org/faostat/en/#home (accessed on 20 September 2019).
2. Goto, T.; Fernandes, A.F.A.; Tsudzuki, M.; Rosa, G.J.M. Causal phenotypic networks for egg traits in an F_2 chicken population. *Mol. Genet. Genom.* **2019**, 1–8. [CrossRef] [PubMed]
3. Hunger Map. 2018. Available online: https://www.wfp.org/content/2018-hunger-map (accessed on 20 September 2019).
4. Kralik, G.; Kralik, Z. Poultry products enriched with nutricines have beneficial effects on human health. *Med. Glas.* **2017**, *14*, 1–7. [CrossRef]
5. Laudodio, V.; Lorusso, V.; Lastella, N.M.B.; Dhama, K.; Karthik, K.; Tiwari, R.; Alam, G.M.; Tufarelli, V. Enhancement of nutraceutical value of table eggs through poultry feeding strategies. *Int. J. Pharm.* **2015**, *11*, 201–212. [CrossRef]
6. Rajasekaran, A.; Kalaivani, M. Designer foods and their benefits: A review. *J. Food Sci. Technol.* **2013**, *50*, 1–16. [CrossRef]
7. Zaheer, K. An updated review on chicken eggs: Production, consumption, management aspects and nutritional benefits to human health. *Food Nutr. Sci.* **2015**, *6*, 1208–1220. [CrossRef]

8. Alagawany, M.; Farag, M.R.; Dhama, K.; Patra, A. Nutritional significance and health benefits of designer eggs. *Worlds Poult. Sci. J.* **2018**, *74*, 317–330. [CrossRef]
9. Roberts, J. Factors affecting egg internal quality and egg shell quality in laying hens. *J. Poult. Sci.* **2004**, *41*, 161–177. [CrossRef]
10. Wilson, P.B. Recent advances in avian egg science: A review. *Poult. Sci.* **2017**, *96*, 3747–3754. [CrossRef] [PubMed]
11. Goto, T.; Tsudzuki, M. Genetic mapping of quantitative trait loci for egg production and egg quality traits in chickens: A review. *J. Poult. Sci.* **2017**, *54*, 0160121. [CrossRef]
12. Wolc, A.; White, I.M.S.; Hill, W.G.; Olori, V.E. Inheritance of hatchability in broiler chickens and its relationship to egg quality traits. *Poult. Sci.* **2010**, *89*, 2334–2340. [CrossRef]
13. Wolc, A.; Arango, J.; Settar, P.; O'Sullivan, N.P.; Olori, V.E.; White, I.M.S.; Hill, W.G.; Dekkers, J.C.M. Genetic parameters of egg defects and egg quality in layer chickens. *Poult. Sci.* **2012**, *91*, 1292–1298. [CrossRef] [PubMed]
14. Zhang, L.C.; Ning, Z.H.; Xu, G.Y.; Hou, Z.C.; Yang, N. Heritabilities and genetic and phenotypic correlations of egg quality traits in brown-egg dwarf layers. *Poult. Sci.* **2005**, *84*, 1209–1213. [CrossRef] [PubMed]
15. Fraeye, I.; Bruneel, C.; Lemahieu, C.; Buyse, J.; Muylaert, K.; Foubert, I. Dietary enrichment of eggs with omega-3 fatty acids: A review. *Food Res. Int.* **2012**, *48*, 961–969. [CrossRef]
16. Surai, P.F.; Sparks, N.H.C. Designer eggs: From improvement of egg composition to functional food. Trends. *Food Sci. Technol.* **2001**, *12*, 7–16. [CrossRef]
17. Yin, J.D.; Shang, X.G.; Li, D.F.; Wang, F.L.; Guan, Y.F.; Wang, Z.Y. Effects of dietary conjugated linoleic acid on the fatty acid profile and cholesterol content of egg yolks from different breeds of layers. *Poult. Sci.* **2008**, *87*, 284–290. [CrossRef] [PubMed]
18. Mori, H.; Takaya, M.; Nishimura, K.; Goto, T. Breed and feed affect amino acid contents of egg yolk and eggshell color in chickens. *Poult. Sci.* **2019**. [CrossRef]
19. Weckwerth, W. Metabolomics in systems biology. *Annu. Rev. Plant Biol.* **2003**, *54*, 669–689. [CrossRef] [PubMed]
20. Griffiths, W.J.; Koal, T.; Wang, Y.; Kohl, M.; Enot, D.P.; Deigner, H.P. Targeted metabolomics for biomarker discovery. *Angew. Chem. Int. Ed. Engl.* **2010**, *26*, 5426–5445. [CrossRef]
21. Li, B.; He, X.; Jia, W.; Li, H. Novel applications of metabolomics in personalized medicine: A mini-review. *Molecules* **2017**, *22*, 1173. [CrossRef]
22. Wishart, D.S. Metabolomics: Applications to food science and nutrition research. *Trends Food Sci. Technol.* **2008**, *19*, 482–493. [CrossRef]
23. Cubero-Leon, E.; Peñalver, R.; Maquet, A. Review on metabolomics for food authentication. *Food Res. Int.* **2014**, *60*, 95–107. [CrossRef]
24. Bundy, J.G.; Davey, M.P.; Viant, M.R. Environmental metabolomics: A critical review and future perspectives. *Metabolomics* **2009**, *5*, 3. [CrossRef]
25. Macel, M.; Van Dam, N.M.; Keurentjes, J.J. Metabolomics: The chemistry between ecology and genetics. *Mol. Ecol. Resour.* **2010**, *10*, 583–593. [CrossRef] [PubMed]
26. Goto, T.; Kubota, Y.; Toyoda, A. Plasma and liver metabolic profiles in mice subjected to subchronic and mild social defeat stress. *J. Proteome Res.* **2015**, *14*, 1025–1032. [CrossRef] [PubMed]
27. Goto, T.; Tomonaga, S.; Toyoda, A. Effects of diet quality and psychosocial stress on the metabolic profiles of mice. *J. Proteome Res.* **2017**, *16*, 1857–1867. [CrossRef] [PubMed]
28. Goldansaz, S.A.; Guo, A.C.; Sajed, T.; Steele, M.A.; Plastow, G.S.; Wishart, D.S. Livestock metabolomics and the livestock metabolome: A systematic review. *PLoS ONE* **2017**, *12*, e0177675. [CrossRef] [PubMed]
29. Takemoto, S.; Tomonaga, S.; Funaba, M.; Matsui, T. Effect of long-distance transportation on serum metabolic profiles of steer calves. *Anim. Sci. J.* **2017**, *88*, 1970–1978. [CrossRef]
30. Tomonaga, S.; Okuyama, H.; Tachibana, T.; Makino, R. Effects of high ambient temperature on plasma metabolomic profiles in chicks. *Anim. Sci. J.* **2018**, *89*, 448–455. [CrossRef]
31. Wishart, D.S.; Feunang, Y.D.; Marcu, A.; Guo, A.C.; Liang, K.; Vázquez-Fresno, R.; Sajed, T.; Johnson, D.; Li, C.; Karu, N.; et al. HMDB 4.0: The human metabolome database for 2018. *Nucleic Acid Res.* **2017**, *46*, D608–D617. [CrossRef]
32. Goto, T.; Ishikawa, A.; Nishibori, M.; Tsudzuki, M. A longitudinal quantitative trait locus mapping of chicken growth traits. *Mol. Genet. Genom.* **2019**, *294*, 243–252. [CrossRef]

33. Goto, T.; Ishikawa, A.; Yoshida, M.; Goto, N.; Umino, T.; Nishibori, M.; Tsudzuki, M. Quantitative trait loci mapping for external egg traits in F_2 chickens. *J. Poult. Sci.* **2014**, *51*, 118–129. [CrossRef]
34. Goto, T.; Shiraishi, J.I.; Bungo, T.; Tsudzuki, M. Characteristics of egg-related traits in the Onagadori (Japanese Extremely Long Tail) breed of chickens. *J. Poult. Sci.* **2015**, *52*, 81–87. [CrossRef]
35. Sirri, F.; Zampiga, M.; Berardinelli, A.; Meluzzi, A. Variability and interaction of some egg physical and eggshell quality attributes during the entire laying hen cycle. *Poult. Sci.* **2018**, *97*, 1818–1823. [CrossRef] [PubMed]
36. Grembecka, M. Sugar alcohols—their role in the modern world of sweeteners: A review. *Eur. Food Res. Technol.* **2015**, *241*, 1–14. [CrossRef]
37. Wisselink, H.W.; Weusthuis, R.A.; Eggink, G.; Hugenholtz, J.; Grobben, G. Mannitol production by lactic acid bacteria: A review. *Int. Dairy J.* **2002**, *12*, 151–161. [CrossRef]
38. Bernt, W.O.; Borzelleca, J.F.; Flamm, G.; Munro, I.C. Erythritol: A review of biological and toxicological studies. *Regul. Toxicol. Pharm.* **1996**, *24*, S191–S197. [CrossRef] [PubMed]
39. Malcicka, M.; Visser, B.; Ellers, J. An evolutionary perspective on linoleic acid synthesis in animals. *Evol. Biol.* **2018**, *45*, 15–26. [CrossRef] [PubMed]
40. Abedi, E.; Sahari, M.A. Long-chain polyunsaturated fatty acid sources and evaluation of their nutritional and functional properties. *Food Sci. Nutr.* **2014**, *2*, 443–463. [CrossRef] [PubMed]
41. Stanislaus, A.; Guo, K.; Li, L. Development of an isotope labeling ultra-high performance liquid chromatography mass spectrometric method for quantification of acylglycines in human urine. *Anal. Chim. Acta* **2012**, *750*, 161–172. [CrossRef]
42. Čermáková, M.; Pelantová, H.; Neprašová, B.; Šedivá, B.; Maletínská, L.; Kuneš, J.; Tomášová, P.; Železná, B.; Kuzma, M. Metabolomic study of obesity and its treatment with palmitoylated prolactin-releasing peptide analog in spontaneously hypertensive and normotensive rats. *J. Proteome Res.* **2019**, *18*, 1735–1750. [CrossRef]
43. Lai, Z.; Tsugawa, H.; Wohlgemuth, G.; Mehta, S.; Mueller, M.; Zheng, Y.; Ogiwara, A.; Meissen, J.; Showalter, M.; Takeuchi, K.; et al. Identifying metabolites by integrating metabolome databases with mass spectrometry cheminformatics. *Nat. Methods* **2018**, *15*, 53–56. [CrossRef] [PubMed]
44. Tsugawa, H.; Kanazawa, M.; Ogiwara, A.; Arita, M. MRMPROBS suite for metabolomics using large-scale MRM assays. *Bioinformatics* **2014**, *30*, 2379–2380. [CrossRef] [PubMed]
45. Olejnik, S.; Algina, J. Generalized eta and omega squared statistics: Measures of effect size for some common research designs. *Psychol. Methods* **2003**, *4*, 434–447. [CrossRef] [PubMed]
46. Franz, V.H.; Loftus, G.R. Standard errors and confidence intervals in within-subjects designs: Generalizing Loftus and Masson (1994) and avoiding the biases of alternative accounts. *Psychon. Bull. Rev.* **2012**, *19*, 395–404. [CrossRef] [PubMed]
47. Nikiforuk, A.; Potasiewicz, A.; Kos, T.; Popik, P. The combination of memantine and galantamine improves cognition in rats: The synergistic role of the α7 nicotinic acetylcholine and NMDA receptors. *Behav. Brain Res.* **2016**, *313*, 214–218. [CrossRef] [PubMed]
48. R Core Team. *R: A language and Environment for Statistical Computing*; R Foundation for Statistical Computing: Vienna, Austria, 2019; Available online: https://www.R-project.org/ (accessed on 20 September 2019).
49. Benjamini, Y.; Hochberg, Y. Controlling the false discovery rate: A practical and powerful approach to multiple testing. *J. R. Stat. Soc. Ser. B* **1995**, *57*, 298–300. [CrossRef]
50. McClay, J.L.; Adkins, D.E.; Vunck, S.A.; Batman, A.M.; Vann, R.E.; Clark, S.L.; Beardsley, P.M.; van den Oord, E.J. Large-scale neurochemical metabolomics analysis identifies multiple compounds associated with methamphetamine exposure. *Metabolomics* **2013**, *9*, 392–402. [CrossRef]
51. Pedersen, H.K.; Gudmundsdottir, V.; Nielsen, H.B.; Hyotylainen, T.; Nielsen, T.; Jensen, B.A.; Forslund, K.; Hildebrand, F.; Prifti, E.; Falony, G.; et al. Human gut microbes impact host serum metabolome and insulin sensitivity. *Nature* **2016**, *535*, 376–381. [CrossRef] [PubMed]

5

MEATabolomics: Muscle and Meat Metabolomics in Domestic Animals

Susumu Muroya [1,*], Shuji Ueda [2], Tomohiko Komatsu [3], Takuya Miyakawa [4] and Per Ertbjerg [5]

1 NARO Institute of Livestock and Grassland Science, Tsukuba, Ibaraki 305-0901, Japan
2 Graduate School of Agricultural Science, Kobe University, Hyogo 657-8501, Japan; uedas@people.kobe-u.ac.jp
3 Livestock Research Institute of Yamagata Integrated Research Center, Shinjo, Yamagata 996-0041, Japan; komatsutom@pref.yamagata.jp
4 Graduate School of Agricultural and Life Sciences, University of Tokyo, Bunkyo-ku, Tokyo 113-8657, Japan; atmiya@mail.ecc.u-tokyo.ac.jp
5 Department of Food and Nutrition, University of Helsinki, 00014 Helsinki, Finland; per.ertbjerg@helsinki.fi
* Correspondence: muros@affrc.go.jp.

Abstract: In the past decades, metabolomics has been used to comprehensively understand a variety of food materials for improvement and assessment of food quality. Farm animal skeletal muscles and meat are one of the major targets of metabolomics for the characterization of meat and the exploration of biomarkers in the production system. For identification of potential biomarkers to control meat quality, studies of animal muscles and meat with metabolomics (MEATabolomics) has been conducted in combination with analyses of meat quality traits, focusing on specific factors associated with animal genetic background and sensory scores, or conditions in feeding system and treatments of meat in the processes such as postmortem storage, processing, and hygiene control. Currently, most of MEATabolomics approaches combine separation techniques (gas or liquid chromatography, and capillary electrophoresis)–mass spectrometry (MS) or nuclear magnetic resonance (NMR) approaches with the downstream multivariate analyses, depending on the polarity and/or hydrophobicity of the targeted metabolites. Studies employing these approaches provide useful information to monitor meat quality traits efficiently and to understand the genetic background and production system of animals behind the meat quality. MEATabolomics is expected to improve the knowledge and methodologies in animal breeding and feeding, meat storage and processing, and prediction of meat quality.

Keywords: authentication; biomarker; breed; feeding; meat quality traits; metabolite; postmortem aging; processing; skeletal muscle

---

1. Introduction

In the last two decades, the metabolomic approach has been employed widely in the research fields of animal and plant nutrition, physiology, breeds, environment, post-harvest storage and processing [1,2] with the advantages of its high-throughput capacity. The target molecules of food metabolomics are animal and plant metabolites, including low molecular-weight hydrophilic and hydrophobic compounds. These compounds include key metabolites such as flavor-associated compounds, nutrients, and functionality-associated compounds in food, some of which have a molecular weight of >1000. Such compounds characterize nutritional and sensory properties, and therefore the metabolome of a food provides important phenotypic information. In this context, metabolomics is a powerful tool to obtain a deeper understanding of the biologically and agriculturally meaningful information in the global metabolome profiles and the changes caused by factors in food production processes [3].

Currently, two major types of platforms have been applied for metabolomic studies so far: mass spectrometry (MS)-based [4] and non-MS-based techniques such as nuclear magnetic resonance (NMR) [2]. Moreover, various types of separation techniques are incorporated in most of the MS-based approaches, depending on the lipophilicity and polarity of the metabolites of interest. Combined with advanced statistical analyses, multivariate analyses, and bioinformatic databases, metabolomics provides clues not only to discover biomarkers for monitoring and assessment of food quality but also to uncover molecular pathways in which enzymatic reaction generates key metabolites [5]. Correlation analysis between sensory evaluation scores and metabolomic profiles potentially leads to key compounds that are associated with eating quality, such as flavor and texture [6–8], which may enable us to predict the palatability of the foods by using the biomarker metabolites in pre- and post-harvest materials. Metabolomic profile data have also been utilized to explore the responsible genes for specific metabolite-featured phenotypes in genome-wide association studies (GWAS) [9,10].

In recent years, studies in muscle and meat science have utilized metabolomics as in other fields [11–13]. Skeletal muscle characteristics are designed by a functionally cooperative set of genes specific to the spatiotemporal requirement in each muscle. The gene expression is further modulated at levels of transcription, post-transcription, translation, and protein modification during development, growth, and maturation stages of the muscle. Accordingly, muscle metabolites determine the physiological muscle characteristics and meat quality traits as the major phenotypic components. The history of the muscle in the developmental and physiological specialization, feeding processes of animals, circumstances in slaughtering, postmortem aging, and processing processes all have influences on transcriptomic and/or proteomic profiles, and finally on the muscle metabolome profile. This makes the understanding of mechanisms behind meat quality traits by the use of molecular markers more difficult. Nevertheless, the metabolome profiles have been successfully distinguished with metabolites of biomarker candidates in comparisons such as between physiologically different types of skeletal muscles [14–16], between at-slaughter and postmortem muscles [14,15,17–19], between meats produced by different animal feeding conditions [20–26], and between different types of meat processing [7,27–30]. Thus, to better understand factors determining muscle characteristics and meat quality, the recent metabolomics approach has great advantages, which potentially supports to capture biomarkers of the phenotype. Since raw and cooked meat are rich in flavor-associated volatile compounds and precursors, MEATabolomics studies in combination with sensory evaluation have been conducted to explore biomarker candidates associated with eating quality of meat.

Especially in studies of postmortem muscle and meat quality, MEATabolomics is expected to provide a clue for drawing maps of the metabolic network in postmortem muscle aging and flavor development during cooking, which cannot be replaced by the other biological or chemical approaches. Muscles, especially beef and pork, are more or less aged after death for the tenderization and flavor component accumulation. During the aging, the muscles experience drastic and irreversible physico-chemical and metabolic changes [31]. These metabolic changes have a large impact on subsequent beef and pork quality, under influences of physiological background before slaughter. Despite the impact on meat quality, the metabolic changes have not easily predicted, because coordinated metabolism in live muscle is no longer maintained due to lack of energy supply and arrest of de novo gene expression after animal death. Postmortem metabolic maps drawn by MEATabolomics could lead to a finding of the metabolic factors responsible for meat quality traits and thereby contribute to the exploration of biomarkers in quality monitoring, processing, and authentication of meat including chicken. Thus, postmortem muscle metabolism and the factors associated with meat quality have been the major and unique challenges to be addressed for MEATabolomics.

This review provides an overview of past applications, recent findings, and topics achieved by MS- and NMR-based MEATabolomics approaches associated with animal muscle physiology and meat quality traits. We especially focus on studies of meat quality traits and the factors influencing meat quality, such as postmortem aging, as well as animal breed and feeding, processing, spoilage, and authentication. Using those keywords, a total of 78 studies were collected by search in databases such as PubMed, ScienceDirect, and Springer Link, from which 56 studies were selected and featured.

2. MEATabolomic Methodologies and Approaches

The MS-based metabolomics need a separation step suitable for target metabolites of interest (Table 1). Of the current separation techniques, capillary electrophoresis (CE)–MS has high performance in the acquisition of polar and charged metabolite information with high resolution and sensitivity, but not in non-ionic molecules [32,33]. CE has an advantage over gas chromatography (GC) and high-performance liquid chromatography (HPLC; LC) for the resolution of ionic compounds, including their isomers, owing to separation in CE by their charge-to-mass ratio. GC–MS has been widely used for decades due to its established high-separation efficiency, selective and sensitive mass detection, and a broad range of target molecules, mainly fatty acids and sugars, although it needs derivatization of sample molecules, especially for non-volatile compounds. Derivatization artifacts may be generated by decomposition of thermolabile molecules in GC, during the analysis after enhancement of volatility of molecules. On the other hand, LC–MS targets compounds with a relatively narrow range but has the flexibility to change the type of targets with replaceable separation columns. As the common stationary columns in LC, C18 reversed-phase is the most frequently used for separation of hydrophobic molecules, while the polar phase, such as silica and amide, is used for hydrophilic molecules. Running time of LC separation can be much reduced to less than 10 min in the case of ultra HPLC (UHPLC, UPLC).

Table 1. Features of the most commonly used separation techniques in mass spectrometry (MS)-based metabolomics.

	CE	GC	LC
Favorable target metabolites	Polar, charged	Mainly volatile	Non-polar, neutral
Sample derivatization	Unnecessary	Required for non-volatile compounds	Unnecessary
Number of theoretical plate	$10^5 \sim 10^6$	$10^4 \sim 10^5$	10^4
Separation of structural isomer	High	High	Low
Running time	>20 min	20–40 min	15–40 min
Downstream ionization	ESI	EI, CI	ESI, APCI

The separated molecules are required to be ionized generally by electrospray ionization (ESI) in CE, electron ionization (EI) or chemical ionization (CI) in GC, and ESI or atmospheric pressure chemical ionization (APCI; API) in LC. MS detection techniques have been developed to various types, and, currently, the most frequently used technique for matching to the upstream CE, GC, and LC separation steps is time-of-flight (TOF) to achieve higher sensitivity, accuracy, rate of measurement, and mass and dynamic range for the acquisition of more metabolite information [4]. Fourier transform (FT) type MS also has a high-throughput performance with high resolution and a broad range of target molecules and is thereby used for metabolomics. Of the FTMS, FT-ion cyclotron resonance (FT-ICR) MS utilizes a magnetic field to detect resonance of cyclotron motion of metabolite ion, while the orbitrap type is based on a system using an electric field. In some cases, tandem MS (MS/MS) is employed to acquire more structural information for the characterization of compounds. These MS-based approaches are utilized not only for targeted metabolomics but also for untargeted metabolomics [26,34,35]. Recently, rapid evaporative ionization mass spectrometry (REIMS) has been introduced in MEATabolomics [26,36]. This method allows direct MS analysis of a biological sample with no preparative steps, under normal atmospheric conditions, due to being based on the ambient ionization [37,38]. A trend of MEATabolomic approach can be seen in the times of employment for each technique in the 56 studies collected in this review: CE–MS, GC–MS, LC–MS, NMR, and REIMS approaches have been employed in 5, 17, 17, 22, and 1 studies, respectively, with some cases combining multiple approaches (Table 2).

Table 2. Overview of MEATabolomics studies cited in this review.

Category of Objective	Species/Meat Type	Factors Analyzed	Methodology	Multivariate Data Analysis	Ref.	Authors
Meat characterization	Cattle	Muscle type	HR-MAS ^{1}H-NMR	PLS-DA, OPLS-DA	[39]	Ritota et al.
	Lamb	Storage time, display time, packaging conditions	HILIC-MS	PCA	[40]	Subbaraj et al.
	Chicken	pHu	^{1}H-NMR	OPLS-DA, MSEA	[41]	Beauclercq et al.
	Beef	Flavor	GC-MS	-	[42]	Takakura et al.
	Beef	Flavor, aging period	HS/SPME GC-MS	-	[43]	Watanabe et al.
	Beef, pork, chicken	Flavor, species, breeds, tissues	GC-MS	OPLS-DA	[44]	Ueda et al.
	Chicken	Age of chicken, muscle type	^{1}H-NMR	PLS-DA	[45]	Xiao et al.
Meat abnormality	Chicken	Dystrophy of breast	HR-MAS ^{1}H-NMR	PCA, OPLS-DA	[46]	Sundekilde et al.
	Chicken	Wooden breast	GC-MS, LC-MS/MS	RF	[47]	Abasht et al.
	Chicken	Wooden breast	^{1}H-NMR	OPLS-DA	[48]	Wang et al.
	Chicken	Wooden breast	^{1}H-NMR	OPLS-DA	[49]	Xing et al.
	Chicken	White striping	GC-MS, LC-MS	PCA, Pathway	[50]	Boerboom et al.
Genetic background	Pig	Crossbreeds	^{1}H-NMR	PLS	[6]	Straadt et al.
	Pig	Drip loss, association with SNP	GC-MS, LC-MS	Pathway, GWAS	[35]	Welzenbach et al.
	Cattle	Genetic parameters for growth and precocity	^{1}H-NMR	PLS-DA	[51]	Consolo et al.
	Cattle	Genetic parameters for chemical traits	GC, LC	-	[52]	Sakuma et al.
	Cattle	NT5E genotype	GC, LC	-	[53]	Komatsu et al.
Animal feeding	Cattle	Grass-fed/grain-fed	GC-MS, LC-MS/MS	PCA, RF	[20]	Carrillo et al.
	Cattle	Dietary amino acid supplementation	^{1}H-NMR	PCA	[21]	Yu et al.
	Cattle	Dietary mate extract supplementation	^{1}H-NMR	PCA	[22]	de Zawadzki et al.
	Pig	Clenbuterol supplementation	GC-MS	PCA, PLS-DA, OPLS-DA	[23]	Li et al.
	Chicken	Lysine supplementation	CE-MS	-	[24]	Watanabe et al.
	Chicken	Age	^{1}H-NMR	PCA, OPLS-DA	[25]	Liu et al.
	Pig	Ractopamine supplementation	REIMS	PCA, LDA, OPLS-DA	[26]	Guitton et al.

Table 2. *Cont.*

Category of Objective	Species/Meat Type	Factors Analyzed	Methodology	Multivariate Data Analysis	Ref.	Authors
Postmortem aging	Pork	Pm. aging period, muscle type	CE–MS	PCA	[14]	Muroya et al.
	Beef	Pm. aging period, muscle type	LC–MS	PCA	[15]	Ma et al.
	Pork	Pm. aging period, muscle type	UPLC–MS/MS	PCA	[16]	Yu et al.
	Beef	Pm. aging period	CE–MS	PCA	[17]	Muroya et al.
	Beef	Pm. aging period	^{1}H–NMR	OPLS–DA	[18]	Kodani et al.
	Beef	Pm. aging period	^{1}H–NMR	PCA	[19]	Graham et al.
	Pork	Pm. aging period, muscle type (on thiamine)	CE–MS	-	[54]	Muroya et al.
	Beef	Pm. aging period	LC–MS	PCA	[55]	Lana et al.
	Beef	Pm. period of dry-aging	^{1}H–NMR	-	[56]	Kim et al.
	Lamb	Fast chilling effect	LC–MS, ^{1}H– and ^{31}P–NMR	PCA	[57]	Warner et al.
	Beef	Pm. aging period (on oxidative stability)	GC–MS	PCA	[58]	Mitacek et al.
Processing	Pork	Marination time	^{1}H–NMR	PCA, OPLS–DA	[30]	Yang et al.
	Pork	Drying/aging period, fermentation of sausage	HR–MAS ^{1}H–NMR	PCA	[59]	García-García et al.
Processing, authentication	Pork	Geographic origin, processing method	CE–MS	PCA	[27]	Sugimoto et al.
	Pork	Geographic origin, processing method	^{1}H–NMR	PCA, OPLS–DA	[28]	Zhang et al.
Processing, Spoilage	Chicken	Marinade type, storage time, microbial load, sensory score	GC–MS	PCA, FDA	[7]	Lytou et al.
	Chicken	Marinade type, marination time and temperature	LC	PCA	[29]	Lytou et al.
Sensory evaluation	Beef	Grinding score, packaging method	LC–MS	PCA, PLS	[60]	Jiang et al.
	Beef	Commercial brands	GC–MS	–	[61]	Suzuki et al.

Table 2. *Cont.*

Category of Objective	Species/Meat Type	Factors Analyzed	Methodology	Multivariate Data Analysis	Ref.	Authors
Spoilage	Pork	Salmonellae contamination, time of microbial exposure	GC–MS	PCA, etc.	[62, 63]	Xu et al.
	Beef	Packaging, temperature	LC–MS	PCA, FDA, PLS–R	[64]	Argyri et al.
	Beef	Packaging, temperature, sensory score, microbial growth	HS/SPME GC–MS	PCA, FDA, PLS–R	[65]	Argyri et al.
Authentication	Beef	Geographic origin	^{1}H–NMR	PCA, OPLS–DA	[66]	Jung et al.
	Beef	Geographic origin	IMS	PCA	[67]	Zaima et al.
	Beef	Production system	^{1}H–NMR	PLS–DA	[68]	Osorio et al.
	Beef, Pork	Species	GC–MS, UPLC–MS	PCA, PLS–DA, Pathway	[69]	Trivedi, et al.
	Beef, Pork	Species	HS/SPME GC–MS	PCA, PLS–DA	[70]	Pavlidis et al.
	Chicken	Live/dead on arrival	LC–MS	PCA	[71]	Sidwick et al.
	Chicken	Live/dead on arrival	LC–MS	PCA, Pathway	[72]	Cao et al.
	Beef	Irradiation doses (on lipids)	^{1}H–NMR	sLDA, ANN	[73]	Zanardi et al.
	Beef	Irradiation doses (on hydrophilic compounds)	^{1}H–NMR	PCA, CT	[74]	Zanardi et al.

ANN: artificial neural networks; CT: classification tree; FDA: factorial discriminant analysis; GWAS: genome-wide association analysis; HILIC: hydrophilic interaction liquid chromatography; HS/SPME: head space–solid phase microextraction; LDA: linear discriminant analysis; MSEA: metabolite set enrichment analysis; OPLS–DA: orthogonal PLS–discrimination analysis; Pathway: pathway enrichment analysis; Pm.: postmoretm; PCA: principal component analysis; PLS: partial least square analysis; PLS–DA: PLS–discrimination analysis; REIMS: rapid Evaporative Ionization Mass Spectrometry; RF: random forest; sLDA: stepwise linear discriminant analysis; UPLC: ultra-performance LC.

After MS measurement, initial putative metabolite can be identified on the basis of the accurate mass–to–charge ratio (m/z) of the mass spectral ion in MS-based metabolomics. This is assisted by the use of metabolite databases such as METLIN (https://metlin.scripps.edu/), the Human Metabolome Database (HMDB; http://www.hmdb.ca/), and MassBank (http://www.massbank.jp/).

NMR is also highlighted as a method for practical use, such as authentication purposes, in an analytical routine. This technique can provide rapid and reproducible measurements in complex mixtures without a time-consuming pretreatment. Although NMR has relatively low ability to profile metabolites compared to the MS-based techniques due to its low resolution and sensitivity, it has the ability to collect distinct information that the other metabolomics cannot access in a non-destructive and non-biased way [2]. Especially, approaches with ^{1}H–high-resolution magic angle spinning (HR–MAS) have also been applied to characterize meat quality [39,46,59]. Unlike the other types of NMR, this NMR enables us to investigate intact tissue specimens (10–50 mg) and allows the spectra to be obtained with a high resolution compatible to that obtained in liquid samples in less than 30 min.

Once the data matrix is produced from the collected raw data, subsequent statistical analyses and data mining are often performed to identify samples or variables (metabolites) that characterize the variations between datasets and may represent biologically meaningful determinants. In most cases, the statistical sample classification is conducted to overview pattern recognition of sample categories, by multivariate analyses such as principal component analysis (PCA), clustering analysis, partial least square analysis (also called projection to latent structures, PLS), PLS–discriminant analysis (PLS–DA), and random forests (RF) [5,75].

PCA is an unsupervised statistical method that reduces dimensions of high dimensional data to visualize sample distribution and grouping on the principal component (PC) plot based on the pattern of the metabolite dataset and thereby is employed by most of the metabolomic studies as the first step of data analysis [75]. Clustering analysis, especially hierarchical clustering analysis (HCA), is also widely used to generate a snapshot profiling of dataset. In HCA, the algorithm divides the measured datasets into subgroups so that datasets with similar metabolomic profiles are placed in each group. PCA and HCA are frequently used for visualization of classification in omics studies, including metabolomics.

PLS has been developed as a supervised extension of PCA [75]. This regression-based method is especially useful when fewer samples are available than measured metabolites. PLS–DA is used to elucidate the metabolites that carry the information of classification, which screens highly contributing metabolites to the classification. PLS and PLS–DA have often been employed to sharpen the separation between the groups, especially when the groups are not sufficiently separated in PCA. Orthogonal PLS (OPLS)–DA is an extension of PLS–DA to cover the defect of PLS–DA, owing to its robustness against noise. RF is a machine learning method used to discriminate two groups. Different from conventional methods such as PCA and PLS–DA, RF allows data structure understood without dimensional reduction with its low bias and low variance. As with other data mining analyses, the application of support vector machine (SVM) and neural networks are attempted for some of MEATabolomic studies [34]. These analytical methods have been applied for sample characterization and determination of biomarker candidates.

Classical statistical analyses such as Student's *t*-test and analysis of variance (ANOVA) are also used to compare metabolite levels between sample groups, with care about the detection of false-positive metabolites in multiple comparisons of the datasets. To reduce false-positive detection caused by the familywise error rate (FWER) in multiple comparisons, procedures of false discovery rate (FDR), Holm, Bonferroni, or Benjamini–Hochberg are applied to adjust the levels of significance detection in metabolomic studies [5]. Obviously, there are advantages and disadvantages to every method and database. Further information and details in these statistical analyses are described for reference in methodological reviews [5,75].

Moreover, recent progress of bioinformatics analytic tools and databases largely contributes to compound annotation, metabolic pathway finding, and biological data interpretation of the extracted compounds (see the websites of The Metabolomics Society, Inc. for more detail information; http://metabolomicssociety.org/). HMDB (http://www.hmdb.ca/) and PubChem (https://pubchem.ncbi.nlm.nih.gov/) are among the most widely used databases, and they play important roles in the annotation of compound signals.

The Kyoto Encyclopedia of Genes and Genomes (KEGG; https://www.genome.jp/kegg/) is frequently used to connect the metabolome profile with genomic, transcriptomic, enzymatic, and pathway information, which supports prediction and pathway mapping of potential molecular factors responsible for the phenotypic events. Software for analyses of metabolic pathways such as MetaboAnalyst (https://www.metaboanalyst.ca/) has also often been used for biological interpretation in recent years. MetaboAnalyst provides a variety of statistical analysis tools, including visualizing programs, multivariate analyses, and metabolite pathway analyses. These metabolomic approaches, especially CE–MS, GC–MS, LC–MS, and NMR, in combination with subsequent data analyses have recently been increasingly used in animal-based food research fields [10–12,76].

3. Metabolomes Associated with Meat Quality Traits

Metabolomics is used for the exploration of key compounds contributing to the physico-chemical properties and sensory evaluation scores, and thereby it contributes to accounting for meat palatability and quality traits such as color and water-holding capacity (WHC). As described below, the contents of skeletal muscle metabolites, including amino acids and sugars, are affected by animal genetic background, feeding, muscle type, postmortem aging, and meat processing. These metabolomic changes are linked with the physico-chemical properties of muscles and meat, as shown in variations in meat color and WHC between different types of muscles or between differently treated muscles. Accordingly, the metabolites are beneficial indicators to predict the physico-chemical meat quality traits, which indicate a result more or less from the influences of animal genetic background, feeding, and postmortem processes (Table 2).

3.1. Meat Color, WHC, and pH Decline

The color of meat is one of the commercially important meat quality traits since the appearance of meat has a great influence on the consumer's desire to purchase [77]. The redness and discoloration are determined by the chemical status of myoglobin, which is affected by multiple factors that are inherent to both live animal and postmortem conditions. With hydrophilic interaction liquid chromatography–mass spectrometry (HILIC–MS)-based metabolomics, influences of postmortem aging time, packaging atmosphere, and display time on lamb meat color was investigated to identify the metabolites affected by those postmortem conditions [40]. Many compounds were found to change with time of aging and display or packaging atmosphere, which indicated the contribution of the compounds, including amino acids, sugars, nucleotides, and organic acids, to the sample discrimination. Furthermore, the abundance of reducing or antioxidant compounds such as L-glutathione and taurine contributed to color stability. Intriguingly, boron complexes of sugar and malate, presumably color stability-related compounds present in plants but not in animals, were detected, although the role of the compounds remains unknown [40]. WHC is also one of the most important traits of meat quality. Various genetic and environmental factors can induce myofibrillar protein denaturation and thereby lower the WHC of meat. Meat with lower WHC generally has higher drip loss, which has a negative impact on juiciness and palatability of the meat. To assess the responsible genes for WHC in pork, metabolomics was applied to characterize pork with high drip loss via metabolic pathway enrichment analysis in an integrative omics approach, as described below (see Section 4.1) [35].

Due to the large impact on meat color, WHC, and other final meat quality traits, postmortem pH decline is also an important factor to be controlled. In a study investigating two chicken lines differing in ultimate pH (pHu), the *pectoralis major* muscles of the two lines were discriminated against by high-resolution NMR metabolomics [41]. Subsequent metabolite set enrichment analysis (MSEA) of the data showed that carbohydrate metabolism in the low pHu line and metabolism of amino acid and protein in the high pHu line were over-representative pathways. The difference in the metabolome profile between high and low pHu chicken might be due to the ability of the muscles of glycogen storage and carbohydrate use.

3.2. Flavor and Palatability

Skeletal muscle metabolites include amino acids and sugars that are precursors of volatile compounds associated with meat aroma. The contents of these compounds are altered, depending on animal feeding, the genetic background, and the postmortem aging process. This could cause variations in meat flavor between meats from different animal production or postmortem conditions. Therefore, muscle and meat metabolites are useful indices to predict or evaluate meat flavor and overall palatability in a comparison between meats of different animal breeds, feeding conditions, and/or postmortem processes. After postmortem aging, amino acids and sugars in muscles make a large contribution to the quality of cooked meat, both as intact forms and as products by the Maillard reaction that occurs between amino compounds and reducing sugars during heat treatment. One of the final volatile products, 2,5-dimethyl-4-hydroxy-3(2H)-furanone (DMHF; furaneol), has been identified as a key aroma compound in beef extract by aroma extract dilution analysis (AEDA) [42]. Even though the aroma of furaneol depends on its concentration, it has great influences on sensory characteristics as a flavor component in meat as well as other foods [78]. In fact, furaneol was detected on Japanese Shorthorn beef during cooking at 180 °C by analysis with headspace solid-phase micro-extraction (HS/SPME) GC–MS, as well as other flavor-associated volatile compounds [43]. Moreover, inosine-5′-monophosphate (5′-IMP) greatly contributes to the enhanced flavor of the meat due to its primary role as an *umami* compound, in a different manner from aroma compounds [79,80]. Thus, the MEATabolomics approach provides tools to access these palatability-associated compounds with the high-throughput analytic systems.

In a beef study combined with flavor-associated evaluation scores, the *longissimus thoracis* (LT) muscles of Japanese Black (JB) and Holstein cattle were discriminated by the metabolome profiles especially with decanoic acid and glutamine [44], indicating that these compounds were associated with the difference in flavor between the two breeds. Metabolites in beef affecting the sensory attributes have also been explored using commercial ground beef differing in grinding and packaging methods [60]. The study resulted in the determination of 33 metabolites differentiating the grinding and 22 metabolites associated with beef flavor and off-flavor. Results from a pork study of differences between breeds indicated that high carnosine content in meat was associated with a low flavor/taste score in pork [6]. Sensory score-associated variation of sugar content and composition of fatty acids, amino acids, and nucleotides were also observed in a comprehensive analysis of the metabolites in seven brands of JB beef [61]. The results of these studies suggest the influences of animal genetic and feeding factors on sensory evaluation scores via the meat metabolites. It is quite likely that altered contents of amino acids and sugars in the animal and meat-producing processes cause changes in the content of volatile flavor-associated compounds, including furaneol in the cooked meat. Even in the presence of complicated biological and manufactural factors, these studies revealed that metabolomic information is expected to provide indices to predict sensory phenotypes of meat.

3.3. Chicken Meat Quality Traits

Chicken meat quality traits are one of the most intensively focused targets in recent MEATabolomics, due to the globally growing market. A Chinese local breed, Wuding chicken meat, was analyzed by the ^{1}H NMR-based approach [45], in which the age of chicken affected the levels of metabolites such as lactate, creatine, IMP, glucose, carnosine, anserine, and taurine. Moreover, the abnormal meat quality of chicken is currently an important issue in the poultry industry. Increasing incidents of muscle abnormalities, including white striping (WS), wooden breast (WB), and spaghetti meat (SM) phenotypes, have been linked to a drastic improvement of breast muscle mass [81]. Metabolomics approaches have been applied to especially elucidate the mechanisms underlying the hardness of the WB phenotype. As WB-related biomarker candidates, Abasht et al. [47] identified compounds associated with protein levels, muscle protein degradation, and altered glucose metabolism, using GC–MS and LC–MS/MS followed by RF procedure analysis. Another study with ^{1}H–NMR approach revealed that WB-affected broiler chicken breasts had a higher content of leucine, valine, alanine, glutamine, lysine, lactate, succinate, taurine, glucose, and IMP, but lower histidine, β-alanine, acetate, creatine, creatinine, and anserine compared to normal fillets [48]. In another attempt to explore WB myopathy biomarkers, muscle exudate was used as a sample. Linked with discrimination of the samples between WB and non-WB phenotypes, amino acids, nucleotides, and organic acids were identified as the WB-associated metabolites by an NMR approach [49], suggesting the usefulness of those metabolites as the WB markers. This result was partly consistent with the previous study using breast samples with/without WB [48]. WS, on the other hand, has been associated with altered metabolism related to carbohydrates, long-chain fatty acids, and carnitine in a study with GC–MS and LC–MS approaches, suggesting the involvement of altered β-oxidation in the WS phenotype [50]. A HR–MAS NMR approach revealed that muscle dystrophy of pectoralis in chicken is associated with low content of anserine and carnosine [46]. Thus, the results of these MEATabolomics studies have suggested that the chicken breast muscle abnormalities could be caused by the altered metabolism of carbohydrates, protein and amino acids, and β-oxidation.

4. Factors Affecting Muscle and Meat Metabolomes

As mentioned above, meat is originally skeletal muscles of livestock; consequently, meat quality is greatly affected by factors in the livestock production system, such as animal genetic background, the feeding and stress that animals experienced, but also the physiological characteristics of live muscle, postmortem aging, processing, and spoilage of meat. All the influences of these factors can be assessed by the muscle or meat metabolome profile that is the phenotype resulting from animal experiences through the expression of gene transcription, translation, and the protein modifications as well.

4.1. Animal Species, Breeds, and Genetic Backgrounds

A variety of characteristics in meat depends on species or breeds of animals to a great extent. The genetic background of animals thus has an influence on the phenotype expression resulting in an inherent metabolome profile. For the purpose of investigating the effect of genetic background, metabolomics has so far been applied to capture muscle and meat phenotypic differences between species (beef, pork, chicken) [44] or breeds in beef cattle [44] and pigs [6]. The meat of those species was discriminated by PCA of GC–MS metabolome profiles with specific compounds being characteristic to each species [44], although different postmortem muscle metabolism might affect the profile. In comparison between Holstein and the highly marbled JB cattle that differ in oily flavor, sweet flavor, wagyu beef aroma, and the overall evaluation in a sensory test, their LT muscles were discriminated by OPLS–DA [44]. A higher level of decanoic acid in JB than in Holstein cattle, and differences between skeletal muscle, inter-, and intramuscular fat tissues of JB cattle were also observed.

Regarding cattle, in the comparison between different strains of Nellore cattle genetically selected by precocity, a genetic effect on *longissimus lumborum* (LL) muscle metabolome was observed by ^{1}H–NMR [51]. The purpose of that study was to investigate changes in metabolites associated with muscle physiology and quality of the LL meat, under the established improving effect of high growth or precocity background on daily gain, carcass characteristics, and other industrial meaningful traits. In the subsequent PLS–DA using the detected compounds (metabolites related to glycolysis and the citric acid cycle, amino acids, organic acids, nucleotides, and sugars), the muscles were discriminated between high and low precocity groups. Pathway analysis in the study highlighted the association of the muscle protein metabolism with the intensity of the selection. Thus, these MEATabolomic studies have revealed their potential to uncover species- or breed-specific differences in metabolites, through which differences in meat quality between breeds can be assessed at the molecular level.

A study comparing ^{1}H–NMR metabolome and sensory evaluation scores between five crossbreds of pigs observed an inter-crossbreed effect on metabolites, such as amino acids, lactate, IMP, glycerol, and choline-containing compounds. Some of those compounds were associated with meat quality, including sensory scores [6]. The results also suggested the association of live muscle metabolism, membrane properties, muscle fiber glycolytic potential, lipolysis, and proteolysis with the metabolomic difference.

Furthermore, in a study on pork investigating the relationship between metabolites and drip loss, metabolomics was applied in an integrative way to understand the association of a high drip loss phenotype with the genetic background [35]. In this single nucleotide polymorphism (SNP)-based genome-wide association study (GWAS), a region of candidate genes was identified on chromosome 18 as one associated with drip loss, and the metabolite glycine. Compared to conventional genetic studies using quantitative trait loci (QTL) and candidate genes for drip loss in pigs, GWAS is expected to improve the efficiency of candidate gene detection and accuracy of genomic prediction by avoiding detection of false-positive associations. In beef, GWAS was conducted on JB cattle to explore genes responsible for the palatability of beef, which revealed that SNPs in the ecto-5′-nucleotidase (NT5E) gene affected the content of IMP in the postmortem aged beef, due to the modulating effects of NT5E enzymatic activity. This is a result of which GC- and LC-based metabolomics contributed to GWAS on meat [52,53]. In these comprehensive omics approaches such as GWAS, metabolomics is useful to capture phenotypic metabolites for the candidate genes, due to its wide range of phenotypic molecular measurements.

4.2. Animal Feeding

Feeding conditions, one of the environmental factors for animals, have great influences on animal physiology, including skeletal muscle growth and maturation, and the final meat yield and quality. As the feeding factors, the nutritional condition and stress originating from the farming system are considered as the most important ones [82,83], due to their influences on downstream meat production. The influence of feeding on animals are expected to be observed as metabolome changes via gene expression and physiological alteration in tissues. For these reasons, metabolomics approaches have attempted to capture metabolite signatures of feeding systems aiming at the improvement of meat production.

In beef cattle, to address an increasing demand by consumers for beef produced in a sustainable farming system, the effect of differences in feeding on the muscle metabolome was assessed by metabolomic approaches with GC–MS and LC–MS/MS [20]. Between grass- and grain-fed Angus steers, significant variations between the cattle groups was observed in a variety of lipids, including polyunsaturated fatty acid (PUFA) in the *latissimus dorsi* muscle, showing higher $\omega3$ and lower $\omega6$ fatty acids in grass-fed cattle than in the grain-fed. The PCA classified these samples by the diet, and RF analysis based on the metabolites resulted in a predictive accuracy of 100% between the dietary conditions, in which lipids and amino acids were potential biomarkers discriminating the two groups of cattle. Intriguingly, the cortisol level was lower in grass-fed cattle, suggesting they might experience less stress than the grain-fed [20]. These results indicated that metabolite signatures could be utilized as indices not only in feeding-originated beef quality but also in animal welfare and authentication. In other studies, the effects of dietary amino acids and mate extract administration on beef were also investigated by NMR approaches followed by PCA [21,22]. Biopsy *semitendinosus* muscle samples of dairy calves fed protein-based milk replacer at 7 weeks of age were classified by amino acid supplementation [21]. Additionally, in a study of feedlot Nellore cattle, the dietary addition of mate extract, one of the antioxidant additives administered to broiler chickens, resulted in increased content of IMP, creatine, carnosine, and conjugated linoleic acid in the LT muscle, and some of these were in a dose-dependent manner [22]. Coupled with redox status analysis, the results further indicated that the beef of mate extract had increased oxidative stability, tenderness, and consumer acceptability. Thus, not only the global feeding conditions but also nutrients and ingredients in the animal feed cause changes in muscle and meat metabolomes, which can be assessed by MEATabolomics. In such studies, metabolites can be screened and be identified as potential biomarkers of the feeding.

Regarding pork, a GC–MS approach was applied to the screening of pigs fed with Clenbuterol, a β-adrenergic agonist [23]. The *biceps femoris* muscles of pigs treated with Clenbuterol were discriminated from the counterparts in the subsequent PLS-DA by compounds associated with fatty acids and amino acids. Recently, a similar study was conducted using a REIMS approach [26], in which carcasses of pigs fed with ractopamine, another β-agonist, were discriminated by the lipid profiles of REIMS with high accuracy of classification in three different types of muscle. Due to the high-throughput and accurate performance, this methodology can be widely applied to fields of practical meat research.

Such nutrimetabolomic approaches have also been attempted in chicken, of which growth performance, breast yield, and meat quality are improved with dietary lysine supplementation [12]. Aiming at an increase in flavor compounds in chicken, the effect of dietary lysine supplementation at a 1.5-fold level of a standard requirement on the breast muscle was investigated by a CE–MS approach. Supplementation resulted in increased levels of lysine degradation products, such as saccharopine, α-amino adipic acid, and glutamate, the latter being one of the most important taste-improving amino acids in meat [24]. On the other hand, in a study investigating the effect of duck aging on the meat metabolome, duck meat samples of different ages were classified in PCA and PLS–DA by NMR metabolite data, showing increased levels of lactate and anserine and decreased levels of fumarate, betaine, taurine, and inosine with increasing age [25].

4.3. Muscle Type

The type of skeletal muscle of pig and cattle is associated with meat quality (WHC, meat color, and tenderness) [84,85] and sensory traits [86,87]. This is due to the differences in physiological characteristics [88,89] and compositions of proteins [90,91] and metabolites [14,54,92,93] between different types of muscles that have distinct compositions of fast and slow type muscle fibers. Due to the different physiological properties of the muscle fibers, muscles with different fast/slow fiber type composition undergo different postmortem aging processes, as shown in protein [94,95] and metabolite [14,16] degradation, which could be the cause of the intermuscular difference in meat quality. Contents of flavor-associated metabolites such as amino acids, sugar, and nucleotides differed between porcine LL and *vastus intermedius* (VI) muscles in a CE–MS metabolomics study [14], and between bovine muscles in an LC–MS [15] or a UPLC–MS/MS study [16]. These results indicate the association of fast and slow type fiber composition with contents of flavor-associated metabolites in aged meat [14,15]. It is noticeable that the rate and extent of the postmortem metabolite changes including glycolysis, metabolism of amino acids and nucleotides are different between the muscle types, as shown in different accumulation of lactate in glycolysis, hypoxanthine in purine metabolism, and aliphatic amino acids and methionine in amino acid metabolism [14]. During postmortem aging of pork, thiamine is accumulated, while thiamine triphosphate level declines, which is another difference between muscle types [54]. In addition, the metabolomic difference between *longissimus* and *semitendinosus* muscles of four cattle breeds was investigated by an HR–MAS NMR approach, which resulted in a good classification of the two muscles in Chianina and Buffalo, but not in Holstein Friesian and Maremmana cattle [39]. Such intermuscular differences may be associated with differences in meat quality and sensory traits between those muscles.

4.4. Postmortem Aging and Storage

Aging and storage of postmortem muscle have great influences on meat quality, as well as animal genetic background and feeding conditions. Even when an animal has muscles of high meat quality potential, the final meat quality could be lowered by inappropriate management, resulting in meat deterioration such as spoilage and discoloration in parallel with abnormal metabolome changes. Inversely, the development of metabolites in meat during postmortem aging under the appropriate condition has a beneficial influence on the final meat quality.

MEATabolomics has contributed to postmortem meat aging studies to account for influences of postmortem aging on meat quality in terms of metabolites. In pork, postmortem changes in the LL and VI muscles were investigated by a CE–MS approach [14]. The result of this study indicated that a variety of postmortem muscle metabolisms were activated, especially within 24 h after slaughter. The PCA classified the samples primarily into LL and VI muscles, and secondarily to at-slaughter and day 14 (168 h) aging points (Figure 1). In the LL samples, intermediate glycolytic products, glucose 6-phosphate and fructose 6-phosphate (F-6P), increased until 24 h postmortem, while the downstream compounds such as fructose 1,6-bisphosphate (F-1,6P) and phosphoenolpyruvate decreased with being almost exhausted at 24 h postmortem, indicating the rate-determining activity of phosphofructokinase (F-6P → F-1,6P) in postmortem LL muscle glycolysis. Additionally, in the LL muscle, most of the amino acids and identified dipeptides increased after day 1 during the aging, suggesting that protein degradation began to be dominant after day 1. Regarding ATP degradation, marked differences in compound levels were observed in the pathways of ADP → AMP → IMP → inosine → hypoxanthine between the LL and VI muscles, suggesting the intermuscular differences in enzymatic activities in the pathways [14]. These pathways in pig LL and VI muscles were overrepresented in the analysis of pathway impact by MetaboAnalyst (Figure 2), indicating a number of biologically meaningful differences in metabolic pathways between the two muscles. Such pathway analyses further afforded new insights: pathways associated with thiamine, pentose phosphate, and NADH metabolism had a high impact on postmortem aging in both muscles. In other studies analyzing bovine LL, *psoas major* (PM), and/or *semimembranosus* muscles, intermuscular metabolomic differences in energy metabolism during 24 h aging [16], and differences in metabolites associated with meat color and lipid oxidation during 23 d aging [15] were also observed by LC–MS-based approaches. Thus, the skeletal muscle type is associated with the muscle metabolome as described above, in which postmortem aging makes the intermuscular metabolome differences inherent to live animals further complicated, as shown in the complexity of the differences in the final meat quality. Nevertheless, MEATabolomics approaches have shown its advantages to discriminate meat samples by muscle type and postmortem aging conditions.

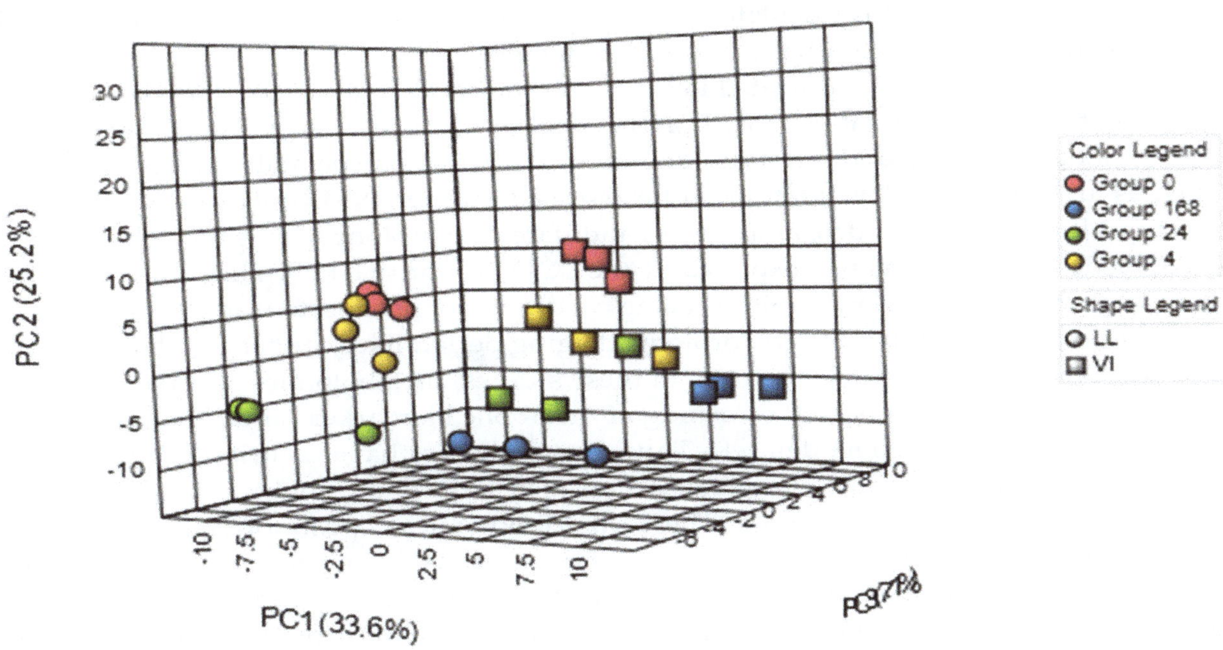

Figure 1. Classification of pig LL and VI muscle samples of different postmortem aging period by PCA. Pig LL and VI muscles were aged during 0, 4, 24, 168 h postmortem.

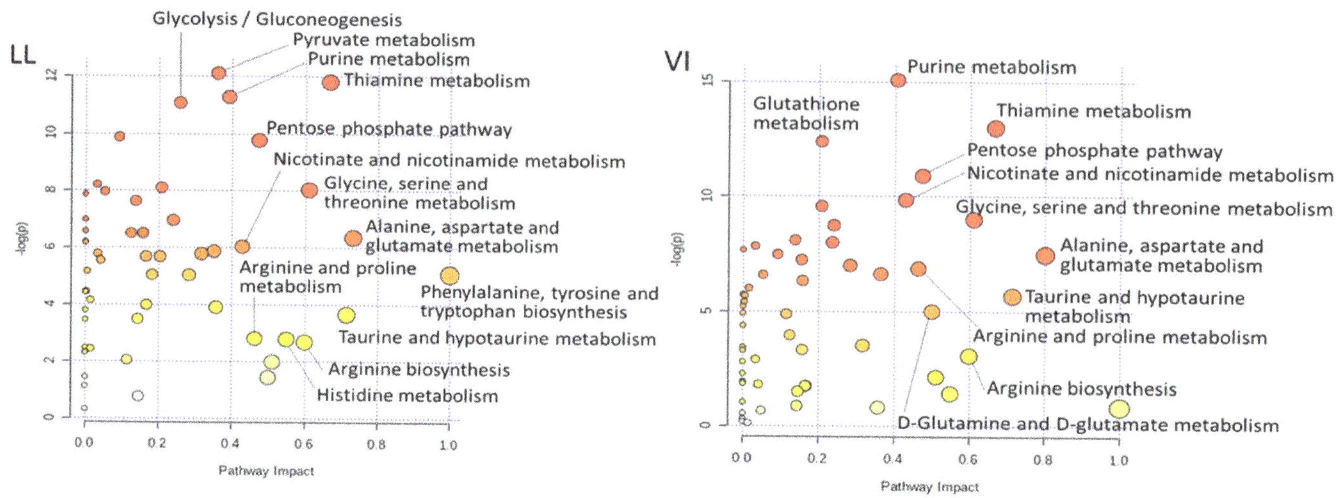

Figure 2. Metabolic pathway impact analysis of pig LL and VI muscles during postmortem aging. Pathway impact values (horizontal axis) were calculated from the pathway topology analysis in MetaboAnalyst. A pathway with small p value and high impact is plotted as a red and large circle.

Regarding beef, metabolomic changes during postmortem aging was also analyzed by CE–MS [17] and NMR approaches [18,19] in *longissimus* muscle. In the CE–MS metabolomics targeting charged metabolites, changes in nutritionally important compounds such as choline, nicotinamide, and thiamine were observed during 14 d aging of JB cattle beef, some of which contributed to classification of the samples by the aging period [17]. On the other hand, NMR studies were conducted for metabolomic profiling during 10 weeks [18] and 21 days [19] postmortem. The former NMR study aimed to explore excellent biomarkers of postmortem JB beef aged up to 10 weeks, and revealed that leucine and creatine are available as biomarkers for long-term aging (>2 weeks) because of positive and negative correlation with aging duration, respectively [18]. The L-value, defined as the ratio of leucine to creatine in the study, was more appropriate to evaluate meat quality during long-term aging than the conventional K-value that is calculated by using the contents of ATP, ADP, AMP, and IMP. In addition, this study also showed that the NMR metabolomics was able to estimate unsaturation degree of triacylglycerol that was correlated to content ratio of unsaturated fatty acids ($R^2 = 0.944$) and melting point of intramuscular fat ($R^2 = 0.871$). This suggests that such an approach can evaluate not only postmortem meat aging and intramuscular fat content, but also palatability-associated quality traits such as juiciness and WHC that might be assessed through NMR-measured pH profile [18] in highly marbled JB beef. Besides, biomarker candidates for prolonged postmortem aging (up to 44 days) of Piedmontese LT beef was explored by the LC–MS approach [55]. The results indicated the possibility of glutamate, serine, and arginine being candidates of postmortem aging indicators due to their constant increase until day 44. The prolonged aging period likely enhanced the accumulation of amino acids and some other metabolites. In these studies, metabolomics approaches combined with statistical analyses have enabled screening of metabolic biomarker candidates of postmortem aging from the meat metabolome dataset. Thus, metabolomics have contributed not only to the understanding of postmortem muscle metabolisms but also to the finding of better biomarkers that the conventional studies have never assessed. These MEATabolomics of postmortem aging of pork and beef re-emphasize the possibility that the amino acid generation during cold storage is primarily due to postmortem protein degradation in the muscles, as shown in previous studies of aging [31,94–101]. Besides, the elucidation of flavor-associated compound generation from muscle metabolites has been accelerated with the accumulating knowledge of the uncovered meat metabolomes through postmortem aging studies on cooked meat [43].

Other optional postmortem storage conditions are also to be investigated due to the influences on a variety of metabolites associated with meat quality traits. A dry-aging condition of 3 °C with

0.2 m/s air-velocity for 3 weeks showed high improvement of beef palatability in a sensory panel evaluation compared to conditions of general wet-aging and other dry conditions, with higher contents of compounds including branched amino acids and IMP in a study by an NMR approach [56]. The air flow of this storage method might simply accelerate the up-concentration of the flavor-associated metabolites, but other undetermined factors are likely also involved. On the other hand, the chilling rate of lamb LT muscle during aging affected the muscle energy metabolism within 24 h postmortem, which was associated with meat tenderness in a study utilizing ^{1}H–, ^{31}P–NMR, and LC–MS [57]. MEATabolomics was also conducted to investigate roles of NADH in discoloration mechanism via mitochondria activity during postmortem wet-aging of bovine LL muscle [58]. The results suggested that discoloration during postmortem aging of beef was due to increased oxidative stress, mitochondrial damages, and decreasing metabolite sources to regenerate NADH.

4.5. Processing

Meat processing is important to improve the microbiological safety, color, flavor, and texture for the development of favorable meat products. Product chemical information in processing is useful to monitor the processing steps for quality management and to design better products for quality development. To date, metabolomics has been used for the characterization of processed products in studies that assess effects of marinating conditions of chicken breast fillets [7,29] and the marination time on pork [30], and processing conditions on dry-cured hams [27,28] and fermented sausages [59], and texture defects in dry-cured ham [102]. Metabolomic profiling of hams from Japan and European countries identified a total of 203 charged metabolites by CE–MS, and revealed that redness and fat whiteness are associated with metabolomic profiles. The result of this study suggested that the metabolome of hams might be affected by country of origin and processing methods such as smoking and use of starter culture [27]. Difference between Chinese dry-cured and other hams were investigated by NMR metabolomics, in which a total of 33 charged metabolites were identified and each ham was characterized by a specific metabolite set [28]. In an application to a traditional Spanish fermented *salchichón* sausage, a HR–MAS NMR approach showed that the metabolome profile changed, depending primarily on the fermentation and secondarily on the drying process [59]. This result suggested that proteolysis and lipolysis attributed to microbial activities could be monitored by NMR metabolomics.

Metabolomics was also used to test the effect of marination and storage conditions on preservation and sensory quality of chicken breast fillets. By comparison between different types of marinade, temperatures, and intervals during marinating process, each marinade was characterized by a distinct organic acid profile [29]. The results of this study also showed that marinating time has influence on the indigenous microorganisms and the sensory characteristics. A subsequent study to test the effect of the processing conditions (marinade type, storage, and microbial load) showed that the profiles of organic acid and volatile compounds were discriminated between pomegranate-based marinated and control samples according to storage time, microbial load, and sensory score [7]. On the other hand, in a Chinese traditional marinated meat product, amino acids, sugars, acetate, succinate, uracil and inosine increased during marinating, while lactate, creatine, IMP and anserine decreased [30]. In this study, combined with sensory test, a negative effect on the taste of marinating meat in soy sauce was observed during the late stage of dry-ripening, accompanying decreases in most of the metabolites, which suggests that shortening the dry-ripening period could be better to improve the taste quality.

4.6. Spoilage

Spoilage developed by microbiological activity has been the greatest concern for consumers in terms of meat safety. Meat spoilage is developed through microbial growth during meat cutting, storage, and distribution processes after the slaughter of animals. Microbial growth causes chemical changes that result from changes in the microorganism itself and the metabolic output on meat. Therefore, chemical information acquired by metabolomics is expected to be utilized to quantify the degree of spoilage and predict the number or activity of microorganisms [103]. As a candidate of such compound set, volatile organic compounds (VOC) have been focused on in metabolomic studies of spoilage. In an attempt targeting microorganism-associated VOC, metabolome profiling, combined with a multivariate analysis utilizing total ion currents, was able to distinguish naturally spoiled pork samples from those artificially contaminated with *Salmonella typhimurium*, a food poisoning pathogen commonly recovered from pork products [62]. Levels of a total of 16 compounds, including phenylethyl alcohol and dimethyl disulfide, differed between the contaminated and non-contaminated pork samples. However, the identified compounds in such approaches to microorganism-associated VOC may depend on the multivariate analysis algorithm or other factors originating from meat samples [63]. Impact of other microorganisms (*Pseudomonas* spp., *Brochothrix thermosphacta*, lactic acid bacteria, *Enterobacteriaceae*, and yeasts/molds) on minced beef under the various temperature and packaging conditions was also investigated by a LC–MS approach [64].

With high sensitivity and good selectivity, HS/SPME GC–MS was applied in a study to assess spoilage of minced beef [65]. In this study, the authors focused on influences of temperature and atmosphere in packaging during storage of minced beef with indigenous microorganisms. According to the result, identified compounds such as 2-pentanone and 2-heptanone significantly contributed to unacceptable sensory scores and were associated with spoiled samples. This study showed that the GC–MS-based approach has a potential to estimate microbial counts of the different microorganisms and sensory scores of a meat samples independently of storage conditions.

4.7. Authentication

To address the growing consumer awareness, the meat industry has a need for systems to guarantee the authenticity of the meat and the products in order to take measures against increasing fraud/adulteration in the growing complexity of global food chains. Regarding beef, concerns about bovine spongiform encephalopathy (BSE) have increased the awareness of safety. In this field, numerous DNA and protein-based target detection techniques, including species-specific PCR, has been proved to be effective. MEATabolomics is expected to provide useful chemical information regarding authenticity as an alternative method. In this context, analysis of geographic origin of beef by metabolomics was attempted with NMR [66] and imaging MS (IMS) [67]. In the NMR approach, extracts of imported beef from 4 countries were discriminated in the results of PCA and OPLS–DA. The major compounds responsible for the discrimination contained succinate and some of amino acids, suggesting that these compounds could be potential biomarkers to discriminate the geographic origin of beef and that NMR has potential to efficiently work for this analysis. IMS-based direct analysis visualizes distribution of target biomolecules such as lipids, glycolipids, and peptides on biological tissue section samples, without complicated pretreatment procedures. In a matrix-associated laser desorption/ionization–IMS analysis of beef from 3 different Japanese regions, the three types of beef were discriminated by subsequent PCA [67]. In these analyses of geographic origin of beef, although geographic difference in metabolome changes may be due to multiple factors of animal breeds, feeds, and environment, including water and climate, the results revealed availability of metabolomics to distinguish production regions. Geographic differences may also contain the production systems of animals. A study aiming to discriminate the pre-slaughter production system was able to distinguish differences between 1-year cattle fed with barley-based concentrate indoors and with pasture feeding outdoors, using NMR metabolomics followed by PLS–DA, analyzing the *longissimus dorsi* (LD) muscle [68]. The results suggested that NMR approach is suitable for authentication of cattle production history.

Challenges to discriminate beef, pork, and mixture samples by metabolomics were also conducted recently for the establishment of measures against meat adulteration. To this end, different grades of minced beef and pork samples were mixed in various ratios and analyzed by GC–MS and reverse-phase LC–MS approaches [69]. The metabolite content and percentages of fat declared on meat product labeling were correlated each other, and the species of meat was identified by chemometrics using differential metabolite sets. Another volatilomic approach using HS/SPME GC–MS was also applied for the similar purpose [70]. In this study, multiple volatile compounds correlated not only to beef and pork but also the mixture was identified. With use of datasets divided 70% for model calibration and 30% for model prediction, the overall correct classification rate was 99% on average in both datasets. As the authors concluded, this volatilomic approach could be developed for robust and reliable off-line discrimination of meat samples. LC–MS approaches were also applied to classification of chicken meats into normally slaughtered and dead on arrival [71,72]. Such applications along with the development of discrimination analysis is expected to solve the current fraud issues related not only to chicken but also beef and pork.

On the other hand, the irradiation of meat is currently highlighted as an issue for its negative impact in cases when used out of appropriate range of strength, despite positive evaluation of the irradiation effect on meat such as disinfestation, growth inhibition, parasite control, reduction of pathogenic bacteria, and shelf-life extension. Metabolomics is expected to be a tool to monitor the negative impact of irradiation, but it needs to be a non-time-consuming, non-invasive, and reproducible method. In this regard, NMR metabolomics seems suitable for such purpose, and, accordingly, has been used in attempts to investigate the influence of irradiation on beef [73,74]. In the analysis targeting lipids, stepwise linear discriminant analysis (sLDA) following the NMR data profiling allowed the classification of 81.9% of the beef samples according to the irradiation dose (0, 2.5, 4.5, and 8 kGy) [73]. Moreover, the NMR analysis targeting hydrophilic compounds, combined with subsequent classification tree (CT) analysis, was able to distinguish between the irradiated and non-irradiated beef samples [74]. In addition, glycerol, lactic acid esters, and tyramine were found to be important biomarkers for the classification. Thus, MEATabolomics has been applied in a variety of research fields associated with meat production methods and the quality traits, including authentication.

5. Conclusions

MEATabolomics has allowed us to better understand skeletal muscle physiology in animals and molecular factors associated with meat quality. Information raised in MEATabolomics can be used as phenotypic indices of muscle properties and meat quality traits, which connects the external phenotype of meat (the quality traits) to regulatory factors in animals or conditions in the production systems. The techniques for metabolomics have been progressing, especially in the process of statistical data analyses, as shown in some examples of attempts to introduce new algorithms and to develop models for phenotype prediction. On the other hand, metabolomics targeting animal blood and meat exudate metabolome, along with use of meat metabolome data, would promote the prediction of meat quality as a non-invasive methodology by utilizing these data. Further studies on the associations of muscles and meat with animal development, stress conditions, welfare and sustainability issues, cooking methods, consumer acceptability, and sensory characteristics would be future challenges for MEATabolomics to be applied to, despite the complex biological processes during meat production and difficulties in chemical identification of secondary metabolites observed in processed and cooked meat. Nevertheless, MEATabolomics is expected to further expand comprehensive association studies with genomics or transcriptomics in an integrative fashion, being supported by the development of new detecting techniques such as REIMS and statistical analytic resources.

Author Contributions: Conceptualization, S.M.; writing—original draft preparation, S.M.; writing—review and editing, T.M., S.U., T.K., and P.E. All authors have read and agreed to the published version of the manuscript.

References

1. Rubert, J.; Zachariasova, M.; Hajslova, J. Advances in high-resolution mass spectrometry based on metabolomics studies for food—A review. *Food Addit. Contam. Part. A* **2015**, *32*, 1685–1708. [CrossRef] [PubMed]
2. Consonni, R.; Cagliani, L.R. The potentiality of NMR-based metabolomics in food science and food authentication assessment. *Magn. Reson. Chem.* **2019**, *57*, 558–578. [CrossRef] [PubMed]

3. García-Cañas, V.; Simó, C.; Herrero, M.; Ibáñez, E.; Cifuentes, A. Present and Future Challenges in Food Analysis: Foodomics. *Anal. Chem.* **2012**, *84*, 10150–10159. [CrossRef] [PubMed]
4. Junot, C.; Fenaille, F.; Colsch, B.; Bécher, F. High resolution mass spectrometry based techniques at the crossroads of metabolic pathways. *Mass Spectrom. Rev.* **2014**, *33*, 471–500. [CrossRef]
5. Sugimoto, M.; Kawakami, M.; Robert, M.; Soga, T.; Tomita, M. Bioinformatics Tools for Mass Spectroscopy-Based Metabolomic Data Processing and Analysis. *Curr. Bioinform.* **2012**, *7*, 96–108. [CrossRef]
6. Straadt, I.K.; Aaslyng, M.D.; Bertram, H.C. An NMR-based metabolomics study of pork from different crossbreeds and relation to sensory perception. *Meat Sci.* **2014**, *96*, 719–728. [CrossRef]
7. Lytou, A.E.; Nychas, G.E.; Panagou, E.Z. Effect of pomegranate based marinades on the microbiological, chemical and sensory quality of chicken meat: A metabolomics approach. *Int. J. Food Microbiol.* **2018**, *267*, 42–53. [CrossRef]
8. Lee, S.M.; Kwon, G.Y.; Kim, K.O.; Kim, Y.S. Metabolomic approach for determination of key volatile compounds related to beef flavor in glutathione-Maillard reaction products. *Anal. Chim. Acta* **2011**, *703*, 204–211. [CrossRef]
9. Adamski, J.; Suhre, K. Metabolomics platforms for genome wide association studies–linking the genome to the metabolome. *Curr. Opin. Biotechnol.* **2013**, *24*, 39–47. [CrossRef]
10. Fontanesi, L. Metabolomics and livestock genomics: Insights into a phenotyping frontier and its applications in animal breeding. *Anim. Front.* **2016**, *6*, 73–79. [CrossRef]
11. Berri, C.; Picard, B.; Lebret, B.; Andueza, D.; Lefevre, F.; Le Bihan-Duval, E.; Beauclercq, S.; Chartrin, P.; Vautier, A.; Legrand, I.; et al. Predicting the Quality of Meat: Myth or Reality? *Foods* **2019**, *8*. [CrossRef] [PubMed]
12. Zampiga, M.; Flees, J.; Meluzzi, A.; Dridi, S.; Sirri, F. Application of omics technologies for a deeper insight into quali-quantitative production traits in broiler chickens: A review. *J. Anim. Sci. Biotechnol.* **2018**, *9*, 61. [CrossRef] [PubMed]
13. Picard, B.; Lebret, B.; Cassar-Malek, I.; Liaubet, L.; Berri, C.; Le Bihan-Duval, E.; Hocquette, J.F.; Renand, G. Recent advances in omic technologies for meat quality management. *Meat Sci.* **2015**, *109*, 18–26. [CrossRef] [PubMed]
14. Muroya, S.; Oe, M.; Nakajima, I.; Ojima, K.; Chikuni, K. CE-TOF MS-based metabolomic profiling revealed characteristic metabolic pathways in postmortem porcine fast and slow type muscles. *Meat Sci.* **2014**, *98*, 726–735. [CrossRef]
15. Ma, D.; Kim, Y.H.B.; Cooper, B.; Oh, J.H.; Chun, H.; Choe, J.H.; Schoonmaker, J.P.; Ajuwon, K.; Min, B. Metabolomics Profiling to Determine the Effect of Postmortem Aging on Color and Lipid Oxidative Stabilities of Different Bovine Muscles. *J. Agric. Food Chem.* **2017**, *65*, 6708–6716. [CrossRef]
16. Yu, Q.; Tian, X.; Shao, L.; Li, X.; Dai, R. Targeted metabolomics to reveal muscle-specific energy metabolism between bovine longissimus lumborum and psoas major during early postmortem periods. *Meat Sci.* **2019**, *156*, 166–173. [CrossRef]
17. Muroya, S.; Oe, M.; Ojima, K.; Watanabe, A. Metabolomic approach to key metabolites characterizing postmortem aged loin muscle of Japanese Black (Wagyu) cattle. *Asian Australas. J. Anim. Sci.* **2019**, *32*, 1172–1185. [CrossRef]
18. Kodani, Y.; Miyakawa, T.; Komatsu, T.; Tanokura, M. NMR-based metabolomics for simultaneously evaluating multiple determinants of primary beef quality in Japanese Black cattle. *Sci. Rep.* **2017**, *7*, 1297. [CrossRef]
19. Graham, S.F.; Kennedy, T.; Chevallier, O.; Gordon, A.; Farmer, L.; Elliott, C.; Moss, B. The application of NMR to study changes in polar metabolite concentrations in beef longissimus dorsi stored for different periods post mortem. *Metabolomics* **2010**, *6*, 395–404. [CrossRef]
20. Carrillo, J.A.; He, Y.; Li, Y.; Liu, J.; Erdman, R.A.; Sonstegard, T.S.; Song, J. Integrated metabolomic and transcriptome analyses reveal finishing forage affects metabolic pathways related to beef quality and animal welfare. *Sci. Rep.* **2016**, *6*, 25948. [CrossRef]
21. Yu, K.; Matzapetakis, M.; Valent, D.; Saco, Y.; De Almeida, A.M.; Terre, M.; Bassols, A. Skeletal muscle metabolomics and blood biochemistry analysis reveal metabolic changes associated with dietary amino acid supplementation in dairy calves. *Sci. Rep.* **2018**, *8*, 13850. [CrossRef] [PubMed]
22. De Zawadzki, A.; Arrivetti, L.O.R.; Vidal, M.P.; Catai, J.R.; Nassu, R.T.; Tullio, R.R.; Berndt, A.; Oliveira, C.R.; Ferreira, A.G.; Neves-Junior, L.F.; et al. Mate extract as feed additive for improvement of beef quality. *Food Res. Int.* **2017**, *99*, 336–347. [CrossRef] [PubMed]

23. Li, G.; Fu, Y.; Han, X.; Li, X.; Li, C. Metabolomic investigation of porcine muscle and fatty tissue after Clenbuterol treatment using gas chromatography/mass spectrometry. *J. Chromatogr. A* **2016**, *1456*, 242–248. [CrossRef] [PubMed]
24. Watanabe, G.; Kobayashi, H.; Shibata, M.; Kubota, M.; Kadowaki, M.; Fujimura, S. Regulation of free glutamate content in meat by dietary lysine in broilers. *Anim. Sci. J.* **2015**, *86*, 435–442. [CrossRef]
25. Liu, C.; Pan, D.; Ye, Y.; Cao, J. ^{1}H NMR and multivariate data analysis of the relationship between the age and quality of duck meat. *Food Chem.* **2013**, *141*, 1281–1286. [CrossRef]
26. Guitton, Y.; Dervilly-Pinel, G.; Jandova, R.; Stead, S.; Takats, Z.; Le Bizec, B. Rapid evaporative ionisation mass spectrometry and chemometrics for high-throughput screening of growth promoters in meat producing animals. *Food Addit. Contam. Part A* **2018**, *35*, 900–910. [CrossRef]
27. Sugimoto, M.; Obiya, S.; Kaneko, M.; Enomoto, A.; Honma, M.; Wakayama, M.; Soga, T.; Tomita, M. Metabolomic Profiling as a Possible Reverse Engineering Tool for Estimating Processing Conditions of Dry-Cured Hams. *J. Agric. Food Chem.* **2017**, *65*, 402–410. [CrossRef]
28. Zhang, J.; Ye, Y.; Sun, Y.; Pan, D.; Ou, C.; Dang, Y.; Wang, Y.; Cao, J.; Wang, D. ^{1}H NMR and multivariate data analysis of the differences of metabolites in five types of dry-cured hams. *Food Res. Int.* **2018**, *113*, 140–148. [CrossRef]
29. Lytou, A.E.; Panagou, E.Z.; Nychas, G.E. Effect of different marinating conditions on the evolution of spoilage microbiota and metabolomic profile of chicken breast fillets. *Food Microbiol.* **2017**, *66*, 141–149. [CrossRef]
30. Yang, Y.; Ye, Y.; Pan, D.; Sun, Y.; Wang, Y.; Cao, J. Metabonomics profiling of marinated meat in soy sauce during processing. *J. Sci. Food Agric.* **2018**, *98*, 1325–1331. [CrossRef]
31. Huff-Lonergan, E.; Zhang, W.; Lonergan, S.M. Biochemistry of postmortem muscle—Lessons on mechanisms of meat tenderization. *Meat Sci.* **2010**, *86*, 184–195. [CrossRef]
32. Ramautar, R.; Somsen, G.W.; De Jong, G.J. CE-MS for metabolomics: Developments and applications in the period 2016–2018. *Electrophoresis* **2019**, *40*, 165–179. [CrossRef]
33. Stolz, A.; Jooss, K.; Hocker, O.; Romer, J.; Schlecht, J.; Neususs, C. Recent advances in capillary electrophoresis-mass spectrometry: Instrumentation, methodology and applications. *Electrophoresis* **2019**, *40*, 79–112. [CrossRef]
34. Welzenbach, J.; Neuhoff, C.; Looft, C.; Schellander, K.; Tholen, E.; Grosse-Brinkhaus, C. Different Statistical Approaches to Investigate Porcine Muscle Metabolome Profiles to Highlight New Biomarkers for Pork Quality Assessment. *PLoS ONE* **2016**, *11*, e0149758. [CrossRef] [PubMed]
35. Welzenbach, J.; Neuhoff, C.; Heidt, H.; Cinar, M.U.; Looft, C.; Schellander, K.; Tholen, E.; Grosse-Brinkhaus, C. Integrative Analysis of Metabolomic, Proteomic and Genomic Data to Reveal Functional Pathways and Candidate Genes for Drip Loss in Pigs. *Int. J. Mol. Sci.* **2016**, *17*. [CrossRef] [PubMed]
36. Gredell, D.A.; Schroeder, A.R.; Belk, K.E.; Broeckling, C.D.; Heuberger, A.L.; Kim, S.-Y.; King, D.A.; Shackelford, S.D.; Sharp, J.L.; Wheeler, T.L.; et al. Comparison of Machine Learning Algorithms for Predictive Modeling of Beef Attributes Using Rapid Evaporative Ionization Mass Spectrometry (REIMS) Data. *Sci. Rep.* **2019**, *9*, 5721. [CrossRef]
37. Cameron, S.J.S.; Alexander, J.L.; Bolt, F.; Burke, A.; Ashrafian, H.; Teare, J.; Marchesi, J.R.; Kinross, J.; Li, J.V.; Takáts, Z. Evaluation of Direct from Sample Metabolomics of Human Feces Using Rapid Evaporative Ionization Mass Spectrometry. *Anal. Chem.* **2019**, *91*, 13448–13457. [CrossRef] [PubMed]
38. Balog, J.; Sasi-Szabó, L.; Kinross, J.; Lewis, M.R.; Muirhead, L.J.; Veselkov, K.; Mirnezami, R.; Dezső, B.; Damjanovich, L.; Darzi, A.; et al. Intraoperative tissue identification using rapid evaporative ionization mass spectrometry. *Sci. Transl. Med.* **2013**, *5*, 194ra193. [CrossRef] [PubMed]
39. Ritota, M.; Casciani, L.; Failla, S.; Valentini, M. HRMAS-NMR spectroscopy and multivariate analysis meat characterisation. *Meat Sci.* **2012**, *92*, 754–761. [CrossRef] [PubMed]
40. Subbaraj, A.K.; Kim, Y.H.; Fraser, K.; Farouk, M.M. A hydrophilic interaction liquid chromatography-mass spectrometry (HILIC-MS) based metabolomics study on colour stability of ovine meat. *Meat Sci.* **2016**, *117*, 163–172. [CrossRef]
41. Beauclercq, S.; Nadal-Desbarats, L.; Hennequet-Antier, C.; Collin, A.; Tesseraud, S.; Bourin, M.; Le Bihan-Duval, E.; Berri, C. Serum and Muscle Metabolomics for the Prediction of Ultimate pH, a Key Factor for Chicken-Meat Quality. *J. Proteom. Res.* **2016**, *15*, 1168–1178. [CrossRef] [PubMed]
42. Takakura, Y.; Sakamoto, T.; Hirai, S.; Masuzawa, T.; Wakabayashi, H.; Nishimura, T. Characterization of

the key aroma compounds in beef extract using aroma extract dilution analysis. *Meat Sci.* **2014**, *97*, 27–31. [CrossRef] [PubMed]
43. Watanabe, A.; Kamada, G.; Imanari, M.; Shiba, N.; Yonai, M.; Muramoto, T. Effect of aging on volatile compounds in cooked beef. *Meat Sci.* **2015**, *107*, 12–19. [CrossRef] [PubMed]
44. Ueda, S.; Iwamoto, E.; Kato, Y.; Shinohara, M.; Shirai, Y.; Yamanoue, M. Comparative metabolomics of Japanese Black cattle beef and other meats using gas chromatography-mass spectrometry. *Biosci. Biotechnol. Biochem.* **2018**. [CrossRef]
45. Xiao, Z.; Ge, C.; Zhou, G.; Zhang, W.; Liao, G. ^{1}H NMR-based metabolic characterization of Chinese Wuding chicken meat. *Food Chem.* **2019**, *274*, 574–582. [CrossRef]
46. Sundekilde, U.K.; Rasmussen, M.K.; Young, J.F.; Bertram, H.C. High resolution magic angle spinning NMR spectroscopy reveals that pectoralis muscle dystrophy in chicken is associated with reduced muscle content of anserine and carnosine. *Food Chem.* **2017**, *217*, 151–154. [CrossRef]
47. Abasht, B.; Mutryn, M.F.; Michalek, R.D.; Lee, W.R. Oxidative Stress and Metabolic Perturbations in Wooden Breast Disorder in Chickens. *PLoS ONE* **2016**, *11*, e0153750. [CrossRef]
48. Wang, Y.; Yang, Y.; Pan, D.; He, J.; Cao, J.; Wang, H.; Ertbjerg, P. Metabolite profile based on ^{1}H NMR of broiler chicken breasts affected by wooden breast myodegeneration. *Food Chem.* **2020**, *310*, 125852. [CrossRef]
49. Xing, T.; Zhao, X.; Xu, X.; Li, J.; Zhang, L.; Gao, F. Physiochemical properties, protein and metabolite profiles of muscle exudate of chicken meat affected by wooden breast myopathy. *Food Chem.* **2020**, *316*, 126271. [CrossRef]
50. Boerboom, G.; Van Kempen, T.; Navarro-Villa, A.; Perez-Bonilla, A. Unraveling the cause of white striping in broilers using metabolomics. *Poult. Sci.* **2018**, *97*, 3977–3986. [CrossRef]
51. Consolo, N.R.B.; Silva, J.D.; Buarque, V.L.M.; Higuera-Padilla, A.; Barbosa, L.; Zawadzki, A.; Colnago, L.A.; Netto, A.S.; Gerrard, D.E.; Silva, S.L. Selection for Growth and Precocity Alters Muscle Metabolism in Nellore Cattle. *Metabolites* **2020**, *10*. [CrossRef] [PubMed]
52. Sakuma, H.; Saito, K.; Kohira, K.; Ohhashi, F.; Shoji, N.; Uemoto, Y. Estimates of genetic parameters for chemical traits of meat quality in Japanese black cattle. *Anim. Sci. J.* **2017**, *88*, 203–212. [CrossRef] [PubMed]
53. Komatsu, T.; Komatsu, M.; Uemoto, Y. The NT5E gene variant strongly affects the degradation rate of inosine 5′-monophosphate under postmortem conditions in Japanese Black beef. *Meat Sci.* **2019**, *158*, 107893. [CrossRef] [PubMed]
54. Muroya, S.; Oe, M.; Ojima, K. Thiamine accumulation and thiamine triphosphate decline occur in parallel with ATP exhaustion during postmortem aging of pork muscles. *Meat Sci.* **2018**, *137*, 228–234. [CrossRef]
55. Lana, A.; Longo, V.; Dalmasso, A.; D'Alessandro, A.; Bottero, M.T.; Zolla, L. Omics integrating physical techniques: Aged Piedmontese meat analysis. *Food Chem.* **2015**, *172*, 731–741. [CrossRef]
56. Kim, Y.H.B.; Kemp, R.; Samuelsson, L.M. Effects of dry-aging on meat quality attributes and metabolite profiles of beef loins. *Meat Sci.* **2016**, *111*, 168–176. [CrossRef]
57. Warner, R.D.; Jacob, R.H.; Rosenvold, K.; Rochfort, S.; Trenerry, C.; Plozza, T.; McDonagh, M.B. Altered post-mortem metabolism identified in very fast chilled lamb M. longissimus thoracis et lumborum using metabolomic analysis. *Meat Sci.* **2015**, *108*, 155–164. [CrossRef]
58. Mitacek, R.M.; Ke, Y.; Prenni, J.E.; Jadeja, R.; VanOverbeke, D.L.; Mafi, G.G.; Ramanathan, R. Mitochondrial Degeneration, Depletion of NADH, and Oxidative Stress Decrease Color Stability of Wet-Aged Beef Longissimus Steaks. *J. Food Sci.* **2019**, *84*, 38–50. [CrossRef]
59. García-García, A.B.; Lamichhane, S.; Castejón, D.; Cambero, M.I.; Bertram, H.C. ^{1}H HR-MAS NMR-based metabolomics analysis for dry-fermented sausage characterization. *Food Chem.* **2018**, *240*, 514–523. [CrossRef]
60. Jiang, T.; Bratcher, C.L. Differentiation of commercial ground beef products and correlation between metabolites and sensory attributes: A metabolomic approach. *Food Res. Int.* **2016**, *90*, 298–306. [CrossRef]
61. Suzuki, K.; Shioura, H.; Yokota, S.; Katoh, K.; Roh, S.-g.; Iida, F.; Komatsu, T.; Syoji, N.; Sakuma, H.; Yamada, S. Search for an index for the taste of Japanese Black cattle beef by panel testing and chemical composition analysis. *Anim. Sci. J.* **2017**, *88*, 421–432. [CrossRef] [PubMed]
62. Xu, Y.; Cheung, W.; Winder, C.L.; Goodacre, R. VOC-based metabolic profiling for food spoilage detection with the application to detecting Salmonella typhimurium-contaminated pork. *Anal. Bioanal. Chem.* **2010**, *397*, 2439–2449. [CrossRef]
63. Xu, Y.; Cheung, W.; Winder, C.L.; Dunn, W.B.; Goodacre, R. Metabolic profiling of meat: Assessment of pork hygiene and contamination with Salmonella typhimurium. *Analyst* **2011**, *136*, 508–514. [CrossRef] [PubMed]

64. Argyri, A.A.; Doulgeraki, A.I.; Blana, V.A.; Panagou, E.Z.; Nychas, G.-J.E. Potential of a simple HPLC-based approach for the identification of the spoilage status of minced beef stored at various temperatures and packaging systems. *Int. J. Food Microbiol.* **2011**, *150*, 25–33. [CrossRef] [PubMed]
65. Argyri, A.A.; Mallouchos, A.; Panagou, E.Z.; Nychas, G.J. The dynamics of the HS/SPME-GC/MS as a tool to assess the spoilage of minced beef stored under different packaging and temperature conditions. *Int. J. Food Microbiol.* **2015**, *193*, 51–58. [CrossRef]
66. Jung, Y.; Lee, J.; Kwon, J.; Lee, K.S.; Ryu, D.H.; Hwang, G.S. Discrimination of the geographical origin of beef by ^{1}H NMR-based metabolomics. *J. Agric. Food Chem.* **2010**, *58*, 10458–10466. [CrossRef]
67. Zaima, N.; Goto-Inoue, N.; Hayasaka, T.; Enomoto, H.; Setou, M. Authenticity assessment of beef origin by principal component analysis of matrix-assisted laser desorption/ionization mass spectrometric data. *Anal. Bioanal. Chem.* **2011**, *400*, 1865–1871. [CrossRef]
68. Osorio, M.T.; Moloney, A.P.; Brennan, L.; Monahan, F.J. Authentication of beef production systems using a metabolomic-based approach. *Animal* **2012**, *6*, 167–172. [CrossRef]
69. Trivedi, D.K.; Hollywood, K.A.; Rattray, N.J.; Ward, H.; Greenwood, J.; Ellis, D.I.; Goodacre, R. Meat, the metabolites: An integrated metabolite profiling and lipidomics approach for the detection of the adulteration of beef with pork. *Analyst* **2016**, *141*, 2155–2164. [CrossRef]
70. Pavlidis, D.E.; Mallouchos, A.; Ercolini, D.; Panagou, E.Z.; Nychas, G.E. A volatilomics approach for off-line discrimination of minced beef and pork meat and their admixture using HS-SPME GC/MS in tandem with multivariate data analysis. *Meat Sci.* **2019**, *151*, 43–53. [CrossRef]
71. Sidwick, K.L.; Johnson, A.E.; Adam, C.D.; Pereira, L.; Thompson, D.F. Use of Liquid Chromatography Quadrupole Time-of-Flight Mass Spectrometry and Metabonomic Profiling To Differentiate between Normally Slaughtered and Dead on Arrival Poultry Meat. *Anal. Chem.* **2017**, *89*, 12131–12136. [CrossRef]
72. Cao, M.; Han, Q.; Zhang, J.; Zhang, R.; Wang, J.; Gu, W.; Kang, W.; Lian, K.; Ai, L. An untargeted and pseudotargeted metabolomic combination approach to identify differential markers to distinguish live from dead pork meat by liquid chromatography-mass spectrometry. *J. Chromatograph. A* **2020**, *1610*, 460553. [CrossRef] [PubMed]
73. Zanardi, E.; Caligiani, A.; Padovani, E.; Mariani, M.; Ghidini, S.; Palla, G.; Ianieri, A. Detection of irradiated beef by nuclear magnetic resonance lipid profiling combined with chemometric techniques. *Meat Sci.* **2013**, *93*, 171–177. [CrossRef]
74. Zanardi, E.; Caligiani, A.; Palla, L.; Mariani, M.; Ghidini, S.; Di Ciccio, P.A.; Palla, G.; Ianieri, A. Metabolic profiling by ^{1}H NMR of ground beef irradiated at different irradiation doses. *Meat Sci.* **2015**, *103*, 83–89. [CrossRef] [PubMed]
75. Ren, S.; Hinzman, A.A.; Kang, E.L.; Szczesniak, R.D.; Lu, L.J. Computational and statistical analysis of metabolomics data. *Metabolomics* **2015**, *11*, 1492–1513. [CrossRef]
76. Hollung, K.; Timperio, A.M.; Olivan, M.; Kemp, C.; Coto-Montes, A.; Sierra, V.; Zolla, L. Systems biology: A new tool for farm animal science. *Curr. Protein Pept. Sci.* **2014**, *15*, 100–117. [CrossRef] [PubMed]
77. Carpenter, C.E.; Cornforth, D.P.; Whittier, D. Consumer preferences for beef color and packaging did not affect eating satisfaction. *Meat Sci.* **2001**, *57*, 359–363. [CrossRef]
78. Arihara, K.; Yokoyama, I.; Ohata, M. DMHF (2,5-dimethyl-4-hydroxy-3(2H)-furanone), a volatile food component with attractive sensory properties, brings physiological functions through inhalation. *Adv. Food Nutr. Res.* **2019**, *89*, 239–258. [PubMed]
79. Nishimura, T. Mechanism Involved in the Improvement of Meat Taste during Postmortem Aging. *Food Sci. Technol. Int. Tokyo* **1998**, *4*, 241–249. [CrossRef]
80. Mottram, D.S. Flavour formation in meat and meat products: A review. *Food Chem.* **1998**, *62*, 415–424. [CrossRef]
81. Kuttappan, V.A.; Hargis, B.M.; Owens, C.M. White striping and woody breast myopathies in the modern poultry industry: A review. *Poult. Sci.* **2016**, *95*, 2724–2733. [CrossRef] [PubMed]
82. Andersen, H.J.; Oksbjerg, N.; Young, J.F.; Therkildsen, M. Feeding and meat quality—A future approach. *Meat Sci.* **2005**, *70*, 543–554. [CrossRef] [PubMed]
83. Xing, T.; Gao, F.; Tume, R.K.; Zhou, G.; Xu, X. Stress Effects on Meat Quality: A Mechanistic Perspective. *Compr. Rev. Food Sci. Food Saf.* **2019**, *18*, 380–401. [CrossRef]

84. Hwang, Y.-H.; Kim, G.-D.; Jeong, J.-Y.; Hur, S.-J.; Joo, S.-T. The relationship between muscle fiber characteristics and meat quality traits of highly marbled Hanwoo (Korean native cattle) steers. *Meat Sci.* **2010**, *86*, 456–461. [CrossRef]
85. Chang, K.C.; Da Costa, N.; Blackley, R.; Southwood, O.; Evans, G.; Plastow, G.; Wood, J.D.; Richardson, R.I. Relationships of myosin heavy chain fibre types to meat quality traits in traditional and modern pigs. *Meat Sci.* **2003**, *64*, 93–103. [CrossRef]
86. Lee, S.H.; Choe, J.H.; Choi, Y.M.; Jung, K.C.; Rhee, M.S.; Hong, K.C.; Lee, S.K.; Ryu, Y.C.; Kim, B.C. The influence of pork quality traits and muscle fiber characteristics on the eating quality of pork from various breeds. *Meat Sci.* **2012**, *90*, 284–291. [CrossRef] [PubMed]
87. Nam, Y.J.; Choi, Y.M.; Lee, S.H.; Choe, J.H.; Jeong, D.W.; Kim, Y.Y.; Kim, B.C. Sensory evaluations of porcine longissimus dorsi muscle: Relationships with postmortem meat quality traits and muscle fiber characteristics. *Meat Sci.* **2009**, *83*, 731–736. [CrossRef]
88. Lefaucheur, L. A second look into fibre typing–relation to meat quality. *Meat Sci.* **2010**, *84*, 257–270. [CrossRef]
89. Schiaffino, S.; Reggiani, C. Fiber types in mammalian skeletal muscles. *Physiol. Rev.* **2011**, *91*, 1447–1531. [CrossRef]
90. Eggert, J.M.; Depreux, F.F.S.; Schinckel, A.P.; Grant, A.L.; Gerrard, D.E. Myosin heavy chain isoforms account for variation in pork quality. *Meat Sci.* **2002**, *61*, 117–126. [CrossRef]
91. Oe, M.; Ohnishi-Kameyama, M.; Nakajima, I.; Muroya, S.; Shibata, M.; Ojima, K.; Kushibiki, S.; Chikuni, K. Proteome analysis of whole and water-soluble proteins in masseter and semitendinosus muscles of Holstein cows. *Anim. Sci. J.* **2011**, *82*, 181–186. [CrossRef] [PubMed]
92. Hatazawa, Y.; Senoo, N.; Tadaishi, M.; Ogawa, Y.; Ezaki, O.; Kamei, Y.; Miura, S. Metabolomic Analysis of the Skeletal Muscle of Mice Overexpressing PGC-1α. *PLoS ONE* **2015**, *10*, e0129084. [CrossRef] [PubMed]
93. Bourdeau Julien, I.; Sephton, C.F.; Dutchak, P.A. Metabolic Networks Influencing Skeletal Muscle Fiber Composition. *Front. Cell Dev. Biol.* **2018**, *6*, 125. [CrossRef]
94. Muroya, S.; Nakajima, I.; Oe, M.; Chikuni, K. Difference in postmortem degradation pattern among troponin T isoforms expressed in bovine longissimus, diaphragm, and masseter muscles. *Meat Sci.* **2006**, *72*, 245–251. [CrossRef] [PubMed]
95. Muroya, S.; Ertbjerg, P.; Pomponio, L.; Christensen, M. Desmin and troponin T are degraded faster in type IIb muscle fibers than in type I fibers during postmortem aging of porcine muscle. *Meat Sci.* **2010**, *86*, 764–769. [CrossRef]
96. Kitamura, S.-I.; Muroya, S.; Tanabe, S.; Okumura, T.; Chikuni, K.; Nishimura, T. Mechanism of production of troponin T fragments during postmortem aging of porcine muscle. *J. Agric. Food Chem.* **2005**, *53*, 4178–4181. [CrossRef] [PubMed]
97. Muroya, S.; Kitamura, S.-I.; Tanabe, S.; Nishimura, T.; Nakajima, I.; Chikuni, K. N-terminal amino acid sequences of troponin T fragments, including 30 kDa one, produced during postmortem aging of bovine longissimus muscle. *Meat Sci.* **2004**, *67*, 19–24. [CrossRef] [PubMed]
98. Muroya, S.; Ohnishi-Kameyama, M.; Oe, M.; Nakajima, I.; Chikuni, K. Postmortem changes in bovine troponin T isoforms on two-dimensional electrophoretic gel analyzed using mass spectrometry and western blotting: The limited fragmentation into basic polypeptides. *Meat Sci.* **2007**, *75*, 506–514. [CrossRef]
99. Lametsch, R.; Karlsson, A.; Rosenvold, K.; Andersen, H.J.; Roepstorff, P.; Bendixen, E. Postmortem proteome changes of porcine muscle related to tenderness. *J. Agric. Food Chem.* **2003**, *51*, 6992–6997. [CrossRef]
100. Marino, R.; Albenzio, M.; Della Malva, A.; Santillo, A.; Loizzo, P.; Sevi, A. Proteolytic pattern of myofibrillar protein and meat tenderness as affected by breed and aging time. *Meat Sci.* **2013**, *95*, 281–287. [CrossRef]
101. Della Malva, A.; De Palo, P.; Lorenzo, J.M.; Maggiolino, A.; Albenzio, M.; Marino, R. Application of proteomic to investigate the post-mortem tenderization rate of different horse muscles. *Meat Sci.* **2019**, *157*, 107885. [CrossRef] [PubMed]
102. López-Pedrouso, M.; Pérez-Santaescolástica, C.; Franco, D.; Fulladosa, E.; Carballo, J.; Zapata, C.; Lorenzo, J.M. Comparative proteomic profiling of myofibrillar proteins in dry-cured ham with different proteolysis indices and adhesiveness. *Food Chem.* **2018**, *244*, 238–245. [CrossRef] [PubMed]
103. Nychas, G.-J.E.; Skandamis, P.N.; Tassou, C.C.; Koutsoumanis, K.P. Meat spoilage during distribution. *Meat Sci.* **2008**, *78*, 77–89. [CrossRef] [PubMed]

6

Differential Metabolomics Profiles Identified by CE-TOFMS between High and Low Intramuscular Fat Amount in Fattening Pigs

Masaaki Taniguchi [1,*], Aisaku Arakawa [1], Motohide Nishio [1], Toshihiro Okamura [1], Chika Ohnishi [2], Kouen Kadowaki [3], Kimiko Kohira [4], Fumika Homma [4], Kazunori Matsumoto [4] and Kazuo Ishii [1]

1 Institute of Livestock and Grassland Science, National Agriculture and Food Research Organization, Tsukuba Ibaraki 305-0901, Japan; aisaku@affrc.go.jp (A.A.); mtnishio@affrc.go.jp (M.N.); okamut@affrc.go.jp (T.O.); kazishi@affrc.go.jp (K.I.)

2 Miyazaki Station, National Livestock Breeding Center, Kobayashi Miyazaki 886-0004, Japan; c0ohnisi@nlbc.go.jp

3 Ibaraki Station, National Livestock Breeding Center, Chikusei Ibaraki 308-0112, Japan; k0kadowk@nlbc.go.jp

4 National Livestock Breeding Center, Nishigo Fukushima 961-8511, Japan; k0kohira@nlbc.go.jp (K.K.); f0honma@nlbc.go.jp (F.H.); k0matsumoto@nlbc.go.jp (K.M.)

* Correspondence: masaakit@affrc.go.jp.

Abstract: The amount of intramuscular fat (IMF) present in the loin eye area is one of the most important characteristics of high-quality pork. IMF measurements are currently impractical without a labor-intensive process. Metabolomic profiling could be used as an IMF indicator to avoid this process; however, no studies have investigated their use during the fattening period of pigs. This study examined the metabolite profiles in the plasma of two groups of pigs derived from the same Duroc genetic line and fed the same diet. Five plasma samples were collected from each individual the day before slaughter. Capillary electrophoresis-time of flight mass spectrometry (CE-TOFMS) was used to analyze the purified plasma from each sample. Principle component analysis (PCA) and partial least squares (PLS) were used to find the semi-quantitative values of the compounds. The results indicate that branched-chain amino acids are significantly associated with high IMF content, while amino acids are associated with low IMF content. These differences were validated using the quantification analyses by high-performance liquid chromatograph, which supported our results. These results suggest that the concentration of branched-chain amino acids in plasma could be an indicative biomarker for the IMF content in the loin eye area.

Keywords: biomarker; CE-TOFMS; intramuscular fat; meat quality; metabolomics; porcine

1. Introduction

Pork consumers depend primarily on meat quality when making purchasing decisions. Intramuscular fat (IMF) content in the loin eye muscle, also known as marbling, is a particularly valuable trait of high-quality pork and is associated with the meat's flavor. IMF content may vary by breed types, muscle types, genders, ages, feeding conditions, and final slaughter weight [1–3]. The heritability of IMF content in Duroc breed pigs has been estimated to be moderate to high (0.39 to 0.69) [4–11]. This has led to improvements in the phenotypic value of livestock in recent decades, despite that IMF is recognized as a polygenic trait [1,12].

Pigs grown on energy restricted diets accumulate significantly more IMF content when their gene expression profiles in relation to protein, glycogen and lipid turnover are altered [13]. Dietary regulation, particularly of lysine, has also resulted high amounts of IMF in pork [14]. A better understanding of the molecular and biological mechanisms underlying adipocytes development within muscle tissue could help to create effective nutritional controls during the fattening period of livestock, however few studies have been conducted to date.

Metabolomics are of particular interest in animal production as postmortem aging of meat can be predicted using time course analysis of capillary electrophoresis-time of flight mass spectrometer (CE-TOFMS) metabolomics. Such findings can contribute to the improvement of pork meat quality checks prior to slaughter [15]. More recently, Muroya et al. [16] proposed a novel concept, "MEATabolomics", which combines the muscle biology and meat metabolomics of domestic animals. Additionally, metabolomics data can be defined as intermediate phenotypes, as metabolites are found between the genome level and the external phenotype level, such as growth rate, fat deposition and other economic traits. Metabolomics could lead to "next-generation phenotyping" approaches that improve the prediction of breeding values and selection schemes [17]. Furthermore, Suravajhala et al. [18] reported that high-throughput omics technologies have contributed to our understanding of complex biological phenomena, disease resistance, and holistic production improvements. Hence, metabolomics data may have a potential to improve prediction accuracy of genomic selection and to enhance the impact of breeding schemes for traits related to animal production by incorporating multiple layers of the high-throughput omics approach. In this context, development of non-invasive measurement technologies to obtain the metabolomics data is key for precise evaluation of target traits.

The objective of this study was to analyze and compare metabolite profiles in high and low IMF content pigs, in order to identify metabolites that could be used as IMF content indicators in pigs. To achieve this goal, we used a CE-TOFMS as our primary metabolomics technology due to its highly-sensitive, broad-range detection ability. We used the CE-TOFMS system to screen for candidate metabolites via a semi-quantitative method, then used an absolute quantification method to identify which candidate metabolites are associated with IMF content. We concluded that branched-chain amino acids (BCAA), glycine (Gly), anserine, and carnosine could be used as potential non-invasive biomarkers to estimate IMF content in the loin eye muscle of pigs before slaughter.

2. Results

2.1. Phenotypic Characteristics

Pigs were separated into low and high IMF groups at two locations of the National Livestock Breeding Center (NLBC), based on IMF (%) content in the loin eye muscle. The mean proportion of IMF content in pigs at the Miyazaki (MIY) station were 3.1% and 8.7%, respectively. The mean proportion of IMF content in pigs at the Ibaraki station (IBR) were 2.9% and 5.4%, respectively (Table 1). The differences were highly significant at both stations. Moisture (%) measured in the muscle tissues of low IMF pigs were significantly higher than in the high IMF pigs at both stations. At the MIY station, protein (%) and loin eye area (cm^2) in low IMF pigs were significantly 8% and 11% higher than in high IMF pigs, respectively. At the IBR station, cooking loss (%) was significantly 15% higher than those of high IMF pigs.

Table 1. Comparison of growth performance and carcass characteristics between high and low intramuscular fat (IMF) groups at two stations.

Growth Performance and Carcass Characteristics	Miyazaki (MIY) Station			Ibaraki (IBR) Station		
	L-IMF MIY	H-IMF MIY	p *1	L-IMF IBR	H-IMF IBR	p *1
IMF (%)	3.101 ± 0.255	8.699 ± 1.441	$p < 0.001$	2.868 ± 0.550	5.447 ± 0.797	$p < 0.001$
Days old at slaughter	149.2 ± 9.99	152.2 ± 13.64	N.S.	154.4 ± 8.357	169.8 ± 6.997	$p < 0.05$
Moisture (%)	73.96 ± 0.339	69.98 ± 0.966	$p < 0.001$	73.81 ± 0.978	71.48 ± 0.444	$p < 0.01$
Crude protein (%)	21.82 ± 0.173	20.23 ± 0.917	$p < 0.01$	22.35 ± 0.149	22.02 ± 0.350	N.S.
Cooking loss (%)	24.68 ± 1.548	26.52 ± 1.823	N.S.	28.04 ± 1.966	24.38 ± 2.269	$p < 0.05$
WHC (%) *2	68.30 ± 5.489	72.30 ± 4.511	N.S.	68.14 ± 5.824	68.94 ± 4.199	N.S.
WBSF (kg) *3	2.392 ± 0.338	2.048 ± 0.489	N.S.	2.963 ± 0.550	2.838 ± 0.478	N.S.
Tenderness (kg/cm^2)	38.09 ± 5.586	34.81 ± 2.010	N.S.	46.45 ± 6.442	44.82 ± 6.710	N.S.
Pliability	1.497 ± 0.0918	1.375 ± 0.0609	N.S.	1.527 ± 0.0442	1.494 ± 0.0679	N.S.
Toughness (kg/cm^2 × cm)	8.163 ± 1.474	8.054 ± 1.082	N.S.	10.27 ± 1.900	10.15 ± 1.888	N.S.
Brittleness	1.631 ± 0.0750	1.662 ± 0.0864	N.S.	1.568 ± 0.0880	1.632 ± 0.110	N.S.
Average daily gain (g/day)	1141.3 ± 62.70	1035.9 ± 188.8	N.S.	1006.7 ± 53.91	1008.8 ± 46.73	N.S.
Loin eye area (cm^2)	35.06 ± 1.318	31.45 ± 2.059	$p < 0.05$	35.10 ± 3.369	34.29 ± 3.219	N.S.
Back fat thickness (cm)	2.597 ± 0.547	3.376 ± 0.548	N.S.	2.044 ± 0.568	2.476 ± 0.661	N.S.
Statue height (cm)	63.16 ± 1.895	61.22 ± 0.722	N.S.	65.48 ± 1.001	64.16 ± 2.330	N.S.
Body length (cm)	107.6 ± 3.855	105.28 ± 2.285	N.S.	89.7 ± 39.35	106.5 ± 5.040	N.S.

*1 Statistical significance is defined when $p < 0.05$. *N.S.* denotes no significance ($p > 0.05$). *2 WHC: Water-holding capacity (%) by centrifugation at 10,000× *g*. *3 WBSF: Warner-Bratzler Shear Force value (Kg) measured after cooling of the carcass.

2.2. *Screening of Differential Metabolomics Profiles with CE-TOFMS*

The total number of metabolites detected in the MIY samples were 201 (144 cations and 57 anions), which were used as the semi-quantitative values (relative area based on the internal standard). Likewise, the IBR metabolites detected were 152 (104 cations and 48 anions). The mean detection rates of metabolites, which are defined as the proportion of metabolites identified among either 201 (MIY) or 152 (IBR), were 82.0% and 88.6%, respectively. The mean detection rate of metabolites at IBR was greater than at MIY, despite the fact that more metabolites were detected at MIY than IBR.

Principal component analysis (PCA) was carried out based on the semi-quantitative values (relative area) of all detected metabolites. At MIY, the high IMF group tended to be in the PC1 positive and PC2 negative regions, while the low IMF group showed higher variability (Figure S1 PCA (A)). The high IMF group at IBR was found in the PC3 and PC4 positive regions, but high and low IMF groups were neither separated by PC1 nor PC2 (Figure S1 PCA (B and C)).

Partial least squares (PLS) was then used to identify possible relationships between IMF content and metabolite profiles at MIY and IBR (Figure 1. Tables S1 and S2). Since the PLS was performed between two groups (low or high IMF), we only considered PLS1 separation. All high IMF pigs and low IMF pigs were clearly in the positive and negative PLS1 scores. Therefore, we considered that positive loading values were associated with high IMF and negative loading values were associated with low IMF (Table 2). Statistically significant metabolites with positive loading values at MIY were leucine (Leu), *O*-acetylhomoserine/2-aminoadipic acid, 1-methylnicotinamide, choline, and phosphorylcholine, while all 10 metabolites detected with negative loading values were significant. Statistically significant metabolites with positive loading values at IBR were urea and gluconic acid, while those of the negative loading values were threonine (Thr) and diethanolamine. At both stations a BCAA, valine (Val) was identified from the positive loading metabolites. At MIY, Leu and isoleucine (Ile) were also identified from the positive loading metabolites. However, as we observed in the PCA analysis, BCAAs (Leu, Ile, and Val), tryptophan (Trp) and amino acid metabolites (*O*-acetylhomoserine/2-aminoadipic acid, *N*,*N*-dimethylglycine) were also detected in the list of positive PLS1 loading metabolites at MIY. Negative PLS1 loading metabolites at MIY included amino acids (Gly and beta-alanine (β-Ala)), amino acid metabolites (*N*-acetylornithine, *N*-acetyllisyne, 5-oxoproline,

N5-ethylglutamine), peptides (anserine-divalent, carnosine) and organic acids (creatine, cis-aconitic acid). At IBR, urea, 3-hydroxybuyric acid, nicotinamide, mucic acid and homocitrulline were commonly observed in the top positive loading metabolites on PCA3 and PCA4, While negative PLS1 loading metabolites consisted of several amino acids (Thr, Asn, Arg, Lys, methionine (Met), and Tyr) and amino acid metabolites (5-hydroxylysine and hydroxyproline).

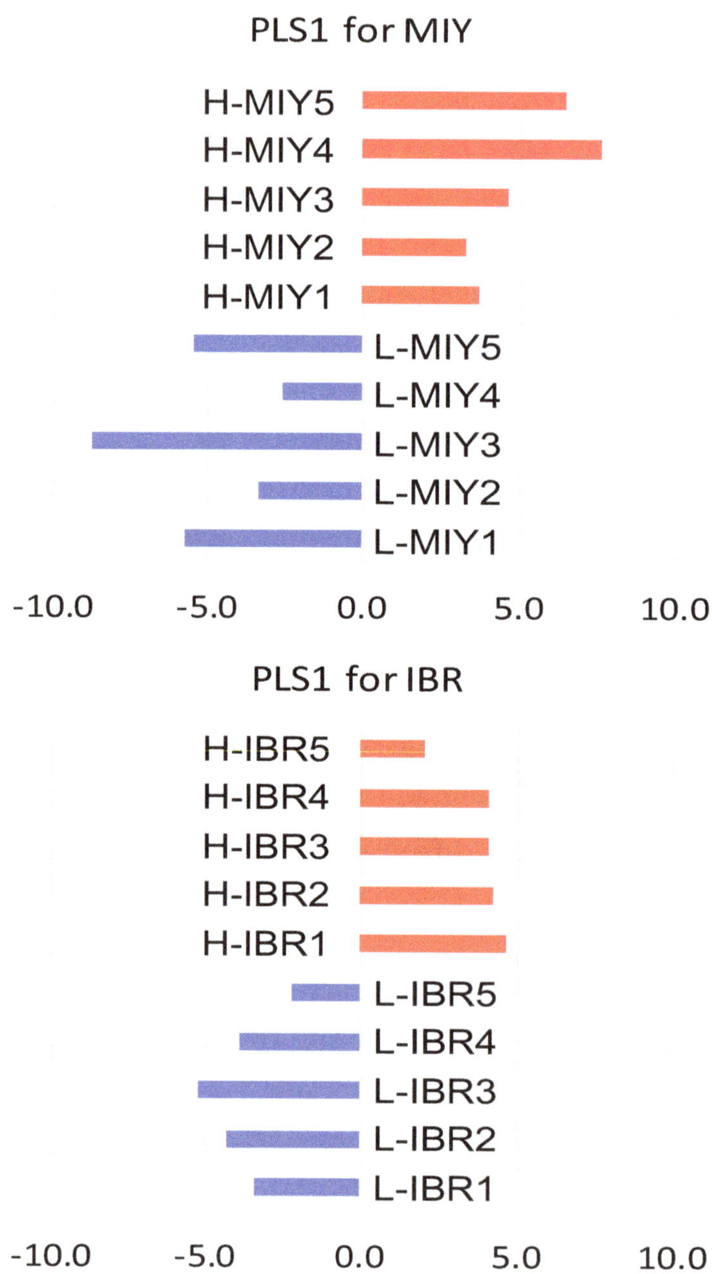

Figure 1. Partial least squares (PLS) analysis by the metabolites detected from pig plasma samples showing the difference in the intramuscular fat contents at the two stations. Blue and red bars indicate low and high IMF pigs, respectively. Top and bottom panes show the results of PLS in MIY and IBR stations, respectively.

Table 2. Top 10 metabolites with positive and negative high loading values in partial least square at two stations.

MIY Station			IBR Station		
Metabolites	PLS1		Metabolites	PLS1	
	R *1	*p*		*R* *1	*p*
Positive loading					
Leu	0.731	1.6×10^{-2}	Urea	0.729	1.7×10^{-2}
O-acetylhomoserine 2-aminoadipic acid	0.727	1.7×10^{-2}	Gluconic acid	0.654	4.0×10^{-2}
1-Methylnicotinamide	0.711	2.1×10^{-2}	3-hydroxybutyric acid	0.544	1.0×10^{-1}
Choline	0.645	4.4×10^{-2}	Isethionic acid	0.505	1.4×10^{-1}
Phosphorylcholine	0.637	4.8×10^{-2}	Nicotinamide	0.494	1.5×10^{-1}
Ile	0.609	6.1×10^{-2}	Val	0.493	1.5×10^{-1}
Creatine	0.603	6.5×10^{-2}	Taurine	0.488	1.5×10^{-1}
Val	0.582	7.8×10^{-2}	Mucic acid	0.476	1.6×10^{-1}
N,N-dimethylglycine	0.569	8.6×10^{-2}	Homocitrulline	0.475	1.7×10^{-1}
Trp	0.470	1.7×10^{-1}	Sarcosine	0.452	1.9×10^{-1}
Negative loading					
Gly	−0.861	1.4×10^{-3}	Thr	−0.716	2.0×10^{-2}
Anserine_divalent	−0.840	2.3×10^{-3}	Diethanolamine	−0.705	2.3×10^{-2}
N-acetylornithine	−0.834	2.7×10^{-3}	Thymidine	−0.611	6.0×10^{-2}
N-acetyllysine	−0.796	5.9×10^{-3}	5-hydroxylysine	−0.589	7.3×10^{-2}
5-oxoproline	−0.753	1.2×10^{-2}	Asn	−0.566	8.8×10^{-2}
Carnosine	−0.741	1.4×10^{-2}	Arg	−0.533	1.1×10^{-1}
Creatinine	−0.707	2.2×10^{-2}	Lys	−0.521	1.2×10^{-1}
N^5-ethylglutamine	−0.705	2.3×10^{-2}	Hydroxyproline	−0.514	1.3×10^{-1}
cis-Aconitic acid	−0.685	2.9×10^{-2}	Met	−0.505	1.4×10^{-1}
β-Ala	−0.683	2.9×10^{-2}	Tyr	−0.459	1.8×10^{-1}

*1 *R* indicates correlation coefficient between PLS score and each metabolite levels. Since *O*-acetylhomoserine/2-aminoadipic acid were detected within the identical single peaks having the consistent *m/z* (molecular mass/electric charge of ions) and MT/RT (migration time/retention time) by the CE-TOFMS system employed in this study, the values in the table indicated the combination of two compounds: The *m/z* and MT/RT for *O*-acetylhomoserine/2-aminoadipic acid is 162.076 and 12.10, respectively.

2.3. Integrated Analysis of Metabolomics Data Using Absolute Quantification

With the CE-TOFMS metabolomics analysis system, the concentration in micro molar (mM) of 110 metabolites was calculated using a standard curve method. We then focused on the differences in these selectively quantified metabolites between the low and high IMF groups at the two stations. Consequently, the total number of metabolites detected in any of the eight samples from MIY and IBR was 59 mM. The mean number of cations detected at MIY was 38.7, and the mean number of anions at MIY was 34.6. At IBR, cations were 13.7, while anions were 11.9. The mean detection ratios of quantified metabolites at MIY were 88.8% and 78.9% at IBR. This indicates that there were a greater number of quantified metabolites identified at MIY than IBR. For the 59 quantified metabolites, we analyzed relationships between IMF content and the metabolites.

We used PLS to further determine which metabolites are definitively associated with IMF content. The PLS score plot showed that the low and high IMF groups at both stations were more separated by PLS2 than PLS1 or PLS3 (Figure 2), while the two stations themselves were clearly separated by PLS1 (Figure 2). We then extracted the top five metabolites with both positively and negatively high loading of PLS2 (Table 3, Table S3). Positively loading PLS2 metabolites included the BCAAs Leu, Ile, and Val, and creatine, which were found to be significant metabolites with relatively moderate correlation coefficient (R). Comparison of these BCAAs and creatine concentrations at MIY indicated that the high IMF group displayed a significantly higher amount or tended to show greater amounts of the metabolites than those in the low IMF group. Basically, these same metabolites, BCAAs and creatine in the high IMF group at IBR also showed a greater amount than those of low IMF, although the difference was not as large as at MIY.

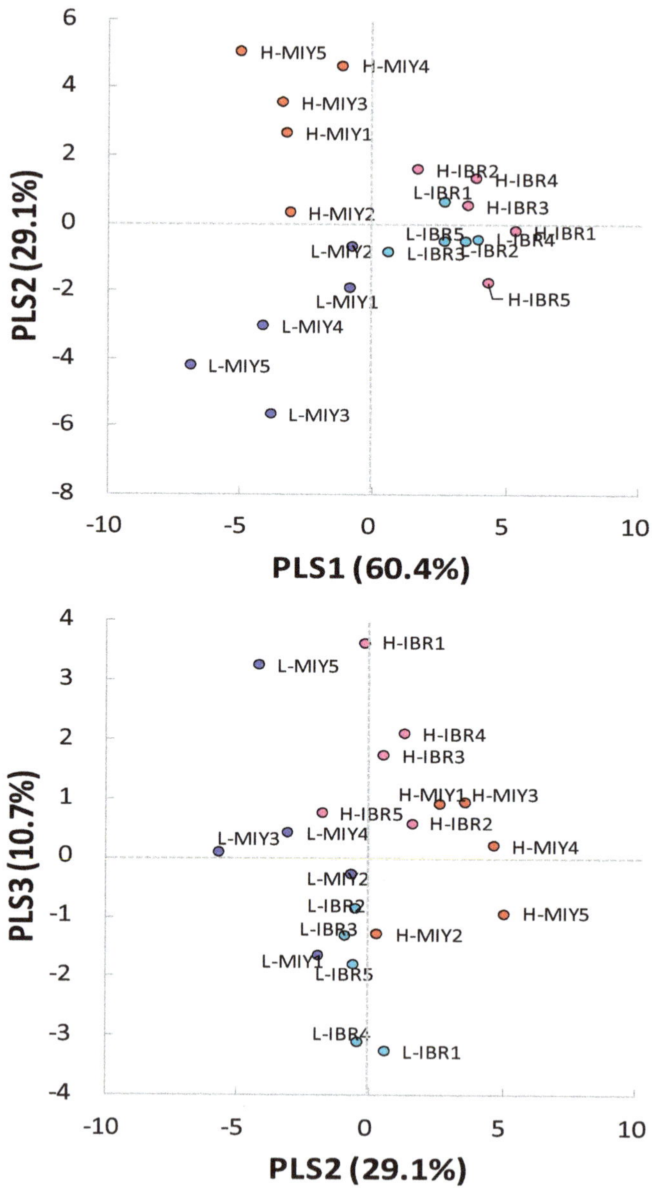

Figure 2. Integrated analyses of partial least squares by the metabolites detected from pig plasma samples showing the difference in the intramuscular fat contents at the two stations. Blue and red dots indicate low and high IMF pigs from MIY station, respectively. Cyan and magenta dots indicate low and high IMF pigs from IBR station, respectively. Top and bottom panes show the results of PLS analyses with PLS1 and 2 and with PLS2 and 3, respectively.

All metabolites from the negatively loading PLS2 metabolites were determined significant with slightly higher correlation coefficients as compared to those in the positively loading metabolites. Similar to the positive loading metabolites, the differences in the negatively loading PLS2 metabolites concentrations between the high and low IMF groups at MIY were clearer and more significant than those at IBR. Consequently, all metabolites identified by PLS2 factor loadings were commonly found in the list of metabolites which were identified with the one-factor PLS analysis at MIY and IBR (Table 2).

Table 3. Top 5 metabolites with positive and negative high loading values of PLS2 in the integrated analysis regarding IMF content at two stations.

Metabolites	R *1	p *2	MIY Station			IBR Station			Total		
			Low IMF	High IMF	p *2	Low IMF	High IMF	p *2	Low IMF	High IMF	p *2
Positively Loading											
Leu	0.610	***	153.33 ± 6.18	180.6 ± 6.56	*	156.41 ± 6.46	165.16 ± 6.13	*N.S.*	154.86 ± 5.81	172.88 ± 7.00	*
Ile	0.577	***	93.43 ± 9.25	120.37 ± 8.14	*tnd.*	107.85 ± 4.29	110.69 ± 2.94	*N.S.*	100.64 ± 7.65	115.53 ± 6.20	*
Val	0.531	*	236.99 ± 19.06	284.70 ± 13.93	*tnd.*	268.24 ± 4.32	285.19 ± 9.64	*N.S.*	252.62 ± 14.94	284.94 ± 11.29	*
Creatine	0.529	*	232.13 ± 34.19	307.37 ± 8.54	*tnd.*	302.38 ± 4.64	317.32 ± 10.72	*N.S.*	267.25 ± 28.34	312.34 ± 9.44	*
Trp	0.370	*tnd.*	60.06 ± 6.08	71.31 ± 4.34	*N.S.*	74.12 ± 2.92	72.079 ± 3.35	*N.S.*	67.08 ± 5.59	71.69 ± 3.66	*N.S.*
Negatively Loading											
Gly	−0.788	***	920.72 ± 48.07	666.51 ± 22.31	**	829.52 ± 14.96	799.15 ± 43.97	*N.S.*	875.12 ± 39.86	732.83 ± 45.36	***
Creatinine	−0.611	***	102.50 ± 5.45	82.42 ± 4.56	*	89.64 ± 3.82	89.99 ± 3.83	*N.S.*	96.07 ± 5.38	86.20 ± 4.35	*tnd.*
Hydroxyproline	−0.573	***	81.09 ± 11.06	51.67 ± 7.05	*tnd.*	66.47 ± 3.25	55.19 ± 5.81	*N.S.*	73.78 ± 8.42	53.43 ± 6.15	*
β-Ala	−0.506	*	6.362 ± 0.755	4.164 ± 0.347	**	6.388 ± 0.323	6.522 ± 0.793	*N.S.*	6.375 ± 0.547	5.343 ± 0.801	*N.S.*
Carnosine	−0.443	*tnd.*	10.97 ± 0.590	8.239 ± 0.646	*	11.52 ± 1.03	11.80 ± 1.01	*N.S.*	11.245 ± 0.800	10.019 ± 1.157	*N.S.*

The unit of metabolites concentration is represented in µM. *1 R indicates correlation coefficient between PLS score and each metabolite levels. *2 Values are denoted with means ± S.E.M. Statistical significance is defined as *tnd.* ($p < 0.10$), * ($p < 0.05$), ** ($p < 0.01$), and *** ($p < 0.001$). *N.S.* indicates no significance.

2.4. Comparison of Amino Acids and Related Metabolites

Since several amino acids and their related metabolites were identified as the differential metabolites between IMF groups, we next focused our analysis on those metabolites derived from BCAAs such as 3-methyl-2-oxovaleric acid (2K3MVA)/4-methyl-2-oxovaleric acid (2-oxoleucine), 2-oxoisovaleric acid (2-KIV) (Figure 3). The combined amount of Ile and Leu in the high IMF group at MIY was significantly higher than in the low IMF group. The amount of Ile and Leu at IBR, and Val at both stations showed a similar tendency, although the difference was not significant. Likewise, the amount of 2K3MVA/2-Oxoleucine and 2-KIV in the high IMF groups at both stations were greater than in the low IMF groups. There was no significant difference in the total free amino acid contents between the low and high IMF groups. The mean concentrations and standard deviations were 3428 ± 240 (μM) and 3258 ± 182 (μM), respectively.

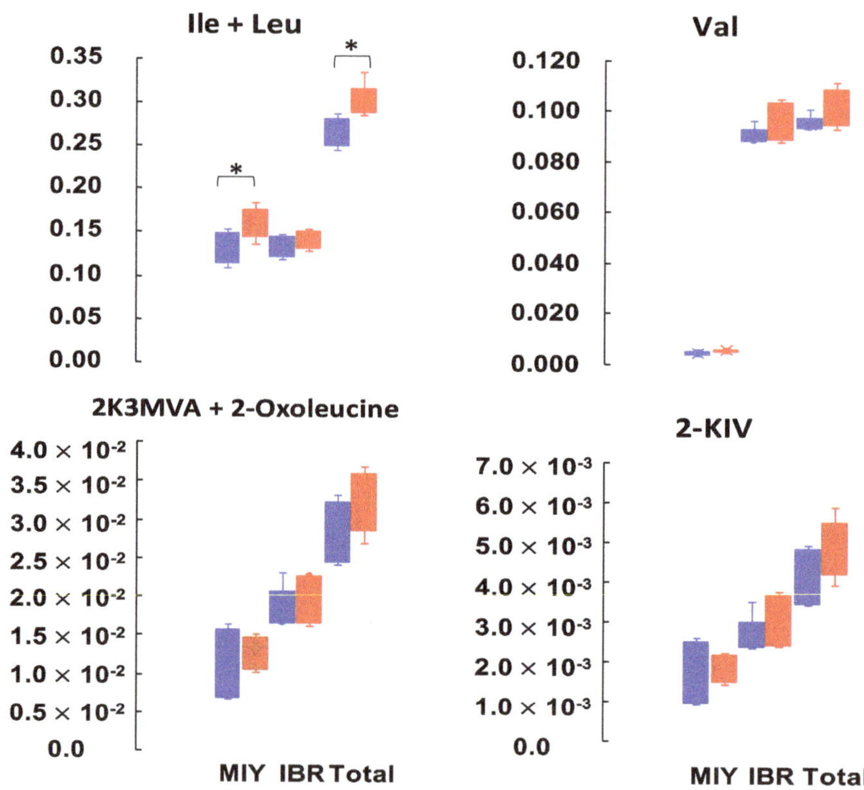

Figure 3. Comparisons of branched-chain amino acids and their degraded metabolites detected from pig plasma samples showing the difference in the intramuscular fat contents at the two stations. Blue and red bars indicate low and high IMF pigs, respectively. Total shows the sum of MIY and IBR. *Y*-axis indicate the relative area of each metabolites normalized by the internal standard. 3-methyl-2-oxovaleric acid (2K3MVA) and 4-methyl-2-oxovaleric acid (2-oxoleucine), which are the degraded metabolites of Ile and Leu, respectively, could not be separately detected by the capillary electrophoresis-time of flight mass spectrometer (CE-TOFMS). Statistical significance is defined as * ($p < 0.05$).

2.5. Amino Acid Content Analysis in the Muscle Tissue of Pig Carcass

Per our CE-TOFMS metabolomics results, we concluded that BCAAs were associated with high IMF content in loin eye muscle. We examined the free amino acid contents of 20 amino acids with taurine, carnosine and anserine (Asn) in the muscle tissues of pig carcasses in order to better understand the mechanism by which high IMF content had either resulted from or was caused by high BCAAs in plasma (Table 4).

At MIY, high IMF pigs had significantly higher concentrations of Asn in loin eye muscle tissue than low IMF pigs, while Gly, carnosine and Asn in high IMF pigs were significantly lower than low

IMF. The results of Gly and Carnosine matched the results shown in Tables 2 and 3. At IBR, low IMF pigs had greater concentrations of Asp, His, Met, Trp, phenylalanine (Phe), Ile, and Leu in loin eye muscle tissues than those of high IMF pigs. The BCAA contents of loin eye muscles in low IMF pigs were significantly higher than those of high IMF pigs. This result is quite opposite from the results obtained from the plasma metabolomics. As seen in the plasma metabolomics, the total free amino acid content of loin eye muscle tissues displayed no difference between the low and high IMF groups.

3. Discussion

This study sought to identify the metabolites associated with IMF content in pork loin eye muscle using CE-TOFMS metabolomics data detected in pig plasma before slaughter. We selected five castrated boars with low IMF content, and five castrated boars with high IMF content in the loin eye muscle from two NLBC stations in Japan (IBR and MIY), for a total of 20 pigs. The difference in IMF content in the high IMF pigs was 2.8 times greater than the low IMF pigs at MIY. At IBR, the difference in IMF content in high IMF pigs was 1.9 times greater than the low IMF pigs. In contrast the moisture content (%), crude protein content (%), and loin eye area (cm^2) in both high IMF groups were either significantly less or tended to be less than those of the low IMF groups. Most of the growth-related traits and the eating-quality traits displayed no differences between the groups. The metabolites differentially identified between the IMF content levels were therefore considered to be primarily associated with IMF content, and secondly with moisture content. Possible partial relationships may have occurred between the candidate metabolites for IMF content, crude protein content, loin eye area or cooking loss (Table 1).

Plasma samples from each pig were profiled using semi-quantification results derived from CE-TOFMS metabolomics. As indicated in the PCA analysis results, large variations were observed, particularly in the low IMF groups at both stations, although the high IMF groups had relatively close PCAs plots. No nutritional controls or genetic factors were used to generate these differences. Since no obvious separation was observed in the IMF content differences from either station, it can also be concluded that none of these factors were involved in samples obtained during this study and it is natural to see such vague separations (Figure S1).

Next, we conducted the PLS analysis. Our results clearly depicted variations in the candidate metabolites in the first PLS1 screening; more specifically, that high IMF groups had positive PLS1 values, while low IMF groups had negative PLS1 values (Figure 1). The metabolites with high positive loading values of PLS1 consistently included BCAAs and amino acid metabolites. This led us to conclude that BCAAs are potential biomarkers for IMF content in pork loin eye muscles. Metabolites with negative loading values of PLS1 included several kinds of amino acids and amino acid metabolites, which suggests that an increased amount of amino acids and amino acid metabolites in plasma may represent lower IMF content (i.e., leaner meat) (Table 2).

We used the quantification results from the CE-TOFMS to validate these findings. We applied the integrated PLS analysis using quantification data from the four groups (Figure 2). Although the PLS1 accounted for 60.4% of the total IMF variability, it appeared to obviously discriminate station differences, which was not the original purpose of this study. The PLS2 showed the comprehensible separation between the IMF content groups. Even though the number of metabolites quantified in the integrated analysis was 59 out of 201 at MIY and 152 at IBR during the screening analysis, BCAAs were commonly extracted from the list of metabolites with high loading PLS2 values (Table 3).

According to the correlation coefficients estimated in Tables 2 and 3, metabolites with higher R values showed clear differences between the high and low IMF groups. Additionally, metabolite concentration differences between the two IMF groups at MIY were more obvious than those at IBR. Nevertheless, the concentration of metabolites in the high IMF groups was consistently greater at both stations, with the opposite holding true for metabolites with negative loading PLS2 values.

Since BCAAs were indicated to correspond to IMF content, we examined the potential BCAA-related metabolites closer (namely 2K3MVA, 2-oxoleucine, and 2-KIV) (Figure 3). Since 2K3MVA and 2-oxoleucine could not be separately identified by the CE-TOFMS system, additional amounts of

2K3MVA and 2-oxoleucine and their original BCAAs (Ile and Leu) were applied (Figure 3). The relative amount of metabolites degraded from BCAAs (2K3MVA, 2-oxoleucine, and 2-KIV) indicated a tendency similar to the original BCAAs. Even if the differences were not statistically significant, all of these metabolites follow the patterns of "low in low IMF", and "high in high IMF". One explanation for this may be that the baseline amount of BCAAs in high IMF pig plasma is greater than in low IMF pigs, which means that the amount of metabolized BCAAs might follow the same pattern. If this is the case, then it is quite important to analyze the amino acid content in tissues that metabolize amino acids.

Post slaughter amino acid analysis was conducted on the loin eye muscle tissues (Table 4). Results of this analysis were inconsistent with the metabolomic plasma analysis, that is, the amino acid tissue analysis showed that high IMF pigs at MIY had significantly lower amounts of Gly, Car, and Ans than low IMF pigs. This suggests that the amount of Gly, Car, and Ans (divalent) detected in the plasma of high IMF pigs might be correlated with the intake of those metabolites into muscle tissues. The amino acid analysis at IBR indicated that the amount of Asp and Met in the low IMF group were significantly greater than those in the high IMF group, which matched our plasma screening results. It is interesting that the amount of BCAA in the muscle tissues of the low IMF group were significantly greater than in the high IMF group. Additionally, the BCAA differences at MIY were not as obvious as in the plasma samples (N.S.), although the same trend was observed in both the plasma and the muscle tissues. These results suggest that the amount of BCAA in plasma is positively associated with high IMF content in the loin eye muscle.

Table 4. Analysis of amino acid contents in the loin eye muscle tissues from two stations.

Amino Acid *1	MIY Station			IBR Station		
	Low IMF	High IMF	*p* *2	Low IMF	High IMF	*p* *2
Asp	$9.6 \times 10^{-2} \pm 8.1 \times 10^{-3}$	$9.5 \times 10^{-2} \pm 2.1 \times 10^{-2}$	*N.S.*	$8.1 \times 10^{-2} \pm 3.3 \times 10^{-2}$	$3.2 \times 10^{-2} \pm 1.0 \times 10^{-2}$	*
Glu	$5.5 \times 10^{-1} \pm 1.0 \times 10^{-1}$	$6.2 \times 10^{-1} \pm 1.2 \times 10^{-1}$	*N.S.*	$6.8 \times 10^{-1} \pm 1.5 \times 10^{-1}$	$5.7 \times 10^{-1} \pm 9.3 \times 10^{-2}$	*N.S.*
Asn	$1.6 \times 10^{-1} \pm 2.8 \times 10^{-2}$	$2.3 \times 10^{-1} \pm 5.0 \times 10^{-2}$	*	$2.6 \times 10^{-1} \pm 6.9 \times 10^{-2}$	$2.4 \times 10^{-1} \pm 1.7 \times 10^{-2}$	*N.S.*
Ser	$4.9 \times 10^{-1} \pm 9.1 \times 10^{-2}$	$4.8 \times 10^{-1} \pm 9.9 \times 10^{-2}$	*N.S.*	$5.5 \times 10^{-1} \pm 1.2 \times 10^{-1}$	$4.5 \times 10^{-1} \pm 5.1 \times 10^{-2}$	*N.S.*
Gln	1.658 ± 0.250	1.490 ± 0.144	*N.S.*	1.485 ± 0.438	$9.9 \times 10^{-1} \pm 1.4 \times 10^{-1}$	*tnd.*
His	$1.7 \times 10^{-1} \pm 2.0 \times 10^{-2}$	$1.7 \times 10^{-1} \pm 2.6 \times 10^{-2}$	*N.S.*	$1.9 \times 10^{-1} \pm 2.3 \times 10^{-2}$	$1.5 \times 10^{-1} \pm 2.3 \times 10^{-2}$	*
Gly	1.62 ± 0.26	1.213 ± 0.0966	*	1.852 ± 0.498	1.789 ± 0.481	*N.S.*
Thr	$3.4 \times 10^{-1} \pm 5.9 \times 10^{-2}$	$3.6 \times 10^{-1} \pm 6.2 \times 10^{-2}$	*N.S.*	$3.9 \times 10^{-1} \pm 7.5 \times 10^{-2}$	$3.3 \times 10^{-1} \pm 3.1 \times 10^{-2}$	*N.S.*
β-Ala	$6.7 \times 10^{-1} \pm 2.3 \times 10^{-1}$	$4.4 \times 10^{-1} \pm 1.0 \times 10^{-1}$	*N.S.*	$5.1 \times 10^{-1} \pm 8.3 \times 10^{-2}$	$5.6 \times 10^{-1} \pm 1.6 \times 10^{-1}$	*N.S.*
Arg	$3.6 \times 10^{-1} \pm 5.9 \times 10^{-2}$	$3.5 \times 10^{-1} \pm 6.6 \times 10^{-2}$	*N.S.*	$4.0 \times 10^{-1} \pm 8.3 \times 10^{-2}$	$2.9 \times 10^{-1} \pm 3.5 \times 10^{-2}$	*tnd.*
Ala	2.143 ± 0.325	2.329 ± 0.397	*N.S.*	2.643 ± 0.472	2.170 ± 0.293	*N.S.*
Tau	2.341 ± 0.441	2.16 ± 0.289	*N.S.*	2.404 ± 0.772	1.829 ± 0.920	*N.S.*
Car	26.41 ±1.56	20.99 ± 2.35	**	28.95 ± 1.65	27.60 ± 2.20	*N.S.*
Ans	$8.5 \times 10^{-1} \pm 6.0 \times 10^{-2}$	$6.6 \times 10^{-1} \pm 8.4 \times 10^{-2}$	**	$7.1 \times 10^{-1} \pm 5.7 \times 10^{-2}$	$7.3 \times 10^{-1} \pm 1.2 \times 10^{-1}$	*N.S.*
Tyr	$2.8 \times 10^{-1} \pm 4.7 \times 10^{-2}$	$2.7 \times 10^{-1} \pm 4.8 \times 10^{-2}$	*N.S.*	$3.0 \times 10^{-1} \pm 5.8 \times 10^{-2}$	$2.5 \times 10^{-1} \pm 2.3 \times 10^{-2}$	*N.S.*
Val	$3.9 \times 10^{-1} \pm 6.1 \times 10^{-2}$	$4.8 \times 10^{-1} \pm 5.7 \times 10^{-2}$	*tnd.*	$5.3 \times 10^{-1} \pm 8.0 \times 10^{-2}$	$4.4 \times 10^{-1} \pm 5.6 \times 10^{-2}$	*N.S.*
Met	$2.5 \times 10^{-1} \pm 4.1 \times 10^{-2}$	$2.7 \times 10^{-1} \pm 4.6 \times 10^{-2}$	*N.S.*	$3.0 \times 10^{-1} \pm 3.4 \times 10^{-2}$	$2.3 \times 10^{-1} \pm 2.1 \times 10^{-2}$	**
Trp	$8.8 \times 10^{-2} \pm 7.8 \times 10^{-3}$	$9.9 \times 10^{-2} \pm 5.9 \times 10^{-3}$	*tnd.*	$1.2 \times 10^{-1} \pm 9.8 \times 10^{-3}$	$1.0 \times 10^{-1} \pm 5.2 \times 10^{-3}$	**
Phe	$3.3 \times 10^{-1} \pm 4.4 \times 10^{-2}$	$3.5 \times 10^{-1} \pm 4.1 \times 10^{-2}$	*N.S.*	$3.9 \times 10^{-1} \pm 4.0 \times 10^{-2}$	$3.4 \times 10^{-1} \pm 1.2 \times 10^{-2}$	*
Ile	$3.1 \times 10^{-1} \pm 5.7 \times 10^{-2}$	$3.4 \times 10^{-1} \pm 5.1 \times 10^{-2}$	*N.S.*	$3.8 \times 10^{-1} \pm 4.9 \times 10^{-2}$	$3.1 \times 10^{-1} \pm 2.5 \times 10^{-2}$	*
Leu	$7.0 \times 10^{-1} \pm 1.2 \times 10^{-1}$	$6.3 \times 10^{-1} \pm 1.0 \times 10^{-2}$	*N.S.*	$7.3 \times 10^{-1} \pm 1.1 \times 10^{-1}$	$5.5 \times 10^{-1} \pm 6.2 \times 10^{-2}$	*
Lys	$4.4 \times 10^{-1} \pm 8.1 \times 10^{-2}$	$4.2 \times 10^{-1} \pm 8.1 \times 10^{-2}$	*N.S.*	$4.5 \times 10^{-1} \pm 8.1 \times 10^{-2}$	$3.5 \times 10^{-1} \pm 3.5 \times 10^{-2}$	*tnd.*
Hyp	3.142 ± 0.253	2.897 ± 1.630	*N.S.*	2.099 ± 0.198	2.054 ± 0.0751	*N.S.*
Pro	$9.6 \times 10^{-1} \pm 8.9 \times 10^{-2}$	2.790 ± 3.31	*N.S.*	$5.4 \times 10^{-1} \pm 1.04 \times 10^{-1}$	$3.7 \times 10^{-1} \pm 1.3 \times 10^{-1}$	*tnd.*
FAA *3	14.42 ± 0.732	15.67 ± 3.60	*N.S.*	15.26 ± 2.76	12.40 ± 1.87	*N.S.*

The unit of amino acid concentration is represented in μmol/g. *1 This list includes analysis result of amino acids together with Taurine, Carnosine and Anserine. *2 Statistical significance is defined as *tnd.* ($p < 0.10$), * ($p < 0.05$), ** ($p < 0.01$). *N.S.* indicates no significance. *3 FAA denotes total amount of free amino acids.

Recently, it was demonstrated that surplus dietary Ile, one of BCAAs, increased the IMF content in skeletal muscle via the upregulation of lipogenic genes by stimulating lipogenesis in skeletal muscle tissue of finishing pigs [19]. In the previous study, Luo et al. clearly showed the effect of extra Ile in the diet on lipogenesis in the muscle tissue, serum cholesterol levels, and fatty acid composition without affecting growth performance. However, it is not possible to further discuss whether Ile (BCAA) intake into the muscle cells affected lipid synthesis due to the lack of Ile level in muscle tissue [19]. In contrast, it has been well known that BCAA are effective amino acids to regulate protein synthesis because of a greater increase in protein synthesis than degradation [20–22].

Therefore, one possible explanation for this is that the availability of BCAA and/or the intake of BCAA in the muscle tissue of low IMF pigs are more effective than in high IMF pigs, which results in the pig building more muscle mass. It is possible, therefore, that the availability of BCAA in the muscle tissue of high IMF pigs is relatively low, which would mean that some kind of molecular mechanism(s) could be triggered that could enhance the preadipocytes near the muscle cells.

The Duroc pigs used for this study were derived from the one previously established to improve IMF content at NLBC. The heritability of IMF in the Duroc genetic line established at NLBC has been previously determined to be 0.52 [23]. Although there were certain differences in the metabolite profiles between MIY and IBR by the PLS1 (Figure 2), these observed differences may be due to environmental factors only, since the genetic Duroc line and the feeding regimes were nearly identical at each location.

Previously, Katsumata et al. [14] reported that a low lysine diet fed to finishing pigs increased IMF content, although the authors recognized that the influence of restrictive amino acid nutrition on IMF content needs to be researched further. Restricted amino acid diets may increase stress sensitivity in pigs, since it has been indicated to do so in other animals [14,24,25]. In the case of beef marbling improvements, vitamin A restriction in the diet of beef steers increased IMF content due to hyperplasia of the adipocytes, although subcutaneous fat depth was not affected [26,27]. This method of feeding beef cattle diets with low vitamin A is well-known and used worldwide. In addition to the vitamin A restriction, it is suggested that a genotype of alcohol dehydrogenase 1C (ADH1C) has an interactive effect with vitamin A on IMF content, because of the differential transcriptional regulation that potentially occurs in this genotype [28]. It is plausible that nutritional control combined with certain genetic backgrounds may increase or decrease the IMF content in meat animals. However, there is still an inevitable phenotypic variation in IMF content, as is the nature of quantitative traits. Therefore, a non-invasive method, such as plasma metabolomics by CE-TOFMS, could be of benefit to determine the physiological status of meat producing animals, since it can measure a broader range of compounds than other metabolomics systems.

In conclusion, in this study we found that pigs with high IMF content in the loin eye muscle also had an increased amount of BCAA in their blood plasma. We also found that pigs with low IMF content in the loin eye muscle had an increased amount of Gly, Ans, and Car in their blood plasma. We suggest, therefore, that BCAA, Gly, Ans, and Car are potential biomarkers that could be used to estimate IMF in the loin eye muscle before slaughter. Further study of these biomarkers is required, since their amounts were not always consistently demonstrated in this study.

4. Materials and Methods

4.1. Animals

This study used castrated boars from the Duroc breed, which has been genetically bred for high IMF content as reported in the previous study [23]. All pigs used in this study were fed identical diets based on the Japan Nutrition Standards (NILGS 2016) and were raised at two locations of the NLBC (IBR and MIY) in Japan.

At each location, five high IMF content pigs and five low IMF content pigs were chosen from among the herd, for a total of 20 animals. Other than IMF content, growth performance and carcass characteristics of pigs were compared (Table 1). Aside from the fact that low IMF pigs took longer to gain slaughter weight, there were no significant phenotypic variables between any of the four IMF groups.

Blood samples were collected in a vacuum blood collection tube with EDTA/2Na from the jugular vein of each pig after fasting overnight. After blood samples were collected, the pigs were slaughtered at the abattoir located at each station. The blood collection tube was kept on ice and centrifuged with 3000× *g* at 4 °C for 10 min. Plasma was removed from the blood collection tube and placed into a new sample tube, then snap frozen, and stored in −80 °C freezer until use. All procedures involving animals were performed in accordance with the National Livestock Breeding Center's guidelines for care and use of laboratory animals.

4.2. Semi-Quantitative Metabolomics (Basic Scan) by Capillary Electrophoresis-Time of Flight Mass Spectrometry (CE-TOFMS)

Preprocessing was initiated by adding 50 μL pig plasma to 450 μL methanol containing 50 μM Internal Standard Solution 1 (H3304-1002, Human Metabolome Technologies (HMT), Tsuruoka, Yamagata, Japan) followed by mixing. To the mixture, 500 μL chloroform and 200 μL ultrapure water were added and then mixed. After centrifugation at 2300× *g*, in 4 °C for 5 min, 400 μL supernatant of the mixture was removed to ultrafiltration using UltraFree MC PLHCC (HMT) and centrifuged at 9100× *g* at 4 °C for 120 min. The filtrate was once frozen-dried and dissolved in 50 μL ultrapure water just before applying to Agilent CE-TOFMS system (Agilent Technologies, Santa Clara, CA, USA). Metabolome analysis was performed by Basic Scan package of HMT using the CE-TOFMS based on the methods described previously [29,30]. Briefly, CE-TOFMS analysis was carried out using an Agilent CE capillary electrophoresis system equipped with an Agilent 6210 time-of-flight mass spectrometer (Agilent Technologies, Santa Clara, CA, USA). CE-TOFMS was operated using Agilent G2201AA ChemStation software version B.03.01 (Agilent Technologies, Santa Clara, CA, USA), compounds were detected using the cation or anion modes. Briefly, the compounds were electrophoresed through fused silica capillaries (50 μM × 80 cm, Agilent Technologies, Santa Clara, CA, USA) with 30 kV under 50 mbar and 10 s pressure injections. The electrophoresed compounds were ionized with either ESI Positive or Negative MS ionization. The scan range for the compounds was between 50 and 1000 *m*/*z*. Detected peaks from the CE-TOFMS system were processed by MasterHands ver2.17.1.11 [31]. Signal peaks corresponding to isotopomers, adduct ions, and other product ions of known metabolites were excluded, and remaining peaks were annotated according to the HMT metabolite database library, which includes 900 metabolites, based on their *m*/*z* values with the MTs determined by TOFMS. Areas of the annotated peaks were then normalized based on internal standard levels and sample amounts to obtain relative levels of each metabolite.

4.3. Absolute Quantification of 110 Target Metabolites for CE-TOFMS Analysis

Based on the same preprocessing and the analysis system, primary 110 metabolites involved in the pathways of glycolytic/gluconeogenesis, TCA cycle, pentose phosphate, lipid metabolites, and nucleic acid metabolites were absolutely quantified based on one-point calibrations using their respective standard compounds.

4.4. Measurement of Free Amino Acids, Peptides, and Carcass Traits in Muscle Tissues

The procedure used to analyze the free amino acids and peptides was carried out using the exact same method described in [32]. Briefly, 0.10 g samples of loin eye muscle tissues were drawn from minced raw meat (about 15 g in total). Samples were homogenized with 4.24 mL of ultrapure water, 4 mL of N-hexane and 0.16 mL of internal standard solution mixed with norvaline (5 nmol/μL) in ultrapure water, and then centrifuged at 1750× *g* for 5 min. The underlayer was mixed with 4 mL of N-hexane and centrifuged at 1750× *g* for 5 min. The resultant underlayer was then mixed with 3.6 mL of acetonitrile and centrifuged at 1750× *g* for 10 min. The resulting supernatant was filtered through a 0.45-μm microfilter (Millex-LH; Merck Millipore, Billerica, MA, USA), and the filtrate was then mixed with 45% acetonitrile solution and analyzed for free amino acids and peptides using an Agilent 1260 infinity high performance liquid chromatograph equipped with an Agilent 1260 Binary Pump, 1260 HiP Degasser, 1260 HiP ALS autosampler, 1290 thermostat, 1260 Thermostatted column compartment control module, 1260 diode array detector and a Poroshell 120 EC-C18 column (3.0 × 100 mm, 2.7 μm; Agilent). The eluents used were: (i) 20 mmol/L disodium hydrogen phosphate (pH 7.6); and (ii) acetonitrile/methanol/water (5:5:1, *v*/*v*/*v*). Amino acids and peptides were identified through the comparison of their retention times with those of established standards. The concentrations of each were calculated using internal and external standard solutions and expressed as μmol per g of meat and μmol per 1% of moisture in meat, respectively. The internal standard solutions were used to account for matter lost during analysis, and the external standard solutions (1, 10, 50, and 100 pmol/μL) were used to plot a calibration curve for each amino acid and peptide. Carcass traits were analyzed according to the methods described previously [32,33]. Moisture content was determined in duplicates by drying 2 g samples of meat drawn from the minced raw meat (approximately 30 g in total) for 24 h at 105 °C. IMF content was determined by Soxhlet extraction of the dried samples with diethyl ether for 16 h. Crude protein content was determined using 1 g samples of meat drawn from the minced raw meat (approximately 30 g in total) by the Kjeldahl method using a nitrogen distillation titration device (2400 Kjeltec auto sampler system, FOSS, Hillerod, Denmark). Cooking loss was determined by the weight difference of meat samples (approximately 50 g) before and after heating at 70 °C for 1 h in a water bath. After cooking loss measurement, the meat sample was then used for WBSF measurement with crosshead speed of 200 mm/min using Instron (model 5542; Instron, University Avenue Norwood, MA, USA). WHC was determined using meat samples (approximately 0.5 g) drawn from the minced raw meat which were centrifuged by 10,000× *g* for 30 min. Determination of tenderness, pliability, toughness, and brittleness were performed using meat samples prepared for the same as the WBSF using Tensipresser (model TTP-50BX II, Takemoto Electric Inc., Tokyo, Japan).

4.5. Data Analysis

Metabolomics data acquired by the CE-TOFMS system were normalized and the semi-quantitative and absolute quantification data were then analyzed by PCA and PLS [34] using MATLAB (The MathWorks, Natick, MA, USA) and R programs [35] developed by HMT Co. Ltd. Differences between high and low IMF groups were examined by Welch's *t*-test. Significant difference was defined by $p < 0.05$.

Supplementary Materials:
Figure S1. Principal component analysis by the metabolites detected from pig plasma samples showing the difference in the intramuscular fat contents at the two stations. Blue and red dots indicate low and high IMF pigs, respectively. (A) The PCA performed with PC1 and PC2 for pigs from MIY station; (B) The PCA performed with PC1 and PC2 for pigs from IBR station; (C) The PCA performed with PC3 and PC4 for pigs from IBR station.

Author Contributions: Conceptualization, M.T. and K.I.; Methodology, M.T., C.O., K.K. (Kouen Kadowaki), K.K. (Kimiko Kohira), F.H., and K.M.; Validation, M.T., A.A., M.N. and T.O.; Formal Analysis, M.T. and A.A.; Investigation, M.T.; Resources, C.O. and K.K. (Kouen Kadowaki); Data Curation, A.A., M.N. and T.O.; Writing—Original Draft Preparation, M.T.; Writing—Review & Editing, M.T., A.A., M.N., T.O., C.O., K.K. (Kouen Kadowaki), K.K. (Kimiko Kohira), F.H., K.M., and K.I.; Visualization, M.T.; Supervision, M.T.; Project Administration, K.I.; Funding Acquisition, K.I. All authors have read and agreed to the published version of the manuscript.

Acknowledgments: We thank Shozo Tomonaga for his constructive comments.

References

1. Hocquette, J.F.; Gondret, F.; Baéza, E.; Médale, F.; Jurie, C.; Pethick, D.W. Intramuscular fat content in meat-producing animals: Development, genetic and nutritional control, and identification of putative markers. *Animal* **2010**, *4*, 303–319. [CrossRef] [PubMed]

2. Apple, J.K. Nutritional effects on pork quality in swine production. In *National Swine Nutrition Guide*; Factsheet No. PIG 12-02-02; U.S. Pork Center of Excellence: Clive, IA, USA, 2015.
3. Schwab, C.R.; Baas, T.J.; Stalder, K.J. Results from six generations of selection for intramuscular fat in Duroc swine using real-time ultrasound. II. Genetic parameters and trends. *J. Anim. Sci.* **2010**, *88*, 69–79. [CrossRef] [PubMed]
4. Fernández, A.; de Pedro, E.; Núñez, N.; Silió, L.; García-Casco, J.; Rodríguez, C. Genetic parameters for meat and fat quality and carcass composition traits in Iberian pigs. *Meat Sci.* **2003**, *64*, 405–410. [CrossRef]
5. Won, S.; Jung, J.; Park, E.; Kim, H. Identification of genes related to intramuscular fat content of pigs using genome-wide association study. *Asian Aust. J. Anim. Sci.* **2018**, *31*, 157–162. [CrossRef]
6. Davoli, R.; Catillo, G.; Serra, A.; Zappaterra, M.; Zambonelli, P.; Meo Zilio, D.; Steri, R.; Mele, M.; Buttazzoni, L.; Russo, V. Genetic parameters of backfat fatty acids and carcass traits in Large White pigs. *Animal* **2019**, *13*, 924–932. [CrossRef]
7. Hovenier, R.; Kanis, E.; van Asseldonk, T.H.; Westerink, N.G. Breeding for pig meat quality in halothane negative populations—A review. *Pig News Inf.* **1993**, *14*, 17N–25N.
8. Sellier, P. Genetics of meat and carcass traits. In *Genetics of the Pig*; Rothchild, M.F., Ruvinsky, A., Eds.; CAB International: Wallingford, Oxon, UK, 1998; pp. 463–510.
9. Newcom, D.W.; Baas, T.J.; Schwab, C.R.; Stalder, K.J. Genetic and phenotypic relationships between individual subcutaneous backfat layers and percentage of longissimus intramuscular fat in Duroc swine. *J. Anim. Sci.* **2005**, *83*, 316–323. [CrossRef]
10. Suzuki, K.; Irie, M.; Kadowaki, H.; Shibata, T.; Kumagai, M.; Nishida, A. Genetic parameter estimates of meat quality traits in Duroc pigs selected for average daily gain, longissimus muscle area, backfat thickness, and intramuscular fat content. *J. Anim. Sci.* **2005**, *83*, 2058–2065. [CrossRef]
11. Solanes, F.X.; Reixach, J.; Tor, M.; Tibau, J.; Estany, J. Genetic correlations and expected response for intramuscular fat content in a Duroc pig line. *Livest. Prod. Sci.* **2009**, *123*, 63–69. [CrossRef]
12. Cesar, A.S.M.; Regitano, L.C.A.; Koltes, J.E.; Fritz-Waters, E.R.; Lanna, D.P.D.; Gasparin, G.; Mourao, G.B.; Oliveira, P.S.N.; Reecy, J.M.; Coutinho, L.L. Putative Regulatory Factors Associated with Intramuscular Fat Content. *PLoS ONE* **2015**, *10*, e0128350. [CrossRef]
13. Da Costa, N.; McGillivray, C.; Bai, Q.; Wood, J.D.; Evans, G.; Chang, K.C. Restriction of dietary energy and protein induces molecular changes in young porcine skeletal muscles. *J. Nutr.* **2004**, *134*, 2191–2199. [CrossRef] [PubMed]
14. Katsumata, M. Promotion of intramuscular fat accumulation in porcine muscle by nutritional regulation. *Anim. Sci. J.* **2011**, *82*, 17–25. [CrossRef] [PubMed]
15. Muroya, S.; Oe, M.; Nakajima, I.; Ojima, K.; Chikuni, K. CE-TOF MS-based metabolomic profiling revealed characteristic metabolic pathways in postmortem porcine fast and slow type muscles. *Meat Sci.* **2014**, *98*, 726–735. [CrossRef] [PubMed]
16. Muroya, S.; Ueda, S.; Komatsu, T.; Miyakawa, T.; Ertbjerg, P. MEATabolomics: Muscle and meat metabolomics in domestic animals. *Metabolites* **2020**, *10*, 188. [CrossRef]
17. Fontanesi, L. Metabolomics and livestock genomics: Insights into a phenotyping frontier andits applications in animal breeding. *Anim. Front.* **2016**, *6*, 73–79. [CrossRef]
18. Suravajhala, P.; Kogelman, L.J.A.; Kadarmideen, H.N. Multi-omic data integration and analysis using systems genomics approaches: Methods and applications in animal production, health and welfare. *Genet. Sel. Evol.* **2016**, *48*, 38. [CrossRef]
19. Luo, Y.; Zhang, X.; Zhu, Z.; Jiao, N.; Qiu, K.; Yin, J. Surplus dietary isoleucine intake enhanced monounsaturated fatty acid synthesis and fat accumulation in skeletal muscle of finishing pigs. *J. Anim. Sci. Biotech.* **2018**, *9*, 88. [CrossRef]
20. Zhang, S.; Chu, L.; Qiao, S.; Mao, X.; Zeng, X. Effects of dietary leucine supplementation in low crude protein diets on performance, nitrogen balance, whole-body protein turnover, carcass characteristics and meat quality of finishing pigs. *Anim. Sci. J.* **2016**, *87*, 911–920. [CrossRef]
21. Zheng, L.; Wei, H.; He, P.; Zhao, S.; Xiang, Q.; Pang, J.; Peng, J. Effects of supplementation of branched-chain amino acids to reduced-protein diet on skeletal muscle protein synthesis and degradation in the fed and fasted states in a piglet model. *Nutrients* **2017**, *9*, 17. [CrossRef]
22. Wang, Y.; Zhou, J.; Wang, G.; Cai, S.; Zeng, X.; Qiao, S. Advances in low-protein diets for swine. *J. Anim. Sci. Biotech.* **2018**, *9*, 60. [CrossRef]

23. Ishii, K.; Arata, S.; Ohnishi, C. Estimates of genetic parameters for meat quality and carcass traits in Duroc pigs. In Proceedings of the World Congress on Genetics Applied to Livestock Production, Auckland, New Zealand, 8 February 2018. Abstract Number 408.
24. Smriga, M.; Kameishi, M.; Uneyama, H.; Torii, K. DietaryL-lysine deficiency increases stress-induced anxiety and fecal excretion in rats. *J. Nutr.* **2002**, *132*, 3744–3746. [CrossRef] [PubMed]
25. Srinongkote, S.; Smrige, M.; Toride, Y. Diet supplied with L-lysine and L-arginine during chronic stress of high stock density normalizes growth of broilers. *Anim. Sci. J.* **2004**, *75*, 339–343. [CrossRef]
26. Harris, C.L.; Wang, B.; Deavila, J.M.; Bushboom, J.R.; Maquivar, M.; Parish, S.M.; McCann, B.; Nelson, M.L.; Du, M. Vitamin A administration at birth promotes calf growth and intramuscular fat development in Angus beef cattle. *J. Anim. Sci. Biotechnol.* **2018**, *9*, 55. [CrossRef] [PubMed]
27. Kruk, Z.A.; Bottema, M.J.; Reyes-Veliz, L.; Forder, R.E.A.; Pitchford, W.S.; Bottema, C.D.K. Vitamin A and Marbling Attributes: Intramuscular Fat Hyperplasia Effects in Cattle. *Meat Sci.* **2018**, *137*, 139–146. [CrossRef]
28. Ward, A.K.; McKinnon, J.J.; Hendrick, S.; Buchanan, F.C. The impact of vitamin A restriction and ADH1C genotype on marbling in feedlot steers. *J. Anim. Sci.* **2015**, *90*, 2476–2483. [CrossRef]
29. Ohashi, Y.; Hirayama, A.; Ishikawa, T.; Nakamura, S.; Shimizu, K.; Soga, T. Depiction of metabolome changes in histidine-starved Escherichia coli by CE-TOFMS. *Mol. Biosyst.* **2008**, *4*, 135–147. [CrossRef]
30. Ooga, T.; Sato, H.; Nagashima, A.; Sasaki, K.; Tomita, M.; Soga, T.; Ohashi, Y. Metabolomic anatomy of an animal model revealing homeostatic imbalances in dyslipidaemia. *Mol. Biosyst.* **2011**, *7*, 1217–1223. [CrossRef]
31. Sugimoto, M.; Goto, H.; Otomo, K.; Ito, M.; Onuma, H.; Suzuki, A.; Sugawara, M.; Abe, S.; Tomita, M.; Soga, T. Metabolomic profiles and sensory attributes of edamame under various storage duration and temperature conditions. *J. Agric. Food Chem.* **2010**, *58*, 8418–8425. [CrossRef]
32. Sakuma, H.; Saito, K.; Kohira, K.; Ohhashi, F.; Shoji, N.; Uemoto, Y. Estimates of genetic parameters for chemical traits of meat quality in Japanese black cattle. *Anim. Sci. J.* **2017**, *88*, 203–212. [CrossRef]
33. Okumura, T.; Saito, K.; Sakuma, H.; Nade, T.; Nakayama, S.; Fujita, K.; Kawamura, T. Intramuscular fat deposition in principal muscles from twenty-four to thirty months of age using identical twins of Japanese Black steers. *J. Anim. Sci.* **2007**, *85*, 1902–1907. [CrossRef]
34. Yamamoto, H. PLS-ROG: Partial least squares with rank order of groups. *J. Chemom.* **2017**, *33*, e2883. [CrossRef]
35. R Core Team. *R: A Language and Environment for Statistical Computing*; R Foundation for Statistical Computing: Vienna, Austria, 2019; Available online: https://www.R-project.org/ (accessed on 29 May 2020).

7

Metabolic Profiling Provides Unique Insights to Accumulation and Biosynthesis of Key Secondary Metabolites in Annual Pasture Legumes of Mediterranean Origin

Sajid Latif [1,2,*], Paul A. Weston [1,3], Russell A. Barrow [1,4], Saliya Gurusinghe [1], John W. Piltz [5] and Leslie A. Weston [1,3]

1 Graham Center for Agricultural Innovation, Locked Bag 588, Wagga Wagga, NSW 2678, Australia; pweston@csu.edu.au (P.A.W.); rubarrow@csu.edu.au (R.A.B.); sgurusinghe@csu.edu.au (S.G.); leweston@csu.edu.au (L.A.W.)
2 School of Animal and Veterinary Sciences, Charles Sturt University, Wagga Wagga, NSW 2678, Australia
3 School of Agriculture and Wine Sciences, Charles Sturt University, Wagga Wagga, NSW 2678, Australia
4 Plus 3 Australia Pty Ltd., P.O. Box 4345, Hawker, ACT 2614, Australia
5 New South Wales Department of Primary Industries, Wagga Wagga, NSW 2678, Australia; john.piltz@dpi.nsw.gov.au
* Correspondence: slatif@csu.edu.au

Abstract: Annual legumes from the Mediterranean region are receiving attention in Australia as alternatives to traditional pasture species. The current study employed novel metabolic profiling approaches to quantify key secondary metabolites including phytoestrogens to better understand their biosynthetic regulation in a range of field-grown annual pasture legumes. In addition, total polyphenol and proanthocyanidins were quantified using Folin–Ciocalteu and vanillin assays, respectively. Metabolic profiling coupled with biochemical assay results demonstrated marked differences in the abundance of coumestans, flavonoids, polyphenols, and proanthocyanidins in annual pasture legume species. Genetically related pasture legumes segregated similarly from a chemotaxonomic perspective. A strong and positive association was observed between the concentration of phytoestrogens and upregulation of the flavonoid biosynthetic pathway in annual pasture legumes. Our findings suggest that evolutionary differences in metabolic dynamics and biosynthetic regulation of secondary metabolites have logically occurred over time in various species of annual pasture legumes resulting in enhanced plant defense.

Keywords: pasture legumes; phytoestrogens; flavonoids; coumestans; polyphenols; proanthocyanidins; metabolic profiling; biosynthesis

---

1. Introduction

Broadacre farming frequently occurs with livestock production throughout southeastern Australia, with the pasture phase of crop rotation sustaining both lamb and cattle enterprises [1]. Lamb and beef production account for the majority of livestock-related income in southeastern Australia (AU$22 billion in 2017) and global demand is projected to dramatically increase over the next decade (Australian Bureau of Statistics, 2018). Legumes are integral to livestock pasture production systems through provision of high quality forage for grazing livestock. Establishment of pasture species that are non-toxic, persistent, and high in nutritional quality is therefore critical for continued improvement of livestock productivity.

Traditionally, subterranean clover (*Trifolium subterraneum* L.) and lucerne (*Medicago sativa* L.) are the most widely utilized pasture species in prime lamb, wool, and cattle producing regions across southeastern Australia. Subterranean clover is compatible across various soil types and is tolerant of pH extremes [2,3]. Lucerne is a deep-rooted perennial species frequently established across diverse rainfall regions and is suitable for neutral or mildly alkaline soils. Recently the establishment of both pasture species has proven challenging to sustain livestock production due to a range of biotic and abiotic factors. For example, lucerne, which contains high protein content, has been shown to undergo rapid fermentation in the rumen resulting in increased incidence of bloat and potential loss of nitrogen due to excretion [4]. In addition, ingestion of lucerne and subterranean clover is associated with metabolic disorders including acute inflammation of both the small and large intestine (red gut), sodium deficiency and pregnancy-related toxemia [5–7].

Increased risk of metabolic disorders, the significant cost of establishment, and low persistence of traditional legume pastures in low-rainfall regions [8,9] has led to the introduction of novel annual pasture legume species originating from Mediterranean regions of Europe and northern Africa to Australia [3]. These include *Biserrula pelecinus* L. (biserrula), *Ornithopus sativus* Brot. (French serradella), *Ornithopus compressus* L. (yellow serradella), *Trifolium glanduliferum* Boiss. (gland clover), *Trifolium spumosum* L. (bladder clover), and *Trifolium vesiculosum* Savi. (arrowleaf clover). These accessions are characterized by their adaptation to deep, acid and sandy soils, drought tolerance, weed suppressive potential, and prolonged availability as feed for livestock [10,11]. While the nutritive value of traditional pasture legumes has been studied in southeastern and Western Australia, a detailed investigation of the phenolic chemistry of annual legumes has not been performed with respect to livestock production [2].

Legumes have evolved chemical defense mechanisms mediated by secondary metabolites, including phytoestrogens, to deter herbivores and plant pests [12,13]. In terms of phenolic chemistry and key secondary metabolites in pasture legumes, flavonoids represent a distinct class of secondary products with both positive and negative impacts on plant–microbial and plant–livestock interactions [14,15]. In general, they are classified by their chemical structure into subgroups including anthocyanidins and anthoxanthins (flavanones, flavans, and flavanonols) [16].

Significant levels of phytoestrogens are produced in pasture legumes including lucerne as well as various clover species and when present at significant levels can seriously reduce reproductive efficiency and livestock fertility [15,17,18]. Coumestans and isoflavone phytoestrogens are stable, non-steroidal secondary metabolites that mimic mammalian estrogen, an endogenous female sex hormone [15,19]. The affinity of these phytoestrogens in binding estrogen receptor-β can result in reproductive abnormalities during embryo development, and infertility in both sexes of grazing livestock [20]. Isoflavone phytoestrogens are typically stored either as glycosides or aglycones in pasture legumes [15,21].

Coumestans are non-flavonoid phytoestrogens, and include coumestrol and 4′-methoxycoumestrol, first isolated from white clover (*Trifolium repens* L.) and lucerne (*Medicago sativa* L.) in 1957 [22]. These polycylic aromatic metabolites are closely related biosynthetically to flavonoids (Figure 1). Elevated concentrations of phytoestrogens, including isoflavones, coumestrol, and related metabolites, from ingestion of fodder or feedstock have been implicated in estrogenic clinical signs in livestock as edematous vaginal and cervical tissue, hypertrophy of mammary glands, and milky secretions from elongated teats. Phytoestrogens in forage may also cause adverse effects in ovarian function resulting in loss of fecundity or early embryonic death [23]. Livestock grazing various *Trifolium* species exhibit varying tolerance to coumestans, which typically range in concentration from 25 to 200 mg kg^{-1} dry matter (DM) [15]. The presence of coumestrol at concentrations greater than $\approx$40 mg kg^{-1} DM in plant tissues is associated with reproductive inefficiency in sheep and cattle through disruption of several endocrine mechanisms [24,25].

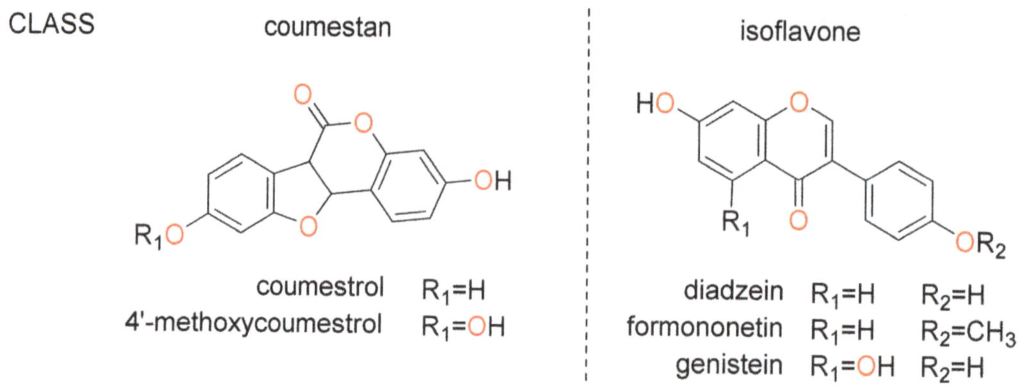

Figure 1. Phytoestrogens found in abundance in lucerne (*Medicago sativa*) and subterranean clover (*Trifolium subterraneum*).

Isoflavonoids are 3-phenylchromen-4-ones (3-phenyl-1,4-benzopyrone) and are important in regulating numerous interactions in higher plants. The isoflavone biosynthetic pathway is one of the most well elaborated pathways in plant secondary metabolism due to the importance of isoflavones as chemoattractants for rhizobia and their involvement in plant defense. Isoflavones are mainly derived from the phenylpropanoid pathway but can be generated through multiple pathways in many plant species [26]. Legumes possess a unique enzyme, isoflavone synthase (IFS), which is a cytochrome P450 monooxygenase that catalyzes the 2, 3 migration of the B-ring of naringenin or liquiritigenin, resulting in the production of various biologically active isoflavonoids [27]. The genes encoding enzymes in this pathway are tissue specific and are regulated both spatially and temporally in legumes [28]. Such catalytic enzymes are induced by various stress factors influencing plant condition, including climate, temperature, soil moisture availability, nutrient deficiency, and herbivory [29].

Polyphenolic compounds, including condensed tannins (proanthocyanidins), are another key group of metabolites possessing a range of biological and nutritional properties in grazing livestock. These compounds vary with regard to chemical structure, plant source, and target animal species [30]. For example, monomeric phenolic acids in forages are associated with enhanced milk production and acid-base imbalance in the rumen following microbial degradation [31]. Traditionally, the proanthocyanidins, which are oligomeric polyphenols, have been considered as anti-nutritional factors leading to reduced palatability of forages, but recent research has shown they provide several potential advantages to grazing livestock, including reduced risk of bloat at moderate intake of 3–4% of dry matter (DM) [32] and increased weight gain, while reducing greenhouse gas emissions [33–35] and parasite burden [36,37]. However, feed composition, age and physiological status of the animal are some of the key factors need to be taken into account while studying the effects of plant secondary metabolites on livestock [38,39].

There is a marked lack of information on the phytochemical profiles of aerial tissues of annual pasture legumes, particularly those recently introduced to the southern hemisphere from the Mediterranean. Recent advances in quadrupole time-of-flight mass spectrometry (MS-QToF) instrumentation have resulted in increasing usage of time-of-flight mass spectrometry (MS-ToF) instruments as quantitation tools as well as prediction of molecular formulae because of their high resolving power and wide dynamic range [40,41]. To broaden our understanding of the secondary chemistry of novel pasture legumes in Australia, we took a metabolomics approach using LC-MS-QToF to investigate (a) the distribution of secondary metabolites that may impact livestock performance, including flavonoids and phytoestrogens (coumestrol, 4′-methoxycoumestrol, formononetin, genistein and daidzein), in above-ground tissues of selected traditional and newly introduced annual pasture legumes grown under field conditions in southern Australia; (b) the total polyphenol and proanthocyanidin content to better delimit their prevalence in these pasture species, and (c) the biosynthetic pathways associated with the production of flavonoids and certain phytoestrogens in these plants.

2. Results

2.1. *Quantification of Phytoestrogens*

Concentration of coumestans in foliar tissues of all species ranged between 0.13 and 48.4 mg kg^{-1} and varied significantly across pasture species (Table 1). Both coumestrol and 4′-methoxycoumestrol accumulated at higher concentrations in leaf and stem compared to inflorescence tissues. Bladder clover possessed significantly higher concentrations of both coumestrol and 4′-methoxycoumestrol in leaf tissue compared to other pasture legumes (48.4 and 24.8 mg kg^{-1}, respectively) while in stem tissue, bladder clover had the highest concentration of coumestrol (39.6 mg kg^{-1}). Lucerne contained the highest concentration of 4′-methoxycoumestrol (27.7 mg kg^{-1}) in any tissue of all species, and lucerne inflorescence tissue contained the highest concentrations of coumestrol and 4′-methoxycoumestrol (0.5 and 0.2 mg kg^{-1}, respectively) of all annual pasture species examined.

Table 1. The concentration of coumestrol and 4′-methoxycoumestrol (mg kg^{-1}) at physiological maturity in selected annual pasture legume species.

Tissue Type	Pasture Species	Coumestrol	4′-Methoxycoumestrol	Total Coumestans
Leaf	Arrowleaf clover	14.7 ± 2.2	5.9 ± 1.5	20.6 ± 5.1
	Biserrula cv.* Casbah	13.1 ± 2.7	5.2 ± 1.7	18.2 ± 6.5
	Biserrula cv. Mauro	14.6 ± 2.4	7.4 ± 5.5	22.0 ± 4.2
	Bladder clover	48.4 ± 10.6	24.8 ± 17.7	73.2 ± 13.6
	French serradella	ND	ND	ND
	Gland clover	27.1 ± 11.1	21.1 ± 5.6	48.2 ± 3.5
	Lucerne	31.8 ± 14.4	20.2 ± 7.5	52.0 ± 6.7
	Subterranean clover	25.1 ± 8.6	12.5 ± 4.5	37.6 ± 7.3
	Yellow serradella	ND	ND	ND
	LSD ($p < 0.05$)	9.3	5.4	13.1
Stem	Arrowleaf clover	24.2 ± 5.3	6.1 ± 4.6	30.3 ± 10.5
	Biserrula cv. Casbah	ND	ND	ND
	Biserrula cv. Mauro	ND	ND	ND
	Bladder clover	39.6 ± 7.7	18.7 ± 3.6	58.3 ± 12.1
	French serradella	ND	ND	ND
	Gland clover	26.2 ± 14.4	3.6 ± 1.9	29.8 ± 13.0
	Lucerne	36.6 ± 16.5	27.7 ± 12.8	64.3 ± 5.1
	Subterranean clover	25.4 ± 7.5	7.1 ± 1.6	32.5 ± 10.6
	Yellow serradella	ND	ND	ND
	LSD ($p < 0.05$)	9.4	14.5	17.5
Inflorescence	Arrowleaf clover	ND	0.1 ± 0.1	<0.1
	Biserrula cv. Casbah	ND	ND	ND
	Biserrula cv. Mauro	ND	0.2 ± 0.1	<0.1
	Bladder clover	0.2 ± 0.2	<0.1	<0.1
	French serradella	ND	ND	ND
	Gland clover	0.1	ND	<0.1
	Lucerne	0.5 ± 0.1	0.2 ± 0.1	0.7 ± 0.1
	Subterranean clover	0.3 ± 0.1	ND	0.3 ± 0.1
	Yellow serradella	ND	ND	ND
	LSD ($p < 0.05$)	0.2	0.1	0.2

ND denotes compounds not detected. Least significant difference (LSD) differentiated between concentration means. Values represent the arithmetic mean of five replicates ± standard deviation. *cv. is acronym for cultivar.

Three isoflavones—daidzein, formononetin and genistein—are commonly found in pasture legumes and were also profiled in this study (Table 2). Isoflavone content in leaf (113.3–443.9 mg kg^{-1}), stem (37.5–968.1 mg kg^{-1}), and inflorescence (29.8–614.1 mg kg^{-1}) varied significantly by both species and tissue (Table 2). Gland clover had the highest concentration of total isoflavones in leaf tissue (443.9 mg kg^{-1}) while bladder clover had the highest concentration in stem and inflorescence tissue

(968.1 and 604.1 mg kg^{-1}, respectively). In terms of individual phytoestrogenic isoflavones in aerial tissues, formononetin concentration was greatest overall, followed by genistein at physiological maturity (Table 2). The production of phytoestrogenic isoflavones was greatest in clover species; specifically, daidzein concentration was highest in leaf tissues of gland clover at 120.2 mg kg^{-1}. Production of these compounds in stems was greatest in bladder clover and subterranean clover (112.0 and 107.8 mg kg^{-1}, respectively) when compared to other species. Subterranean clover contained the highest levels of daidzein (128.9 mg kg^{-1}) in inflorescence tissue when compared to other annual pasture species.

Table 2. The concentration of daidzein, formononetin and genistein (mg kg^{-1}) in annual pasture legume species at physiological maturity, post-flowering.

Tissue Type	Pasture Species	Daidzein	Formononetin	Genistein	Total
Leaf	Arrowleaf clover	55.1 ± 26.4	222.9 ± 84.4	28.3 ± 6.8	306.3 ± 94.3
	Biserrula cv. Casbah	11.5 ± 3.5	73.6 ± 51.3	116.8 ± 2.8	192.5 ± 18.7
	Biserrula cv. Mauro	6.1 ± 0.7	58.9 ± 17.3	124.7 ± 12.5	189.7 ± 24.0
	Bladder clover	95.0 ± 2.1	152.3 ± 18.9	184.1 ± 26.3	431.4 ± 32.8
	French serradella	47.7 ± 12.5	35.5 ± 48.6	130.1 ± 7.6	113.3 ± 8.1
	Gland clover	120.3 ± 16.1	226.1 ± 20.5	97.5 ± 10.8	443.9 ± 61.4
	Lucerne	48.2 ± 13.3	329.4 ± 192.7	61.5 ± 10.1	439.1 ± 141.9
	Subterranean clover	109.1 ± 14.8	157.6 ± 47.3	72.1 ± 244.3	318.8 ± 40.5
	Yellow serradella	89.0 ± 13.	91.6 ± 16.6	118.4 ± 5.4	219.1 ± 43.0
	LSD ($p < 0.05$)	24.9	37.5	34.8	—
Stem	Arrowleaf clover	9.8 ± 3.2	147.7 ± 58.2	27.6 ± 5.2	185.1 ± 67.1
	Biserrula cv. Casbah	7.5 ± 0.9	23.9 ± 12.2	6.1 ± 1.9	37.5 ± 8.9
	Biserrula cv. Mauro	5.1 ± 1.4	60.4 ± 18.4	9.9 ± 3.7	75.4 ± 27.4
	Bladder clover	112.0 ± 33.5	829.8 ± 144.9	126.1 ± 5.4	968.1 ± 394.7
	French serradella	75.2 ± 23.3	96.7 ± 34.4	104.1 ± 18.9	276.0 ± 13.4
	Gland clover	3.4 ± 0.5	239.5 ± 18.4	88.1 ± 37.4	331.0 ± 107.0
	Lucerne	7.5 ± 2.3	120.6 ± 35.3	15.9 ± 5.1	144.0 ± 56.4
	Subterranean clover	107.8 ± 18.9	36.2 ± 6.4	18.2 ± 8.3	162.2 ± 42.4
	Yellow serradella	42.5 ± 7.2	77.7 ± 14.2	20.3 ± 6.9	140.5 ± 25.9
	LSD ($p < 0.05$)	20.8	47.9	19.8	—
Inflorescence	Arrowleaf clover	1.3 ± 0.3	98.2 ± 24.5	18.4 ± 4.6	117.9 ± 45.3
	Biserrula cv. Casbah	2.2 ± 0.5	39.6 ± 9.9	10.2 ± 2.5	52.0 ± 17.6
	Biserrula cv. Mauro	1.7 ± 0.4	18.5 ± 4.6	9.6 ± 4.1	29.8 ± 7.5
	Bladder clover	19.3 ± 4.8	435.8 ± 108.9	158.9 ± 37.2	614.1 ± 190.6
	French serradella	10.5 ± 2.6	98.8 ± 24.7	19.9 ± 4.9	129.2 ± 43.4
	Gland clover	68.3 ± 17.1	184.8 ± 46.2	68.3 ± 17.1	321.0 ± 60.1
	Lucerne	25.5 ± 6.4	182.6 ± 45.6	25.7 ± 6.4	233.8 ± 81.1
	Subterranean clover	128.9 ± 32.2	187.6 ± 46.9	148.7 ± 39.8	465.2 ± 26.3
	Yellow serradella	14.6 ± 3.6	17.6 ± 4.4	25.8 ± 6.6	58.00 ± 5.2
	LSD ($p < 0.05$)	16.8	39.2	24.5	—

Least significant difference (LSD) was used to differentiate between concentration means. Values represent the arithmetic mean of five replicates ± standard deviations.

Formononetin concentration in leaf tissue was significantly higher in the perennial legume lucerne (329.4 mg kg^{-1}) than the other species, while for stem tissue, formononetin concentration was greatest in bladder clover (829.8 mg kg^{-1}). Interestingly, genistein levels were greatest in all three tissues types in bladder clover compared to other species.

2.2. *Effect of Biserrula Cultivar and Growth Stage on Phytoestrogen Levels*

Given the strong potential of biserrula to produce large quantities of biomass and suppress weeds successfully over time, further evaluation was performed to examine temporal effects on the accumulation of phytoestrogens in aerial tissues in both of the commercially available cultivars of biserrula in Australia. Analysis of tissues of Casbah and Mauro analyzed at five different growth stages showed that total concentrations of the phytoestrogens differed significantly between cultivars (Figure 2a,b). Phytoestrogen concentrations reached their maxima at either 50% bloom or full bloom while the lowest concentrations were observed at crop senescence. Significant differences in the concentration of coumestans between the two cultivars were limited to coumestrol at pre-bloom and 50% bloom stages; cv. Mauro was observed to produce greater levels of coumestans than cv. Casbah. The phytoestrogenic isoflavones formononetin and daidzein were more abundant in tissues of Casbah than Mauro, while the opposite was true for genistein.

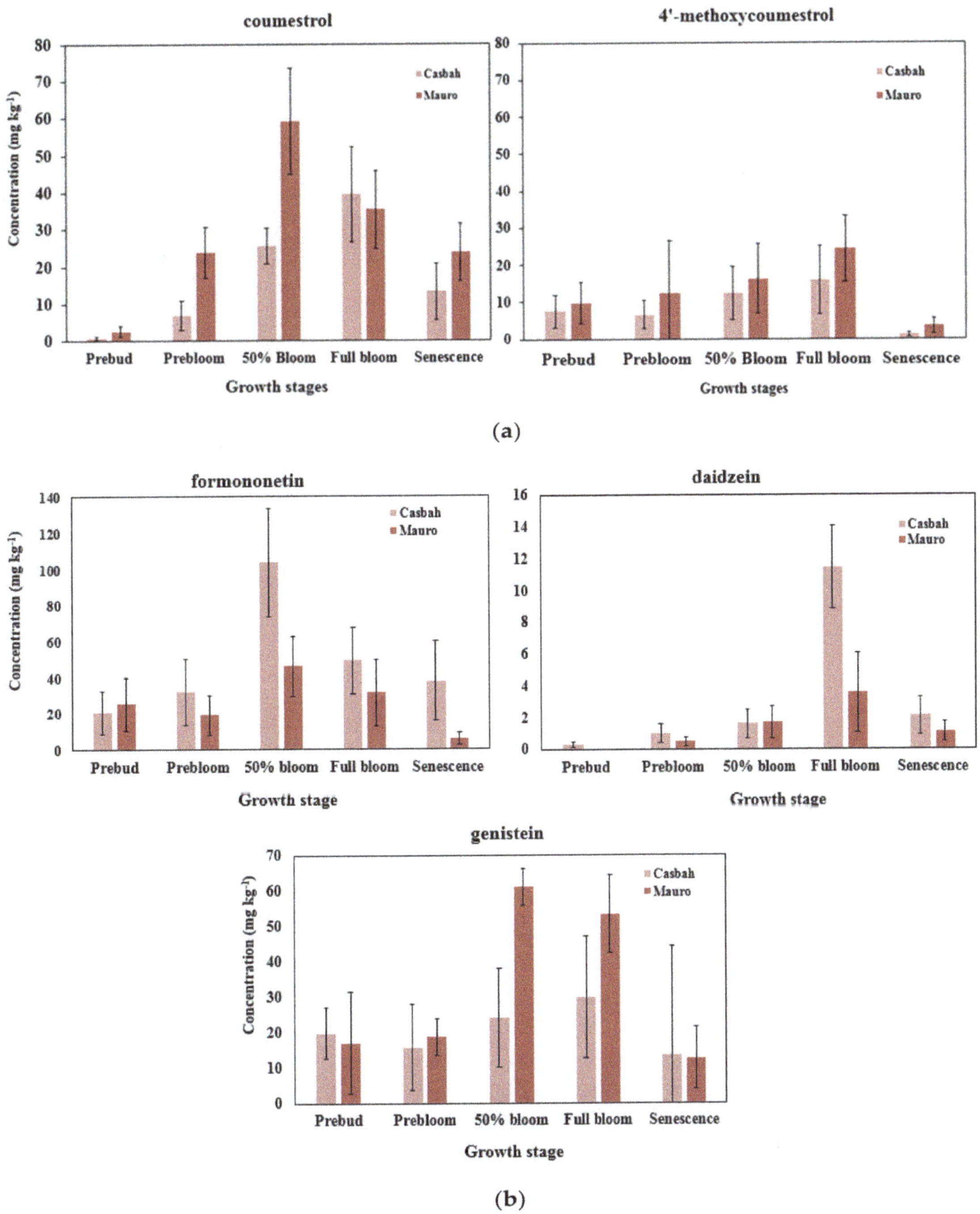

Figure 2. (**a**) The concentration of coumestrol and 4′-methoxycoumestrol at five growth stages i.e., pre bud, pre bloom, 50% bloom, full bloom, and senescence of field-grown biserrula cv. Casbah and cv. Mauro averaged over two years. Error bars indicate standard deviation; (**b**) the concentration of selected isoflavones at five growth stages i.e., pre bud, pre bloom, 50% bloom, full bloom, and senescence in biserrula cv. Casbah and cv. Mauro samples averaged over two years. Error bars indicate standard deviation.

2.3. *Quantification of Total Polyphenols and Proanthocyanidins*

Extractable TPC (total polyphenol content) ranged from 4.40 to 13.84 GAE and TPAC (total proanthocyanidins) ranged from 1.73 to 6.49 mg 100^{-1} g CE (Table 3). Gland clover contained significantly higher extractable TPC levels (13.84 mg 100^{-1} g) compared to all other pasture species. Extractable and bound TPAC was significantly higher in biserrula cv. Casbah compared to other species, while bound TPAC was only detected in the two biserrula cultivars (Casbah and Mauro) and the perennial legume, lucerne.

Table 3. Concentration of extractable and bound total polyphenol and total proanthocyanidins in foliar tissues of pasture legumes.

Pasture Species	TPC		TPAC	
	Extractable	Bound	Extractable	Bound
Arrowleaf clover	10.50^{b}	0.24^{de}	3.71^{c}	ND
Biserrula cv. Casbah	7.69^{cde}	0.43^{cd}	6.49^{a}	8.87^{a}
Biserrula cv. Mauro	6.24^{def}	0.11^{e}	4.29^{b}	4.06^{b}
Bladder clover	6.67^{def}	0.45^{bc}	1.73^{c}	ND
French serradella	8.29^{bcd}	0.35^{cd}	1.94^{c}	ND
Gland clover	13.84^{a}	0.57^{bc}	2.12^{c}	ND
Lucerne	4.40^{f}	0.11^{de}	2.05^{c}	0.33^{c}
Subterranean clover	9.39^{bc}	0.62^{bc}	4.52^{ab}	ND
Yellow serradella	5.17^{ef}	0.72^{b}	1.94^{c}	ND
LSD	2.62	0.28	2.16	1.95

Values represent arithmetic mean of five replicates. Values followed by the same superscript in each column are not significantly different ($p < 0.05$) as determined by ANOVA.

2.4. *Abundance of Flavonoids and Their Glycosides*

The clustering of molecular features profiled through non-targeted metabolic profiling revealed that molecular entities varied between species but were similar in species of the same genera (Supplementary Materials Figure S1). Over 5000 molecular features in total were successfully profiled in legume leaf tissues, with 1727 in stem and 1503 in inflorescence tissues (Figure S2). Interestingly, flavonoids and their glycosides accounted for the majority of constituents among all annotated major classes of secondary metabolites, based on verification with analytical standards as well as METLIN database comparisons. The relative abundance of various flavonoids and their glycosides in selected pasture legumes is presented in Figure 3. Interestingly, the total number of molecular features characterized in this study was highest in biserrula followed by French and yellow serradella, while gland clover exhibited the fewest total number of molecular features (Supplementary Materials Figure S1). However, gland clover, a recent introduction to Australia obtained from native pastures in the Mediterranean, exhibited a considerably higher abundance of secondary metabolites including flavonoids and their glycosides profiled in this study compared to other species (Figure 3). Both Casbah and Mauro cultivars of biserrula presented chemically similar profiles and exhibited relatively low abundance of flavonoids and their glycosides, in contrast to subterranean and gland clover.

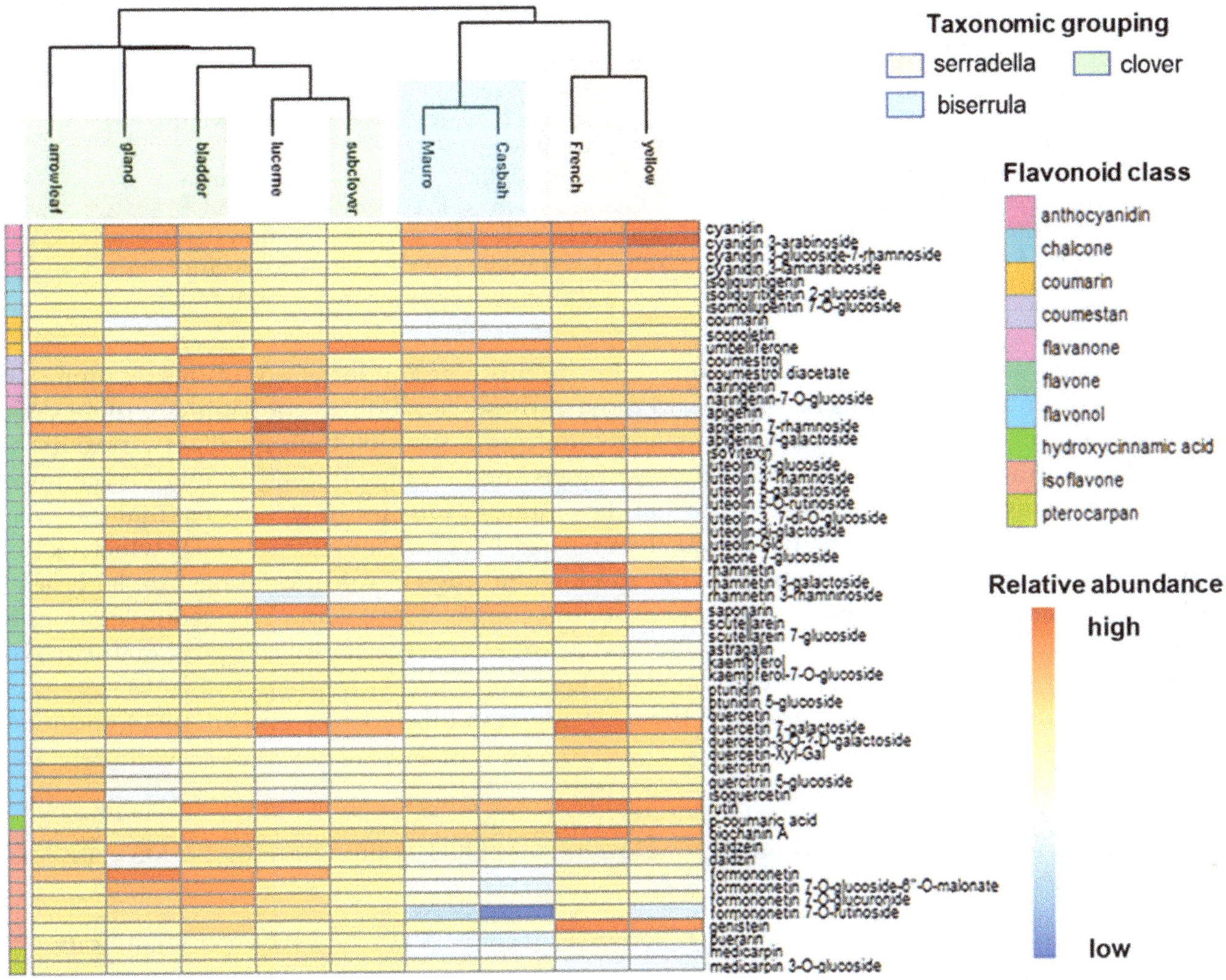

Figure 3. Hierarchal clustering of relative abundance of flavonoids, their glycosides, and coumestrol in leaf tissue in pasture legumes collected in 2016. Hierarchical clustering algorithm and Euclidean distance metric were used on normalized abundance using Mass Profiler Professional MPP (ver. 14.5 Agilent Santa Clara, CA, USA).

3. Discussion

This study represents the first published report on quantification and distribution of secondary products, including flavonoids, phenolics, proanthocyanidins, and phytoestrogens, in aerial tissues of several annual pasture species (Tables 1–5, Figure 3). The metabolic profiling and metabolomics analyses employed in this study provided a comprehensive overview of the specific metabolites pertaining to the biosynthesis and regulation of flavonoids and related metabolites, particularly those with adverse impacts on grazing livestock. The accurate identification of secondary metabolites was facilitated by the use of quantitative liquid chromatography quadrupole time-of-flight mass spectrometry (LC-MS-QToF), allowing their detection and quantification at low concentrations (ng kg^{-1}), with compounds identified by comparison with authentic standards or annotated using the METLIN library of secondary metabolites.

Table 4. Scientific, common, and cultivar names of annual pasture legumes evaluated in this study.

Scientific Name	Common Name	Cultivar
Trifolium vesiculosum Savi.	Arrowleaf clover	Cefalu
Biserrula pelecinus L.	Biserrula	Casbah, Mauro
Trifolium spumosum L.	Bladder clover	Bartolo
Ornithopus sativus Brot.	French serradella	Margarita
Trifolium glanduliferum Boiss.	Gland clover	Prima
Trifolium subterraneum L.	Subterranean clover	Seaton Park
Ornithopus compressus L.	Yellow serradella	Santorini
Medicago sativa L.	Lucerne	Aurora

Along with various isoflavonoids, coumestans are produced in large quantities by members of the *Fabaceae*, commonly known as legumes, and most of these metabolites contribute to plant defense. Plant-produced coumestans are known to be associated with various biological activities, many of which can be attributed to their function as phytoestrogens and polyphenols [42]. 4- methoxycoumestrol has quantified using coumestan as surrogate standard [43,44]. Our findings suggest that the greatest concentration of coumestans occurs in leaf and stem tissues, in contrast to floral tissues which exhibited only trace quantities. This is in agreement with previous findings in traditional pasture legumes [45]. In addition, our results suggest that consumption of pure stands of novel annual pasture legumes such as biserrula, French serradella, yellow serradella, and arrowleaf clover at physiological maturity would likely pose no threat to herd fertility given the lower levels of coumestans than threshold suggested in the literature. In direct contrast, gland and bladder clover exhibited concentrations of coumestans at levels above the suggested tolerance limit of ≈40 mg kg^{-1} in leaf tissues at approximately 48 and 70 mg kg^{-1}, respectively.

Table 5. Phytoestrogens from pasture legume species identified in this study by LC-MS-QToF in positive ionization mode.

Name	Molecular Formula	M + H	Basis for Identification [a]
Isoflavones			
Daidzein	$C_{15}H_{10}O_4$	255.0652	STD
Formononetin	$C_{16}H_{12}O_4$	269.0808	STD
Genistein	$C_{15}H_{10}O_5$	271.0601	STD
Coumestans			
Coumestrol	$C_{15}H_8O_5$	269.0444	STD
4′-methoxycoumestrol	$C_{16}H_{10}O_5$	283.0579	AM

[a] Basis for identification codes: AM [43,44]—match to accurate mass/molecular formula and MS/MS spectra; STD—match to accurate mass and retention time of analytical standards.

The most prevalent traditional pasture legumes in Australia, lucerne and subterranean clover, have been reported to consistently accumulate higher concentrations of coumestans at flowering in both leaf and stem tissues, but recurrent selection by plant breeders has recently resulted in reduced levels in many commercial cultivars. Our results indicated that both lucerne and subterranean clover cultivars selected in this study produced substantial concentrations of both metabolites, leading to a total concentration of 52 and 38 mg kg^{-1}, respectively; levels which may exceed livestock tolerance limits (≈40 mg kg^{-1} DM). Considerably higher concentrations of coumestans in leaf and stem tissues of gland and bladder clover suggest that these cultivars have experienced relatively limited genetic improvement through breeding at this stage (Angelo Loi, personal communication), and produce significant levels of secondary products that may adversely impact livestock health. This observation further supports the hypothesis that coumestans play a role in plant defense against herbivory when established in natural settings [46]. Our findings also imply that these cultivars should be avoided as sole sources of forage for grazing livestock.

Naturally occurring isoflavones found predominately in the *Fabaceae* have been well described, largely based on their pharmacological activities, including estrogenic effects, in humans and animals. The isoflavonoids profiled in the present study demonstrated clear intra-species variation as well as variation associated with tissue type. Other researchers have also noted some chemical variance within the genus *Trifolium* [47]. The ability of plants to regulate production of secondary metabolites in response to biotic and abiotic stressors, including climate change, has been well documented [48–51] and recent studies have shown that these abilities extend to production of flavonoids [12,13].

Qualitative and quantitative variation in flavonoids and associated phytoestrogens profiled in our study is an indicator of both species and tissue specific adaptations to the environment, resulting in modulation in the expression of associated genes [52]. Although recent pasture studies have suggested that isoflavones are highly abundant in *Medicago* and *Trifolium spp.* and are frequently highest before full-bloom [53], considerable variation in total concentration has been reported in mature *Trifolium pratense* pastures, with those surveyed ranging between 9000 and 27,000 mg kg^{-1} DM [48,54,55]. Variations in isoflavone concentration may have been due to environmentally induced responses, time of sampling, extraction efficiency and recovery, and previous analytical workflows employed to profile individual isoflavones [21]. We specifically attempted to address these issues noted in previous published works by conducting experimentation in uniform and replicated field sites, creating composite replicated samples for each pasture plot, and performing sampling at similar stages of physiological maturity in both 2016 and 2017 growing seasons. In addition, we utilized an automated high-pressure extraction device to rapidly and uniformly extract all samples, thereby performing uniform extraction across treatments and replicates, and reducing the possibility of plant to plant variation or inefficiencies associated with manual extraction protocols.

Several phytoestrogenic flavonoids, such as formononetin, the most abundant isoflavone detected followed by and genistein, were prevalent at physiological maturity (50% flowering) in all pasture legumes. These observations are in agreement with previous reports pertaining to various *Trifolium* species [48,56–58]. Recent findings also suggested a greater abundance of phytoestrogens and isoflavonoids in leaf tissues when compared to stems and inflorescence [53]. Biosynthesis and subsequent distribution of isoflavones among leaf, stem, and floral tissues was impacted by cultivar, physiological growth stage, and climatic conditions under which plants were maintained [59]. In *Trifolium pratense* (perennial red clover), isoflavone concentrations were also found to be impacted by growth stage, with inflorescence tissue containing equivalent concentrations of isoflavones to those in leaf tissue in initial bloom stages, with subsequent declines in floral tissue as the crop entered full bloom stage [48]. The disproportionately high levels of formononetin in bladder clover observed in our study suggests that modulation of the phenylpropanoid pathway may, in fact, be species specific and is important in determination of terminal isoflavone accumulation in the pathway.

Only arrowleaf clover, gland clover, bladder clover, and subterranean clover exhibited total isoflavone concentrations above the livestock-safe threshold level of ≈280 mg kg^{-1} DM (Table 4) and therefore, all four of these *Trifolium* spp., could theoretically adversely impact the reproductive performance of livestock. Hashem et al., 2016 similarly reported that cattle fed solely on *Trifolium alexandrinum* (berseem clover) containing isoflavones at high concentrations of ≈280 kg^{-1} DM exhibited hormonal disruption and reduced fertility, comparable to similar cattle grazing other species [18]. Interestingly, lambs grazing *Trifolium pratense* with high levels of isoflavones exhibited weight gain compared to cultivars low in isoflavones, but macromolecular interactions impacting plant nutrition and subsequent weight gain were not fully explored in that study [60]. Future studies investigating the production of phytoestrogens, including coumestans and isoflavones at various growth stages will be important to optimize seasonal grazing.

Isoflavones are synthesized as part of the phenylpropanoid pathway (Figure 4), which has multiple branches common to both legumes and non-legumes. The phenylpropanoid pathway leads to the generation of lignins, anthocyanins, phytoalexins and flavonoids, including isoflavones, as a means of plant protection against stress or predation [29]. Encoding enzymes in this pathway are both developmentally and tissue-specifically regulated, and environmental stressors such as exposure to UV light, drought, prolonged cold, pathogen attack, and nutrient deficiency may also influence end products. Isoflavone synthase (IFS) is a key enzyme involved in the production of an array of isoflavones from naringenin, a common phenylpropanoid pathway intermediate [27].

Legumes can produce both daidzein from the intermediate compound liquiritigenin, and genistein from the intermediate naringenin chalcone in alternative branches of the phenylpropanoid pathway (Figure 4). The presence of multiple copies of enzymes such as chalcone synthase and IFS in species within the *Fabaceae* allows for the differential regulation of isoflavone biosynthesis in response to both developmental and environmental stimuli [61,62].

The flavonoid precursors and various isoflavones profiled in this study varied qualitatively and quantitatively among the annual pasture species investigated, suggesting that these species differ with respect to their metabolic dynamics and ability to regulate flavonoid biosynthesis. Daidzein and formononetin are produced from one branch of the phenylpropanoid pathway while genistein is produced from another branch. Both branches originate from *p*-coumaroyl-CoA, but the bias of the pathway towards one branch or the other is typically determined by the equilibrium between the enzymes chalcone synthase (CS) and chalcone reductase (CR) [27]. Greater production of formononetin in the pasture legume extracts profiled in this study, in contrast to the other isoflavones (daidzein and genistein) (Table 2), suggests that flavonoid biosynthesis in some pasture legumes is biased towards the branch terminating in production of formononetin as opposed to the alternate pathway ending in genistein. These findings are also consistent with previous observations [27,63].

Of note, close metabolite clustering and a higher concentration of genistein in both cultivars of biserrula and serradella species (as opposed to the more typical daidzein and formononetin) suggested potential overexpression of the gene encoding CS for the conversion of *p*-coumaroyl-CoA to naringenin chalcone in biserrula and serradella (Figure 4). This observation is consistent with recent results of a study investigating isoflavone profiles in food-related species in the *Fabaceae* family where higher genistein concentrations discriminated the genera *Biserrula* and *Ornithopus* from other members of the family [52].

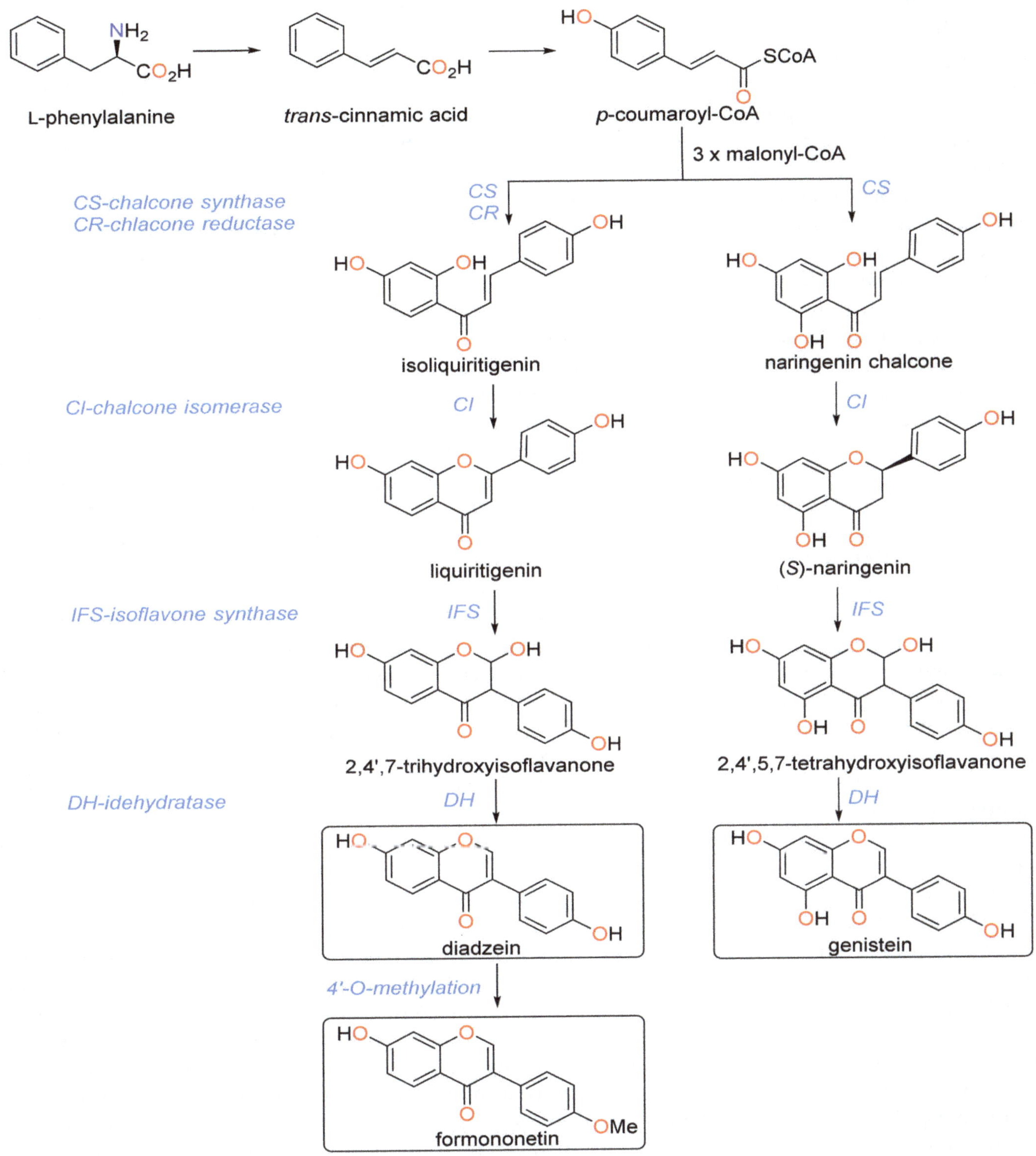

Figure 4. Biosynthetic pathway leading to the formation of various isoflavones including phytoestrogenic isoflavones (boxed) [29,64].

Isoflavones tend to accumulate at highest concentrations at physiological maturity of the plant in *Trifolium* spp. as they commence and complete flowering [65]. To assess the impact of growth stage on the abundance of key phytoestrogens, we further profiled phytoestrogens in two common biserrula cultivars, first commercialized in Australia in 2001, at five different growth stages. Interestingly, we observed no significant differences in coumestan concentrations between the two cultivars, despite the fact that the cultivars were initially isolated from geographically separate locations in the Mediterranean. This suggests that the phenylproponoid pathway was highly conserved in this species, particularly in early growth stages, and was not affected by genotypic differences. However, the production of phytoestrogenic isoflavones varied significantly at 50% bloom and full bloom stages (Figure 2a,b). A previous study also described a significant difference in the accumulation of isoflavonoids (formononetin and biochanin A) in different cultivars of red clover, peaking at 50% bloom and full bloom stage [48]. Our results demonstrated that concentrations of all phytoestrogens assessed were lowest at senescence, and are in agreement with earlier observations in red clover [65]. As noted previously, the genes encoding enzymes for the biosynthesis of isoflavones are developmentally regulated and are frequently influenced by environmental stressors and various biotic factors [29,49,62], some of which are likely experienced in southeastern Australia during a typical growing season. Breeding programs in annual pasture legumes can therefore exploit the bias of the phenylpropanoid pathway by targeted manipulation of genes involved in the biosynthesis of genistein and daidzein through ectopic expression of specific transcription factors over various growth stages [66].

Pasture legume samples collected from multi-year field trials were also subjected to quantification of extractable and bound total polyphenol content (TPC) and total proanthocyanidin content (TPAC) (Table 3). The important role of TPAC in pasture legumes in reducing herbivory by impacting palatability attributes has been elucidated previously [46]. This is the first study to report on both TPC and TPAC in annual self-regenerating pasture legumes at physiological maturity and our findings support previous reports suggesting a similar range for other related pasture legumes [33,67,68]. The range of TPAC in *Ornithopus* reported in a previous study [32] was between 2 g 100 g^{-1} and 2.5 g 100 g^{-1}, which is also in agreement with the current study. However, proanthocyanidins in *Onobrychis viciifolia* (sainfoin) were reported to be up to 10 times higher than measured in this study and this discrepancy could be associated with species, methodology, or growth stage of sampling [69]. Of note is that previous studies using similar assays did not detect measurable TPAC in lucerne [70,71]. This could be due to differences in cultivar genetics and growth conditions; however, we also observed that lucerne exhibited the lowest extractable TPC and TPAC levels of all legumes surveyed. A high concentration of extractable TPAC was noted in biserrula cv. Casbah; both cultivars also exhibited high levels of bound TPAC in their cell walls. High TPAC levels suggest the potential for biserrula to limit pathogen and herbivore attack or reduced palatability to grazing livestock [72], but may potentially offer a cost-effective opportunity for this species to be integrated into multi-species mixtures to reduce parasite burden, an outcome suggested by recent replicated trials with biserrula on grazing sheep. At this time, specific bioactive proanthocyanidins in biserrula remain unidentified.

This study employed metabolic profiling approaches for identification and quantification of a number of more common secondary plant metabolites, particularly flavonoids. Our results clearly demonstrate that (a) this collection of annual pasture legumes produces a diverse array of flavonoids and other phytochemicals associated with plant defense, and in some cases less desirable phytoestrogens or proanthocyanidins and (b) the regulation of the biosynthesis of flavonoids and related metabolites such as coumestans through the phenylpropanoid pathway occurs at multiple branch sites and as reported could be impacted by elicitation in response to various biotic factors including predation and herbivory. Plants have thus developed various evolutionary adaptations for making regulatory decisions to produce an array of secondary metabolites which can provide them with multiple ecological benefits. In some cases, a single compound or related family of compounds can exhibit multiple biological functions in plants [64,73]. Despite significant differences in biological functions of various plant metabolites which are frequently concentration dependent, related compounds often share common biosynthetic pathways while others, including flavonoids, can originate from diverse biosynthetic pathways and precursors [51,74]. Evolutionarily, this chemical diversity or flexibility provides higher plants with a cost-effective strategy for further resource allocation or reallocation. As an example, our results suggest that both lucerne and gland clover have the ability to upregulate the synthesis of formononetin from daidzein while bladder clover evolved to upregulate biosynthesis of genistein from 2′,4′,5,7 tetrahydroxy-isoflavone (Figure 4).

Our study results also support the hypothesis that high concentrations of key phytoestrogens, TPC and TPAC in annual pasture legumes are associated with flavonoid abundance. Interestingly, the relative abundance of total molecular features profiled through non-targeted metabolic profiling was highest in biserrula leaf tissues followed by serradella, while gland clover leaf tissue exhibited the fewest molecular features. In contrast, gland clover exhibited a higher concentration of flavonoids and their glycosides, suggesting that trade-offs in metabolite production and regulation occur in pasture legume species; i.e., if flavonoid synthesis is upregulated, then the expression of TPAC, TPC, and other polyphenols may be downregulated.

4. Materials and Methods

4.1. Chemicals

Analytical grade acetone, acetic acid, anhydrous sodium acetate, ethanol, hexane, hydrochloric acid, potassium chloride, sodium carbonate, and sulfuric acid were obtained from Chem Supply (Port Adelaide, South Australia, Australia). High pressure liquid chromatography (HPLC) grade methanol, acetonitrile, formic acid, and ammonium acetate as well as analytical standards (Folin–Ciocalteu reagent, vanillin, gallic acid, cyanidin-3-*O*-glucoside, (+)-catechin, coumestrol, formononetin, genistein, and naringenin) were purchased from Sigma Aldrich (Australia).

4.2. Plant Material

Monocultures of selected pasture legumes listed in Table 4 were established in 2016 and 2017 (Experiment 1) at the Charles Sturt University research farm in Wagga Wagga, NSW, Australia (35.04° S, 147.36° E) on a red sodosol soil [75]. Each planting was arranged as a randomized complete block design with five replications. Individual pasture legumes were established in plots by direct-seeding with a drill adapted for small-seeded legumes in late May to early June of each year, with plots measuring approximately 4 × 20 m. Following on from experimentation to assess legume establishment and performance [75], above-ground plant tissues (leaf, stem, and flower) were collected from established plots in the third week of October in 2016 and 2017 after >50% flowering was achieved in each species. This corresponded to the growth stage in which peak concentrations of phytoestrogen have been reported in vegetative tissue [76]. An additional trial was also established at Charles Sturt University research farm in 2016 and 2017 which focused on chemical composition in two newly released biserrula cultivars (cv. Casbah and Mauro) using four replicates per cultivar (Experiment 2). In this case, composite fresh plant samples were collected from each plot (two sub-samples per cultivar) at pre-bud, pre-bloom, 50% bloom, full bloom, and senescence between mid-July to mid-November in 2016 and 2017, approximately every three weeks. All plant material was placed on ice at collection and subsequently stored at −20 °C until extraction.

4.3. LC-QToF-MS Analysis

Metabolic profiling of plant tissues was performed using an Agilent 1290 Infinity LC system equipped with a quaternary pump, diode array detector (DAD), degasser, temperature-controlled column (25 °C), and cooled auto-sampler compartment (4 °C) which was coupled to an Agilent 6530 quadrupole time-of-flight (QToF) mass spectrometer (MS) with an Agilent Dual Jet Stream ionization source (Agilent Technologies, Melbourne, Australia). Full-scan mass spectra were acquired over an *m/z* range of 100–1700 Da at a rate of two spectra/second in both positive and negative ion modes. Chromatographic separation was achieved using a reverse-phase C_{18} Poroshell column (2.1 × 100 mm, 2.7 μm particle size) (Agilent Technologies, Santa Clara, CA, USA) equipped with a C_{18} guard column (2.1 × 12.5 mm, 5 μm particle size) (Agilent Technologies, CA, USA) using a flow rate of 0.3 mL min^{-1}. The column was equilibrated for 40 min prior to analysis. Separation was obtained with a gradient of solvent A, [water (Milli-Q, TKA-GenPure, Germany) + 0.1% formic acid (LC-MS grade, LiChropur®, 98–100%, Sigma-Aldrich, MO, USA)] and solvent B [95% HPLC-grade acetonitrile (RCI Labscan, Bangkok, Thailand) + 0.1% formic acid]. The solvent gradient was as follows: 5% B for 0.5 min increasing to 100% B over the next 16.5 min, then holding at 100% B for 23 min and returning to 5% B from 23.1 min to 29 min. The DAD monitored absorbance across a range of wavelengths from 210 to 635 nm. Injection volume was 10 µL for each sample. Nitrogen was used as the drying gas at 250 °C and a flow rate of 9 L min^{-1}. Five biological replicates for each treatment were analyzed. Phytoestrogens (isoflavones and coumestans) identified in annual pasture legumes in the current study are summarized in Table 5.

4.4. Extraction of Polyphenols

Foliar and inflorescence samples (1 g) from each pasture legume were freeze-dried and ground manually to a fine powder (≈1–2 mm size) using a mortar and pestle. Samples were extracted as described previously [77] with minor modifications. Briefly, ground foliar tissue was defatted with *n*-hexane, and residues were extracted three times with 20 mL of solvent (acetone:water:acetic acid 70:29.5:0.5 *v*/*v*/*v*). The three aliquots were combined, and solvent was removed using a rotary evaporator. The residue was lyophilized and reconstituted in 50% aqueous methanol to a final concentration of 1 g mL^{-1} for storage at −20 °C until required.

4.5. Quantification of Total Polyphenol Content

Total free phenolic content was determined as described previously [78] with minor alterations. Briefly, 125 µL of the extract was mixed with 125 µL of Folin–Ciocalteu reagent and 500 µL of deionized water and incubated in the dark for 6 min. The solution was neutralized by adding 1.5 mL 7% aqueous sodium carbonate solution and further incubated in the dark for 90 min, after which absorbance was measured at 725 nm using a UV/Vis spectrophotometer (FLUOstar Omega, BMG Labtech, Offenburg, Germany) against a methanol control. Total phenolic content was expressed as mg 100^{-1} g of gallic acid equivalent (GAE). The experiment was repeated twice with three technical replicates.

4.6. Quantification of Total Proanthocyanidin Content (TPAC)

TPAC was quantified using a vanillin assay as described previously [79,80]. Briefly, 0.2 mL of extract was added to 0.5 mL of 1% (*w*/*v*) vanillin in methanol and 0.5 mL of 25% sulfuric acid in methanol. The extract was thoroughly blended using a vortex mixer and placed in a water bath at 37 °C for 15 min. Absorbance was measured at room temperature at a wavelength of 500 nm. The total proanthocyanidin content was determined as mg 100^{-1} g of (+)-catechin equivalent (CE). The experiment was performed with three technical replicates and the experiment was repeated twice.

4.7. Statistical Analysis

A matrix of molecular features characterized by mass to charge ratio (*m/z*) and retention time (RT) was generated using MassHunter Workstation Qualitative software version B07.00, MassHunter Profinder (version B.08.00), Mass Profiler Professional (MPP version 14.5) and Personal Compound Database Library (PCDL) (Agilent Technologies, CA, USA). Molecular features were extracted and binned/aligned using parameters as follows: peak height ≥ 10,000 counts, compound ion count threshold of two or more ions, compound alignment tolerances 0.00% + 0.15 min for RT and 20.00 mg kg^{-1} ± 2.00 mDa for mass using Profinder. Molecular features which were present only in three samples out of five were included in the analysis. Compounds were tentatively identified by matching molecular entities with PCDL entries having similar accurate mass, RT, and mass spectra (generated from analytical standards) where possible and the METLIN metabolomics database (version B 07.00, Agilent Technologies, CA, USA) otherwise. All descriptive statistical analyses were performed using Statistix (STATISTIX software, version 4.1; Analytical Software, Tallahassee, FL, USA) and standard deviations were calculated and reported where possible.

5. Conclusions

In conclusion, this study has extended our understanding of secondary metabolism associated with the biosynthesis of phytoestrogens and proanthocyanidins in annual and perennial legumes. Metabolic profiling performed with a variety of hardseeded annual pasture legumes demonstrated that phytoestrogens quantified in this study were at concentrations insufficient to pose negative impact on livestock production with the exception of gland clover, bladder clover, and lucerne. Results of the study also suggest that annual hardseeded pasture legumes of Mediterranean origin offer viable and sustainable alternative pasture options for mixed farming systems of southeastern New South Wales. In the future, the use of an integrated experimental approach including multi-omics platforms could also potentially provide deeper insights into pathway dynamics and regulation of associated genes important in the production of secondary metabolites in pasture legumes. Our findings along with those of Butkut et al., 2018 suggest strong potential to improve legume-based forage quality through recurrent selection or engineering for species- or cultivar-specific phytochemicals [53]. In addition, optimization of livestock management to reduce health and reproductive issues by selective grazing through timing of animal movement, manipulation of plant growth stage at harvest, and appropriate selection of pasture species mixtures is also warranted.

Supplementary Materials:
Figure S1: Hierarchal clustering of molecular features in leaf tissue of annual pasture legumes acquired using UHPLC-QToF-MS in positive and negative mode. Hierarchical clustering algorithm and Euclidean distance metric were used on normalized abundance using MPP (ver. 14.5 Agilent Santa Clara, CA, USA), Figure S2: Hierarchal clustering of relative abundance of flavonoids, their glycosides, and coumestrol in inflorescence tissue in pasture legumes collected in 2016. Hierarchical clustering algorithm and Euclidean distance metric were used on normalized abundance using R package.

Author Contributions: Conceptualization, S.L., P.A.W. and L.A.W.; Methodology, S.L., J.W.P., P.A.W. and L.A.W.; Formal analysis, S.L., P.A.W. and L.A.W.; Writing, S.L., S.G., R.A.B., P.A.W. and L.A.W.; Original draft preparation, S.L.; Review and editing, S.G., R.A.B., P.A.W., J.W.P. and L.A.W. Funding acquisition, J.W.P. and L.A.W. All authors have read and agreed to the published version of the manuscript.

Acknowledgments: The authors acknowledge funding support from Meat and Livestock Australia (MLA) (Grant: B.WEF.0146 awarded to L.A. Weston and J. Piltz) for field trials and the Australian Center for International Agricultural Research (ACIAR) for sponsoring the PhD fellowship awarded to S. Latif for studies associated with this project. The authors also wish to acknowledge Ulrike Mathesius of The Australian National University (Acton, ACT, Australia) for providing analytical standards of flavonoids used in this study, and Willian Brown, Graeme Heath, and Simon Flinn for their support in data collection.

References

1. Whitbread, A.M.; Hall, C.A.; Pengelly, B.C. A novel approach to planting grasslegume pastures in the mixed farming zone of southern inland Queensland, Australia. *Crop Pasture Sci.* **2009**, *60*, 1147–1155. [CrossRef]
2. Hackney, B.; Dear, B.; Li, G.; Rodham, C.; Tidd, J. Current and future use of pasture legumes in central and southern NSW—Results of a farmer and advisor survey. In Proceedings of the 14th Australian Society of Agronomy Conference, Adelaide, Australia, 21–25 September 2008.
3. Loi, A.; Howieson, J.G.; Nutt, B.J.; Carr, S.J. A second generation of annual pasture legumes and their potential for inclusion in Mediterranean-type farming systems. *Aust. J. Exp. Agric.* **2005**, *45*, 289–299. [CrossRef]
4. Burggraaf, V.; Waghorn, G.; Woodward, S.; Thom, E. Effects of condensed tannins in white clover flowers on their digestion in vitro. *Anim. Feed Sci. Technol.* **2008**, *142*, 44–58. [CrossRef]
5. Hall, D.; Wolfe, E.; Cullis, B. Performance of breeding ewes on lucerne-subterranean clover pastures. *Aust. J. Exp. Agric.* **1985**, *25*, 758–765. [CrossRef]
6. Humphries, A. Future applications of lucerne for efficient livestock production in southern Australia. *Crop Pasture Sci.* **2013**, *63*, 909–917. [CrossRef]

7. Mulholland, J. Animal production from lucerne-based pastures. In *Realising the Potential of Pastures, the Proceedings of the 16th Riverina Outlook Conference, Wagga Wagga, Australia, 15 July, 1987*; Riverina-Murray Institute of Higher Education: Wagga Wagga, Australia, 1987; pp. 57–66.
8. Loi, A.; Howieson, J.H.; Carr, S.J. Register of Australian herbage plant cultivars. *Biserrula pelecinus* L. (biserrula) cv. Casbah. *Aust. J. Exp. Agric.* **2001**, *41*, 841–842. [CrossRef]
9. Hackney, B.; Nutt, B.; Loi, A.; Yates, R.; Quinn, J.; Piltz, J.; Jenkins, J.; Weston, L.; O'Hare, M.; Butcher, A. "On-demand" hardseeded pasture legumes–a paradigm shift in crop-pasture rotations for southern Australian mixed farming systems. In *Building Productive, Diverse and Sustainable Landscapes, Proceedings of the 17th Australian Agronomy Conference, Hobart, Australia, 20–24 September 2015*; Acuña, T., Moeller, C., Parsons, D., Harrison, M., Eds.; Australian Agronomy society: Hobart, Australia, 2015.
10. Dear, B.; Ewing, M.A. The search for new pasture plants to achieve more sustainable production systems in southern Australia. *Aust. J. Exp. Agric.* **2008**, *48*, 387–396. [CrossRef]
11. Latif, S.; Gurusinghe, S.; Weston, P.A.; Brown, W.B.; Quinn, J.C.; Piltz, J.W.; Weston, L.A. Performance and weed-suppressive potential of selected pasture legumes against annual weeds in south-eastern Australia. *Crop Pasture Sci.* **2019**, *70*, 147–158. [CrossRef]
12. Mathesius, U. Flavonoid Functions in Plants and Their Interactions with Other Organisms. *Plants* **2018**, *7*, 30. [CrossRef]
13. Weston, L.A.; Mathesius, U. Flavonoids: Their structure, biosynthesis and role in the rhizosphere, including allelopathy. *J. Chem. Ecol.* **2013**, *39*, 283–297. [CrossRef]
14. Hassan, S.; Mathesius, U. The role of flavonoids in root–rhizosphere signalling: Opportunities and challenges for improving plant–microbe interactions. *J. Exp. Bot.* **2012**, *63*, 3429–3444. [CrossRef]
15. Reed, K.F.M. Fertility of herbivores consuming phytoestrogen-containing *Medicago* and *Trifolium* species. *Agriculture* **2016**, *6*, 35. [CrossRef]
16. Tiwari, B.K.; Brunton, N.P.; Brennan, C. *Handbook of Plant Food Phytochemicals: Sources, Stability and Extraction*; John Wiley & Sons: Hoboken, NJ, USA, 2013.
17. Wocławek-Potocka, I.; Mannelli, C.; Boruszewska, D.; Kowalczyk-Zieba, I.; Waśniewski, T.; Skarżyński, D.J. Diverse effects of phytoestrogens on the reproductive performance: Cow as a model. *Int. J. Endocrinol.* **2013**, *2013*, 650984. [CrossRef] [PubMed]
18. Hashem, N.; El-Azrak, K.; Sallam, S. Hormonal concentrations and reproductive performance of Holstein heifers fed *Trifolium alexandrinum* as a phytoestrogenic roughage. *Anim. Reprod. Sci.* **2016**, *170*, 121–127. [CrossRef] [PubMed]
19. Hloucalová, P.; Skládanka, J.; Horký, P.; Klejdus, B.; Pelikán, J.; Knotová, D. Determination of Phytoestrogen Content in Fresh-Cut Legume Forage. *Animals* **2016**, *6*, 43. [CrossRef]
20. Burton, J.; Wells, M. The effect of phytoestrogens on the female genital tract. *J. Clin. Pathol.* **2002**, *55*, 401–407. [CrossRef] [PubMed]
21. Cvejić, J.; Bursać, M.; Atanacković, M. Phytoestrogens: "Estrogene-Like" Phytochemicals. In *Studies in Natural Products Chemistry*; Elsevier: Amsterdam, The Netherlands, 2012; Volume 38, pp. 1–35.
22. Bickoff, E.; Booth, A.; Lyman, R.; Livingston, A.; Thompson, C.; Deeds, F. Coumestrol, a new estrogen isolated from forage crops. *Science* **1957**, *126*, 969–970. [CrossRef] [PubMed]
23. Mostrom, M.; Evans, T.J. *Phytoestrogens*; Elsevier: Amsterdam, The Netherlands, 2011; pp. 707–722.
24. Woclawek-Potocka, I.; Okuda, K.; Acosta, T.; Korzekwa, A.; Pilawski, W.; Skarzynski, D. Phytoestrogen metabolites are much more active than phytoestrogens themselves in increasing prostaglandin F2α synthesis via prostaglanin F2α synthase-like 2 stimulation in bovine endometrium. *Prostaglandins Other Lipid Mediat.* **2005**, *78*, 202–217. [CrossRef]
25. Lookhart, G. Analysis of coumestrol, a plant estrogen, in animal feeds by high-performance liquid chromatography. *J. Agric. Food Chem.* **1980**, *28*, 666–667. [CrossRef]
26. Dixon, R.A.; Achnine, L.; Kota, P.; Liu, C.J.; Reddy, M.S.; Wang, L. The phenylpropanoid pathway and plant defence—A genomics perspective. *Mol. Plant Pathol.* **2002**, *3*, 371–390. [CrossRef]
27. Yu, O.; Jung, W.; Shi, J.; Croes, R.A.; Fader, G.M.; McGonigle, B.; Odell, J.T. Production of the Isoflavones Genistein and Daidzein in Non-Legume Dicot and Monocot Tissues. *Plant Physiol.* **2000**, *124*, 781–794. [CrossRef] [PubMed]

28. Gupta, O.P.; Dahuja, A.; Sachdev, A.; Jain, P.K.; Kumari, S.; Praveen, S. Cytosine Methylation of Isoflavone Synthase Gene in the Genic Region Positively Regulates Its Expression and Isoflavone Biosynthesis in Soybean Seeds. *DNA Cell Biol.* **2019**, *38*, 510–520. [CrossRef] [PubMed]
29. Dixon, R.A.; Harrison, M.J.; Paiva, N.L. The isoflavonoid phytoalexin pathway from enzymes to genes to transcription factors. *Physiol. Plant.* **1995**, *93*, 385–392. [CrossRef]
30. Piluzza, G.; Sulas, L.; Bullitta, S. Tannins in forage plants and their role in animal husbandry and environmental sustainability: A review. *Grass Forage Sci.* **2014**, *69*, 32–48. [CrossRef]
31. Salami, S.A.; Luciano, G.; O'Grady, M.N.; Biondi, L.; Newbold, C.J.; Kerry, J.P.; Priolo, A. Sustainability of feeding plant by-products: A review of the implications for ruminant meat production. *Anim. Feed Sci. Technol.* **2019**, *251*, 37–55. [CrossRef]
32. Aerts, R.J.; Barry, T.N.; McNabb, W.C. Polyphenols and agriculture: Beneficial effects of proanthocyanidins in forages. *Agric. Ecosyst. Environ.* **1999**, *75*, 1–12. [CrossRef]
33. Douglas, G.; Stienezen, M.; Waghorn, G.; Foote, A.; Purchas, R. Effect of condensed tannins in birdsfoot trefoil (*Lotus corniculatus*) and sulla (*Hedysarum coronarium*) on body weight, carcass fat depth, and wool growth of lambs in New Zealand. *N. Z. J. Agric. Res.* **1999**, *42*, 55–64. [CrossRef]
34. Ramírez-Restrepo, C.; Barry, T. Alternative temperate forages containing secondary compounds for improving sustainable productivity in grazing ruminants. *Anim. Feed Sci. Technol.* **2005**, *120*, 179–201. [CrossRef]
35. Kingston-Smith, A.H.; Edwards, J.E.; Huws, S.A.; Kim, E.J.; Abberton, M. Plant-based strategies towards minimising 'livestock's long shadow'. *Proc. Nutr. Soc.* **2010**, *69*, 613–620. [CrossRef]
36. Tzamaloukas, O.; Athanasiadou, S.; Kyriazakis, I.; Huntley, J.; Jackson, F. The effect of chicory (*Cichorium intybus*) and sulla (*Hedysarum coronarium*) on larval development and mucosal cell responses of growing lambs challenged with *Teladorsagia circumcincta*. *Parasitology* **2006**, *132*, 419–426. [CrossRef]
37. Hoste, H.; Torres-Acosta, J.; Quijada, J.; Chan-Perez, I.; Dakheel, M.; Kommuru, D.; Mueller-Harvey, I.; Terrill, T. Interactions between nutrition and infections with *Haemonchus contortus* and related gastrointestinal nematodes in small ruminants. In *Advances in Parasitology*; Elsevier: Amsterdam, The Netherlands, 2016; Volume 93, pp. 239–351.
38. Min, B.R.; Barry, T.N.; Attwood, G.T.; McNabb, W.C. The effect of condensed tannins on the nutrition and health of ruminants fed fresh temperate forages: A review. *Anim. Feed Sci. Technol.* **2003**, *106*, 3–19. [CrossRef]
39. Mueller-Harvey, I.; Bee, G.; Dohme-Meier, F.; Hoste, H.; Karonen, M.; Kölliker, R.; Lüscher, A.; Niderkorn, V.; Pellikaan, W.F.; Salminen, J.-P.; et al. Benefits of condensed tannins in forage legumes fed to ruminants: Importance of structure, concentration, and diet composition. *Crop. Sci.* **2019**, *59*, 861. [CrossRef]
40. Weston, L.A.; Skoneczny, D.; Weston, P.A.; Weidenhamer, J.D. Metabolic profiling: An overview—New approaches for the detection and functional analysis of biologically active secondary plant products. *J. Allelochem. Interact.* **2015**, *1*, 15–27.
41. Williamson, L.N.; Bartlett, M.G. Quantitative liquid chromatography/time-of-flight mass spectrometry. *Biomed. Chromatogr.* **2007**, *21*, 567–576. [CrossRef]
42. Nehybova, T.; Smarda, J.; Benes, P. Plant coumestans: Recent advances and future perspectives in cancer therapy. *Anti Cancer Agents Med. Chem.* **2014**, *14*, 1351–1362. [CrossRef] [PubMed]
43. Christ, B.; Hauenstein, M.; Hörtensteiner, S. A liquid chromatography–mass spectrometry platform for the analysis of phyllobilins, the major degradation products of chlorophyll in *Arabidopsis thaliana*. *Plant. J.* **2016**, *88*, 505–518. [CrossRef]
44. Jiang, H.; Liao, X.; Wood, C.M.; Xiao, C.-W.; Feng, Y.-L. A robust analytical method for measurement of phytoestrogens and related metabolites in serum with liquid chromatography tandem mass spectrometry. *J. Chromatogr. B* **2016**, *1012–1013*, 106–112. [CrossRef]
45. Smith, J.; Jagusch, K.; Brunswick, L.; Kelly, R. Coumestans in lucerne and ovulation in ewes. *N. Z. J. Agric. Res.* **1979**, *22*, 411–416. [CrossRef]
46. Molyneux, R.J.; Ralphs, M.H. Plant toxins and palatability to herbivores. *Rangel. Ecol. Manag. J. Range Manag. Arch.* **1992**, *45*, 13–18. [CrossRef]
47. Oleszek, W.; Stochmal, A.; Janda, B. Concentration of isoflavones and other phenolics in the aerial parts of Trifolium species. *J. Agric. Food Chem.* **2007**, *55*, 8095–8100. [CrossRef]
48. Sivesind, E.; Seguin, P. Effects of the environment, cultivar, maturity, and preservation method on red clover isoflavone concentration. *J. Agric. Food Chem.* **2005**, *53*, 6397–6402. [CrossRef] [PubMed]

49. Mahmood, K.; Khan, M.B.; Song, Y.Y.; Ijaz, M.; Luo, S.M.; Zeng, R.S. UV-irradiation enhances rice allelopathic potential in rhizosphere soil. *Plant Growth Regul.* **2013**, *71*, 21–29. [CrossRef]
50. Iannucci, A.; Fragasso, M.; Platani, C.; Papa, R. Plant growth and phenolic compounds in the rhizosphere soil of wild oat (*Avena fatua* L.). *Front. Plant Sci.* **2013**, *4*. [CrossRef] [PubMed]
51. Einhellig, F.A. Allelopathy—A natural protection, allelochemicals. In *Handbook of Natural Pesticides: Methods*; CRC Press: Boca Raton, FL, USA, 2018; pp. 161–200.
52. Visnevschi-Necrasov, T.; Barreira, J.C.; Cunha, S.C.; Pereira, G.; Nunes, E.; Oliveira, M.B.P. Phylogenetic insights on the isoflavone profile variations in *Fabaceae* spp.: Assessment through PCA and LDA. *Food Res. Int.* **2015**, *76*, 51–57. [CrossRef]
53. Butkutė, B.; Padarauskas, A.; Cesevičienė, J.; Taujenis, L.; Norkevičienė, E. Phytochemical composition of temperate perennial legumes. *Crop. Pasture Sci.* **2018**, *69*, 1020–1030. [CrossRef]
54. Tsao, R.; Papadopoulos, Y.; Yang, R.; Young, J.C.; McRae, K. Isoflavone profiles of red clovers and their distribution in different parts harvested at different growing stages. *J. Agric. Food Chem.* **2006**, *54*, 5797–5805. [CrossRef]
55. Wu, Q.; Wang, M.; Simon, J.E. Determination of isoflavones in red clover and related species by high-performance liquid chromatography combined with ultraviolet and mass spectrometric detection. *J. Chromatogr.* **2003**, *1016*, 195–209. [CrossRef]
56. Butkutė, B.; Lemežienė, N.; Dabkevičienė, G.; Jakštas, V.; Vilčinskas, E.; Janulis, V. Source of variation of isoflavone concentrations in perennial clover species. *Pharmacogn. Mag.* **2014**, *10*, S181. [CrossRef]
57. Zgórka, G. Ultrasound-assisted solid-phase extraction coupled with photodiode-array and fluorescence detection for chemotaxonomy of isoflavone phytoestrogens in *Trifolium* L. (Clover) species. *J. Sep. Sci.* **2009**, *32*, 965–972.
58. Ramos, G.P.; Dias, P.M.; Morais, C.B.; Fröehlich, P.E.; Dall'Agnol, M.; Zuanazzi, J.A. LC determination of four isoflavone aglycones in red clover (*Trifolium pratense* L.). *Chromatographia* **2008**, *67*, 125–129. [CrossRef]
59. Saviranta, N.M.; Anttonen, M.J.; von Wright, A.; Karjalainen, R.O. Red clover (*Trifolium pratense* L.) isoflavones: Determination of concentrations by plant stage, flower colour, plant part and cultivar. *J. Sci. Food Agric.* **2008**, *88*, 125–132. [CrossRef]
60. Bennetau-Pelissero, C. Risks and benefits of phytoestrogens: Where are we now? *Curr. Opin. Clin. Nutr. Metab. Care* **2016**, *19*, 477–483. [CrossRef] [PubMed]
61. Dubery, I.A.; Mienie, C. Tissue-specific expression of the chalcone synthase multigene family in *Phaseolus vulgaris*: Development of a RT-PCR method for the expression profiling of the chs isogenes. *J. Plant Physiol.* **2001**, *158*, 115–120. [CrossRef]
62. Martin, C. Structure, function, and regulation of the chalcone synthase. In *International Review of Cytology*; Elsevier: Amsterdam, The Netherlands, 1993; Volume 147, pp. 233–284.
63. Jung, W.; Yu, O.; Lau, S.-M.C.; O'Keefe, D.P.; Odell, J.; Fader, G.; McGonigle, B. Identification and expression of isoflavone synthase, the key enzyme for biosynthesis of isoflavones in legumes. *Nat. Biotechnol.* **2000**, *18*, 208. [CrossRef] [PubMed]
64. Dixon, R.A.; Liu, C.; Jun, J.H. Metabolic engineering of anthocyanins and condensed tannins in plants. *Curr. Opin. Biotechnol.* **2013**, *24*, 329–335. [CrossRef]
65. McMurray, C.H.; Laidlaw, A.S.; McElroy, M. The effect of plant development and environment on formononetin concentration in red clover (*Trifolium pratense* L.). *J. Sci. Food Agric.* **1986**, *37*, 333–340. [CrossRef]
66. Kim, H.-K.; Jang, Y.-H.; Baek, I.-S.; Lee, J.-H.; Park, M.J.; Chung, Y.-S.; Chung, J.-I.; Kim, J.-K. Polymorphism and expression of isoflavone synthase genes from soybean cultivars. *Mol. Cells* **2005**, *19*, 67–73.
67. Jonker, A.; Peiqiang, Y. The Role of Proanthocyanidins Complex in Structure and Nutrition Interaction in Alfalfa Forage. *Int. J. Mol. Sci.* **2016**, *17*, 793. [CrossRef]
68. Wang, Y.; McAllister, T.A.; Acharya, S. Condensed tannins in sainfoin: Composition, concentration, and effects on nutritive and feeding value of sainfoin forage. *Crop. Sci.* **2015**, *55*, 13–22. [CrossRef]
69. Azuhnwi, B.N.; Boller, B.; Dohme-Meier, F.; Hess, H.D.; Kreuzer, M.; Stringano, E.; Mueller-Harvey, I. Exploring variation in proanthocyanidin composition and content of sainfoin (*Onobrychis viciifolia*). *J. Sci. Food Agric.* **2013**, *93*, 2102–2109. [CrossRef]

70. Terrill, T.H.; Rowan, A.M.; Douglas, G.B.; Barry, T.N. Determination of extractable and bound condensed tannin concentrations in forage plants, protein concentrate meals and cereal grains. *J. Sci. Food Agric.* **1992**, *58*, 321–329. [CrossRef]
71. Jackson, F.S.; McNabb, W.C.; Barry, T.N.; Foo, Y.L.; Peters, J.S. The Condensed Tannin Content of a Range of Subtropical and Temperate Forages and the Reactivity of Condensed Tannin with Ribulose- 1,5-bis-phosphate Carboxylase (Rubisco) Protein. *J. Sci. Food Agric.* **1996**, *72*, 483–492. [CrossRef]
72. Johnson, M.T.; Smith, S.D.; Rausher, M.D. Plant sex and the evolution of plant defenses against herbivores. *Proc. Natl. Acad. Sci. USA* **2009**, *106*, 18079–18084. [CrossRef] [PubMed]
73. Sumner, L.; Paiva, N.; Dixon, R.; Geno, P. HPLC-Continuous-flow liquid secondary ion mass spectrometry of flavonoid glycosides in leguminous plant extracts. *J. Mass Spectrom.* **1996**, *31*, 472–485. [CrossRef]
74. Neilson, E.H.; Goodger, J.Q.D.; Woodrow, I.E.; Møller, B.L. Plant chemical defense: At what cost? *Trends Plant Sci.* **2013**, *18*, 250–258. [CrossRef]
75. Latif, S.; Gurusinghe, S.; Weston, P.A.; Quinn, J.C.; Piltz, J.W.; Weston, L.A. Metabolomic approaches for the identification of flavonoids associated with weed suppression in selected Hardseeded annual pasture legumes. *Plant Soil* **2019**, *447*, 199–218. [CrossRef]
76. Dedio, W.; Clark, K. Biochanin A and formononetin content in red clover varieties at several maturity stages. *Can. J. Plant Sci.* **1968**, *48*, 175–181. [CrossRef]
77. Zhou, Z.; Chen, X.; Zhang, M.; Blanchard, C. Phenolics, flavonoids, proanthocyanidin and antioxidant activity of brown rice with different pericarp colors following storage. *J. Stored Prod. Res.* **2014**, *59*, 120–125. [CrossRef]
78. Qiu, Y.; Liu, Q.; Beta, T. Antioxidant properties of commercial wild rice and analysis of soluble and insoluble phenolic acids. *Food Chem.* **2010**, *121*, 140–147. [CrossRef]
79. Min, B.; McClung, A.M.; Chen, M.H. Phytochemicals and antioxidant capacities in rice brans of different color. *J. Food Sci.* **2011**, *76*, C117–C126. [CrossRef]
80. Sun, B.; Ricardo-da-Silva, J.M.; Spranger, I. Critical factors of vanillin assay for catechins and proanthocyanidins. *J. Agric. Food Chem.* **1998**, *46*, 4267–4274. [CrossRef]

8

Exercise Induced Changes in Salivary and Serum Metabolome in Trained Standardbred, Assessed by ^{1}H-NMR

Marilena Bazzano [1], Luca Laghi [2,*], Chenglin Zhu [2], Enrica Lotito [1], Stefano Sgariglia [3], Beniamino Tesei [1] and Fulvio Laus [1]

1 School of Biosciences and Veterinary Medicine, University of Camerino, Via Circonvallazione, 93/95, 62024 Matelica, Italy; marilena.bazzano@unicam.it (M.B.); lotitoenrica@gmail.com (E.L.); beniamino.tesei@unicam.it (B.T.); fulvio.laus@unicam.it (F.L.)

2 Department of Agricultural and Food Sciences, University of Bologna, 47521 Cesena, Italy; chenglin.zhu2@unibo.it

3 Practitioner, 63900 Fermo, Italy; stefano.sgariglia@libero.it

* Correspondence: l.laghi@unibo.it

Abstract: In the present study, data related to the metabolomics of saliva and serum in trained standardbred horses are provided for the first time. Metabolomic analysis allows to analyze all the metabolites within selected biofluids, providing a better understanding of biochemistry modifications related to exercise. On the basis of the current advances observed in metabolomic research on human athletes, we aimed to investigate the metabolites' profile of serum and saliva samples collected from healthy standardbred horses and the relationship with physical exercise. Twelve trained standardbred horses were sampled for blood and saliva before (T_0) and immediately after (T_1) standardized exercise. Metabolomic analysis of both samples was performed by ^{1}H-NMR spectroscopy. Forty-six metabolites in serum and 62 metabolites in saliva were detected, including alcohols, amino acids, organic acids, carbohydrates and purine derivatives. Twenty-six and 14 metabolites resulted to be significantly changed between T_0 and T_1 in serum and saliva, respectively. The findings of 2-hydroxyisobutyrate and 3-hydroxybutyrate in serum and GABA in equine saliva, as well as their modifications following exercise, provide new insights about the physiology of exercise in athletic horses. Glycerol might represent a novel biomarker for fitness evaluation in sport horses.

Keywords: horse; metabolomic; metabolism; exercise; saliva

1. Introduction

Metabolic changes related to energy production and cell growth have been observed in response to exercise in humans [1–4]. Although these studies have contributed to the understanding of the metabolic response after physical exercise, the limited number of metabolites taken into account does not represent the broader metabolic response caused by exercise [2,4,5]. Human serum accounts for about 4000 metabolites [6] that interact in a large and complex network [7]. Therefore, utilizing a more efficient and comprehensive method to analyze all the metabolites within selected biofluids could provide a better understanding of the metabolic response in the context of biochemistry modifications related to exercise.

In this context, the employment of metabolomic analysis can be essential, as demonstrated in a study by Berton and colleagues (2017) [8] that focused on the possibility of using serum as a valuable biofluid to investigate metabolomic modifications induced by exercise in sports players [8].

However, serum is only one of the biofluids suitable for scientific research in the field of sport medicine. A special interest was recently pointed towards the use of saliva as a diagnostic fluid both in human and veterinary medicine [9,10]. It is well known that saliva composition can be affected by systemic disorders and may reflect general metabolic changes [11]. When compared with other biological samples, saliva has the advantage of being easily collected by non-invasive and non-stressful procedures, which is extremely important when sampling animals [9]. The composition of saliva is 99% water but it also contains several compounds such as hormones, glucose, lactate, fatty acids, triglycerides, cholesterol, urea, uric acid and phosphorus [9]. Metabolites, enzymes, proteins and minerals have been found in saliva samples collected from humans, pigs, sheep and horses [9,12–17]. Some of these metabolites have been found to change as a result of pathological conditions like lameness, stress, abdominal pain, inflammation and kidney diseases, and could be eligible as biomarkers in humans and animals [9,12,13,16,18,19]. A proteomic approach was also applied to saliva samples from humans, horses, cattle, dogs, sheep, rabbits and rats with the aim to establish specific proteome signatures of mammals' saliva [20]. Recent studies focused on the possibility of using saliva as a valuable biofluid to investigate enzymatic [17] and metabolomic [21] modifications induced by exercise in soccer players. Exercise testing and monitoring of training sessions have an important value in the assessment of poor performance, fitness and performance potential in athletic horses [22–24]. However, despite a continuous search for novel methods to evaluate the equine athlete, little is known about its metabolomic profile [25,26], and no information is available on the metabolome of standardbred horses

On the basis of the current advances observed in metabolomic research, we aimed to investigate the metabolomic profile of serum and saliva samples collected from healthy standardbred horses and the relationship with physical exercise. Closely monitoring fitness, workload and injuries in sport horses is a major matter to better understand the effects of training methods, so to reduce injuries [24]. That is why the evaluation of indices of fitness (starting from blood parameter, velocity and/or heart rate) is important to adapt training programs with the double aim to improve performance and preserve horses' welfare [24]. Metabolomics applied to athletes could provide new insight in the adaption of horses to exercise, evaluating in detail which metabolites are altered, trying to give an explanation and apply any practical corrective measures. Among the high-throughput platforms employed for metabolomics, we decided to employ proton nuclear magnetic resonance (^{1}H-NMR) spectroscopy, that ensures data traceability, reproducibility and interoperability, because the only variables modulating an NMR spectrum are magnetic field, solvent and pulse sequence [27].

2. Results

All the horses included in the study were accustomed to the training program and showed no clinical sign of disease during the experimental period. The saliva collection method herein used was non-invasive and well tolerated by all horses.

^{1}H-NMR spectroscopy allowed to quantify 46 metabolites in serum and 62 metabolites in saliva, including, among others, alcohols, amino acids, organic acids, carbohydrates and purine derivatives (Figures 1 and 2), with a spectrum of substances broadly in line with those found in the human metabolome [8,28].

Despite the detection of 30 shared metabolites, serum and salivary metabolomes resulted to be different, with 16 metabolites found only in serum, and 32 metabolites found only in saliva.

The concentrations of the metabolites that resulted to be statistically different between the samples collected before (T_0) and after (T_1) exercise in serum are reported in Table 1. *p*-values for all non-significant metabolites are reported in Supplementary Materials Table S1.

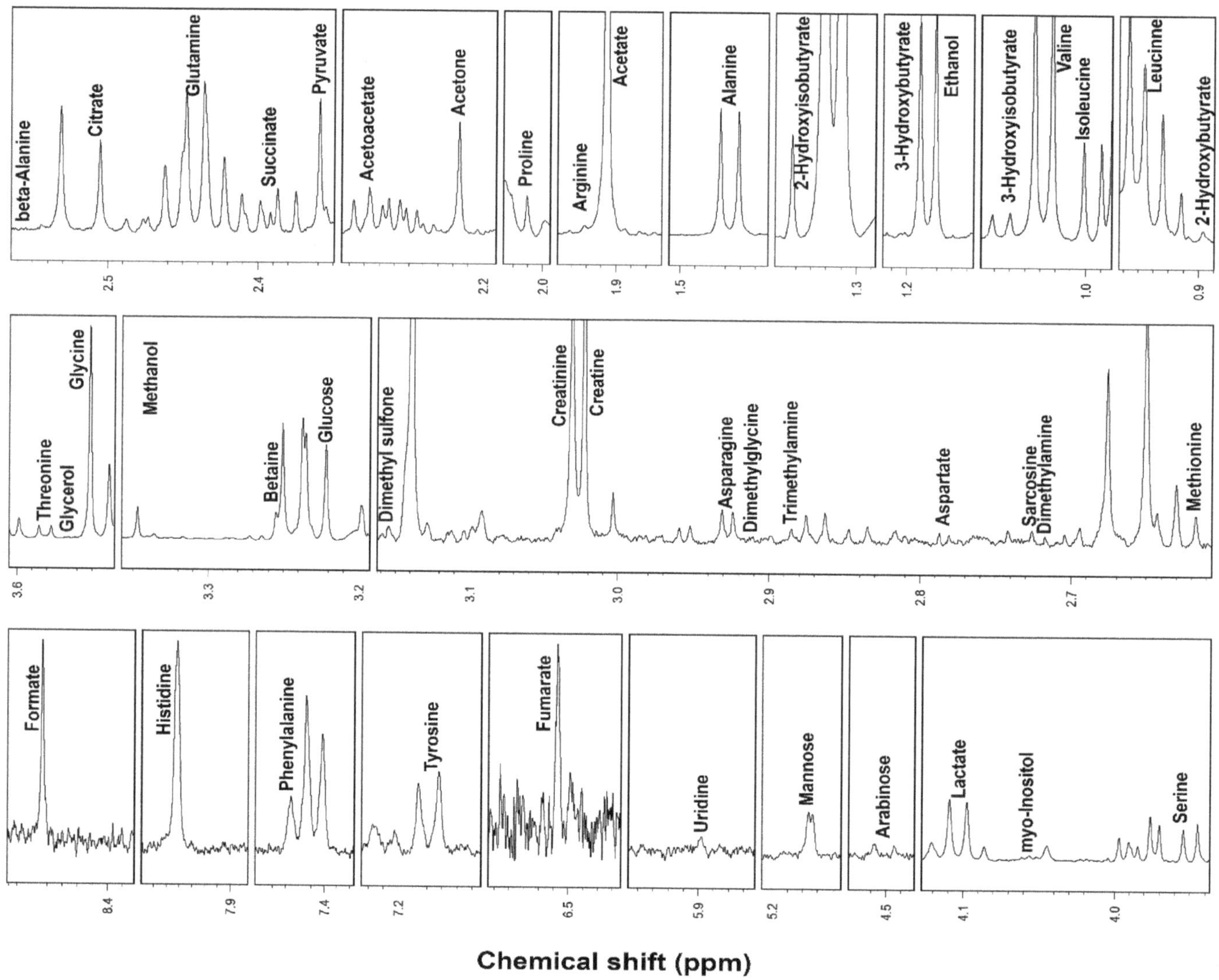

Figure 1. Portions of ^{1}H-NMR spectra from typical serum samples. Assignments appear on the signals used for molecules quantification. The vertical scale of each portion is conveniently set to ease the signals observation.

To observe the overall variations intrinsic to the samples in the space constituted by this restricted group of metabolites, we calculated on their concentrations a robust Principal Component Analysis (rPCA) model (Figure 3). Three principal components (PCs) were accepted, the first of which accounted for 57.5% of the samples' variance represented by the model. Such a PC nicely accounted for the differences among the samples connected to exercise and showed that lactate, pyruvate, succinate, glycerol, fumarate and alanine mostly characterized the samples collected after exercise, while myo-inositol, histidine, proline, asparagine, glutamine and mannose mostly characterized the samples collected before exercise.

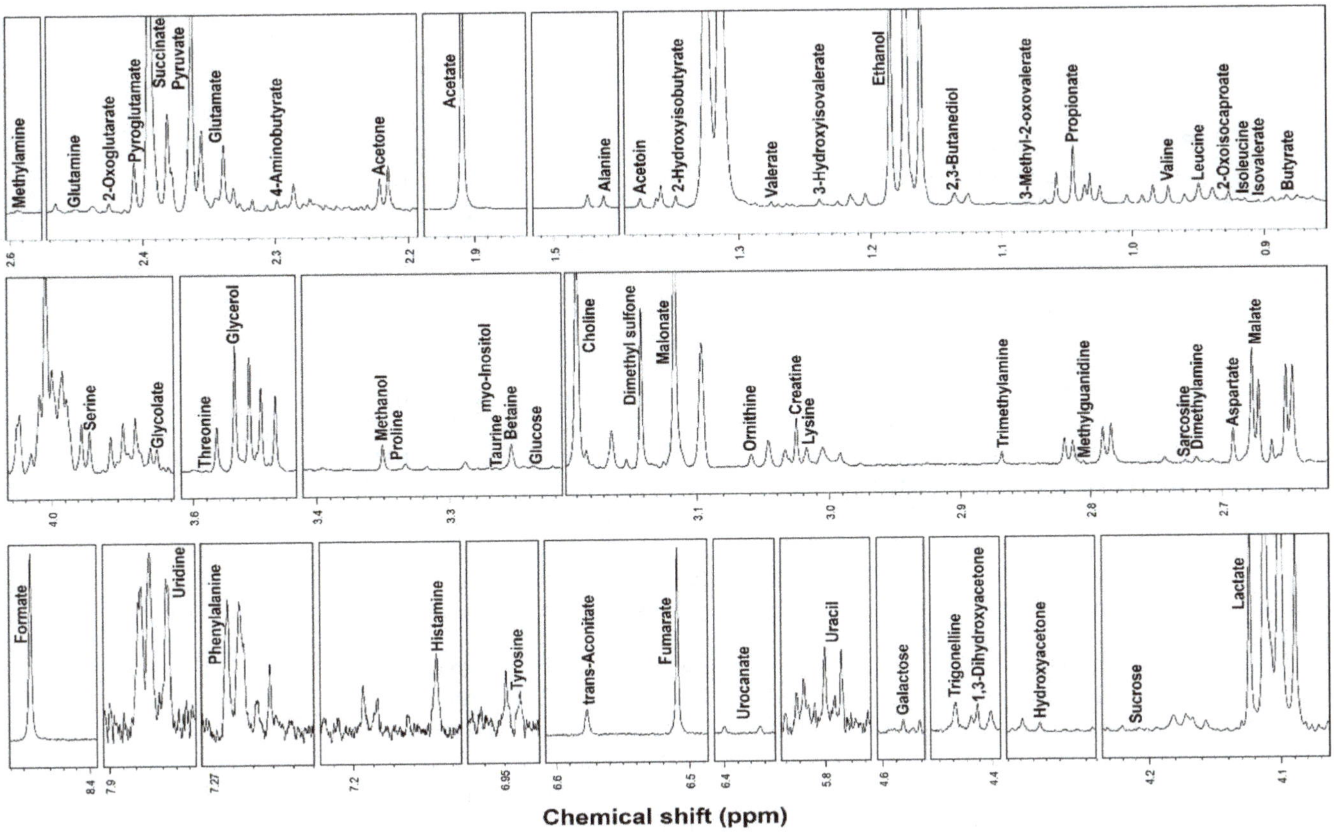

Figure 2. Portions of ^{1}H-NMR spectra from typical saliva samples. Assignments appear on the signals used for molecules quantification. The vertical scale of each portion is conveniently set to ease the signals observation.

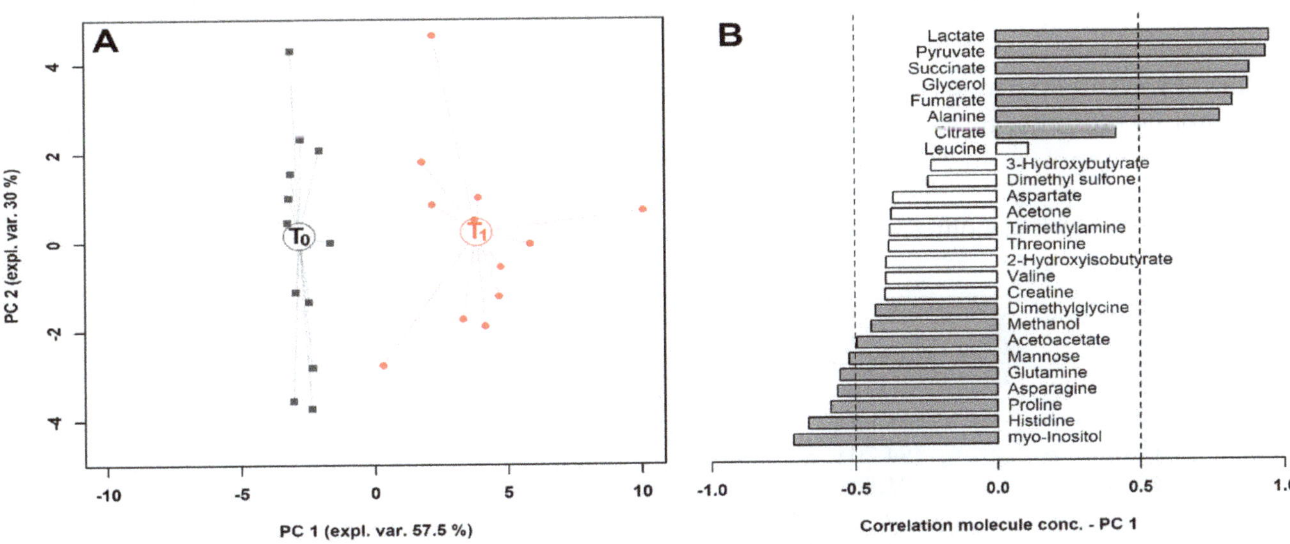

Figure 3. rPCA model built on the space constituted by the concentration of the molecules significantly different in serum, listed in Table 1. (**A**) In the score plot, samples collected at T_0 and T_1 are represented with black squares and red circles, respectively. The wide, empty circles represent the median of the groups. (**B**) The loading plot reports the correlation between the concentration of each substance and its importance over principal component (PC) 1. Significant correlations ($p < 0.05$) are highlighted with gray bars.

Table 1. Serum metabolites with significantly different concentration (μmol/L; mean ± SD) before (T_0) and after (T_1) exercise.

Molecule	T_0	T_1	p	Trend *
3-Hydroxybutyrate	2.09 ± 1.57	1.52 ± 1.33	4.58×10^{-2}	↓
2-Hydroxyisobutyrate	1.64 ± 0.96	0.95 ± 1.19	6.19×10^{-5}	↓
Acetoacetate	3.73 ± 0.73	3.21 ± 0.58	3.20×10^{-4}	↓
Acetone	3.31 ± 1.42	2.56 ± 0.85	1.45×10^{-2}	↓
Alanine	66.34 ± 10.37	104.87 ± 11.85	4.87×10^{-8}	↑
Asparagine	10.36 ± 1.93	7.44 ± 1.82	7.92×10^{-5}	↓
Aspartate	2.17 ± 1.17	1.53 ± 0.61	2.23×10^{-2}	↓
Citrate	11.24 ± 5.88	16.40 ± 7.51	1.79×10^{-2}	↑
Creatine	22.76 ± 7.67	17.94 ± 5.91	2.95×10^{-4}	↓
Dimethyl sulfone	14.74 ± 8.56	11.09 ± 5.96	4.32×10^{-3}	↓
Dimethylglycine	0.36 ± 0.13	0.28 ± 0.09	3.91×10^{-3}	↓
Fumarate	0.84 ± 0.21	1.50 ± 0.73	8.82×10^{-3}	↑
Glutamine	71.51 ± 11.48	58.14 ± 6.78	6.18×10^{-5}	↓
Glycerol	4.03 ± 3.01	45.51 ± 22.46	1.84×10^{-5}	↑
Histidine	29.91 ± 4.34	23.46 ± 3.35	1.52×10^{-3}	↓
Lactate	141.26 ± 50.20	1143.45 ± 582.63	1.03×10^{-4}	↑
Leucine	35.68 ± 6.37	39.22 ± 6.87	9.62×10^{-3}	↑
Mannose	11.20 ± 2.71	7.85 ± 2.09	1.33×10^{-5}	↓
Methanol	16.98 ± 10.21	10.26 ± 8.25	1.49×10^{-3}	↓
myo-Inositol	10.35 ± 1.68	7.37 ± 2.06	8.38×10^{-4}	↓
Proline	10.57 ± 2.45	8.02 ± 1.54	4.19×10^{-4}	↓
Pyruvate	8.72 ± 2.37	26.61 ± 7.59	4.38×10^{-6}	↑
Succinate	1.12 ± 0.21	4.24 ± 1.79	6.11×10^{-5}	↑
Threonine	37.30 ± 5.45	33.26 ± 6.98	4.93×10^{-3}	↓
Trimethylamine	0.16 ± 0.05	0.14 ± 0.05	2.73×10^{-3}	↓
Valine	54.09 ± 6.79	50.40 ± 9.04	1.85×10^{-2}	↓

* Increasing (↑) or decreasing (↓) trends from T_0 to T_1.

About the salivary metabolome, the concentrations of five molecules significantly increased after exercise, while nine significantly decreased between T_0 and T_1 (Table 2). p-values for all non-significant metabolites are reported in Supplementary Materials Table S2.

Table 2. Salivary metabolites with different concentrations (μmol/L; mean ± SD) before (T_0) and after (T_1) exercise.

Molecule	T_0	T_1	p	Trend *
4-Aminobutyrate	30.09 ± 15.39	10.87 ± 16.55	4.35×10^{-3}	↓
Betaine	143.58 ± 86.06	67.19 ± 52.51	1.23×10^{-2}	↓
Creatine	36.68 ± 19.11	62.11 ± 26.59	1.28×10^{-2}	↑
Fumarate	94.70 ± 58.46	23.20 ± 29.32	8.35×10^{-4}	↓
Galactose	21.68 ± 10.18	12.55 ± 12.66	1.98×10^{-2}	↓
Malate	550.72 ± 395.87	138.69 ± 166.50	2.11×10^{-3}	↓
Malonate	193.93 ± 107.17	46.84 ± 35.06	5.10×10^{-4}	↓
Methanol	131.26 ± 73.70	73.05 ± 50.01	5.07×10^{-3}	↓
Ornithine	22.36 ± 12.53	46.91 ± 21.66	3.24×10^{-3}	↑
Phenylalanine	28.22 ± 11.81	53.82 ± 23.36	6.10×10^{-3}	↑
Pyruvate	70.77 ± 32.43	49.34 ± 24.69	7.41×10^{-2}	↓
Sarcosine	2.65 ± 1.64	4.48 ± 2.94	2.48×10^{-2}	↑
Succinate	210.91 ± 78.35	120.47 ± 80.66	3.59×10^{-2}	↓
Tyrosine	33.32 ± 18.13	66.07 ± 46.48	3.29×10^{-2}	↑

* Increasing (↑) or decreasing (↓) trends from T_0 to T_1.

To observe the overall trends associated with the samples, we calculated on their concentrations the rPCA model outlined in Figure 4. Three principal components (PCs) were accepted, the first of which accounted for 84.2% of the samples' variance represented by the model. Such PCs summarized the differences among the samples connected to exercise. The salivary metabolome of horses after exercise was mainly characterized by creatine, ornithine, phenylalanine and tyrosine, while horses before exercise were mainly characterized by fumarate, malate, malonate, 4-aminobutyrate, betaine and galactose. In Figure 5, the metabolic pathways overrepresented by the molecules significantly affected by exercise in serum and saliva are reported.

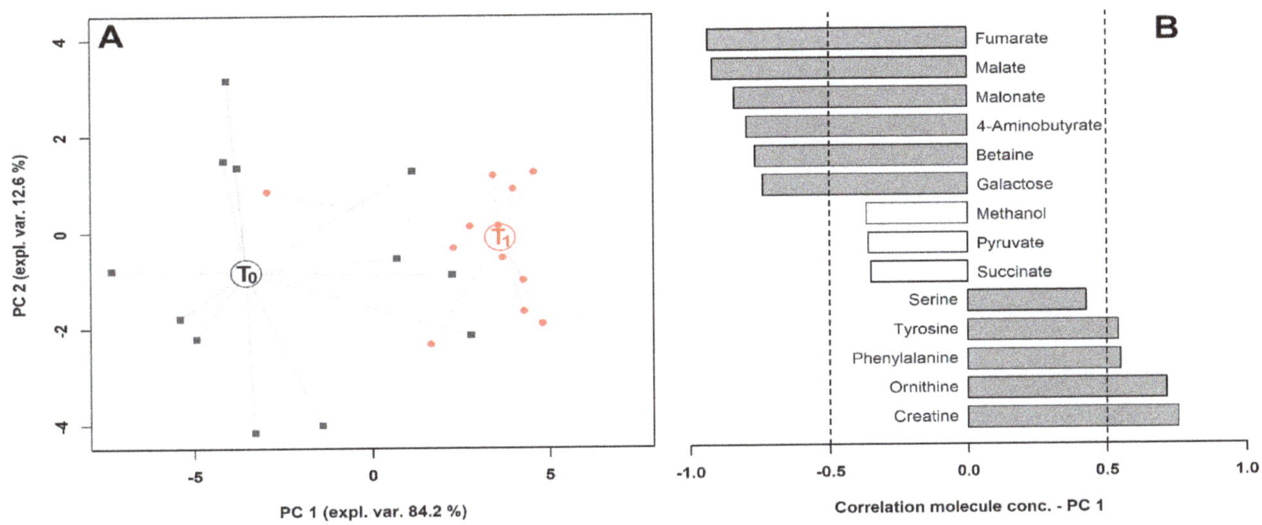

Figure 4. rPCA model built on the space constituted by the concentration of the molecules significantly different in saliva, listed in Table 2. (**A**) In the score plot, samples collected at T_0 and T_1 are represented with black squares and red circles, respectively. The wide, empty circles represent the median of the groups. (**B**) The loading plot reports the correlation between the concentration of each substance and its importance over PC 1. Significant correlations ($p < 0.05$) are highlighted with gray bars.

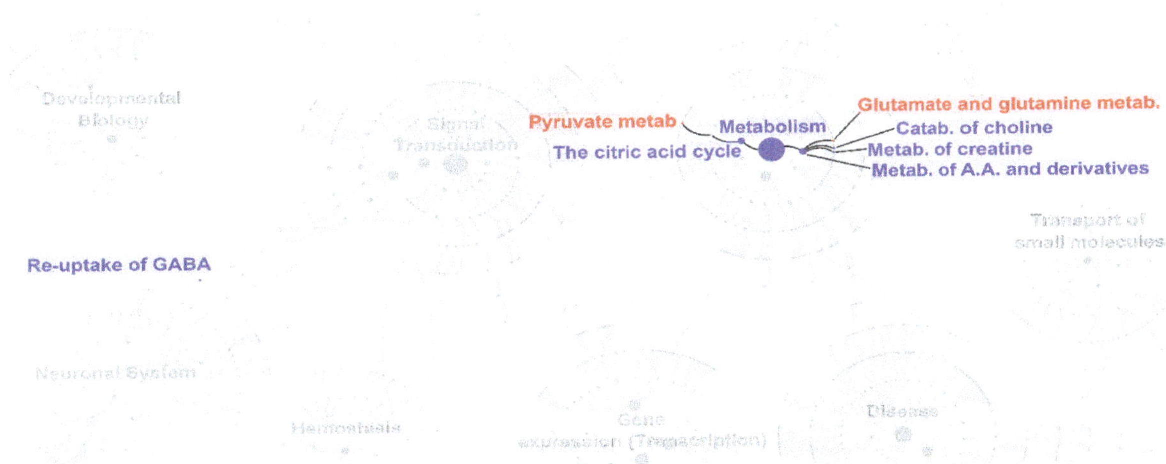

Figure 5. Biomolecular pathways overrepresentation analysis performed on the molecules listed in Table 1 for serum (red) and Table 2 for saliva (blue). The figure replicates a convenient portion of the biomolecular pathways overview according to Reactome, modified to show the main pathways and sub-pathways overrepresented as a consequence of exercise.

3. Discussion

The results showed in the present paper highlight for the first time the changes occurring in the metabolomic profile of trained standardbred horses following exercise in both serum and saliva specimens.

Biochemical constituents of the tricarboxylic acid cycle (TCA or Krebs cycle) including succinate and fumarate were significantly increased in serum after exercise. This is in agreement with results obtained by other authors in human and equine athletes [25,29,30] and could be explained by the need to maintain and/or increase the Krebs cycle flow during exercise [8]. On the contrary, succinate and fumarate, together with malate, another constituent of the TCA cycle, have been found to decrease in saliva after exercise. Among salivary metabolites, malonate was found to have a very strong inhibitory effect on the TCA cycle in complex metabolic systems. Malonate is not normally present in cells [31] and the reason for its presence in saliva, its decrease after exercise and its inhibitory effect on the TCA cycle in horses during exercise need further investigations. Notwithstanding the paucity of data on salivary metabolomics during exercise, we can speculate that the decreased salivary level of succinate, fumarate and malate might reflect the imbalance of the aerobic pathways in favor of the anaerobic production of energy during exercise.

The increase in pyruvate and lactate serum concentrations was expected, since it reflects the anaerobic pathway activation during exercise [8,25]. The high energy demand required during exercise can overload the mitochondria's ability to oxidize pyruvate that is converted to lactate, supplying energy by the Cori cycle [8]. In our study, serum lactate increased up to 11.6-fold following exercise, and this result was beyond the physiological range usually reported for human athletes of a 9.3–9.6-fold increase [8,32]. Probably, the greater muscle mass of horses with respect to humans can account for this difference. Although not significant, pyruvate and lactate slightly decreased in saliva after exercise (Supplemental Materials Table S1), demonstrating once again that salivary metabolic changes do not follow the same metabolic changes observed in serum, at least regarding the energy supply.

Among the metabolites detected in plasma, there are also two carbohydrates: mannose and arabinose.

The concentration of mannose resulted more concentrated in post-exercise equine plasma. This monosaccharide is found in small amounts in the diet, and after conversion to fructose-6-phosphate, it can be used in both glycolysis and gluconeogenesis [33]. Unlike mannose, the concentration of arabinose in serum appeared as decreased by exercise. Arabinose is known to have an inhibitory effect on intestinal sucrase [34], but its role in horse during exercise in not clear.

Galactose was the only carbohydrate with a changed concentration in saliva, as it decreased at T_1. In a study on soccer players, Pitti et al. [21] found an increased level of galactose in blood after exercise. Galactose is an important source of glucose and its role in liver glycogen restoration after exercise has been well recognized in cyclists [33,35]. The reason for galactose reduction in the saliva of horses could reflect a different energetic role of this metabolite compared with humans.

Among amino acids, alanine and histidine increased significantly after exercise. Serum alanine increased 1.6-fold soon after exercise, in agreement with the results obtained in humans by Berton et al. [8] but in contrast with Nieman et al. [29]. Alanine in muscle derives from pyruvate, in turn originated from glucose breakdown [8]. This metabolic mechanism is in line with the rise in serum pyruvate found in our study. Furthermore, alanine has a detoxing function by transporting to the liver the large amount of ammonia typically produced during short and high-intensity exercise derived from branched chain amino acids [8,36]. Similarly, the increase in gluconeogenic amino acid histidine could be linked to the higher energetic demand during acute exercise.

The decreased levels of glutamine, asparagine and proline observed in horses after training were similar to the findings in human athletes [29,30] thus supporting the hypothesis of an enhanced muscle amino acid oxidation during exercise to sustain the energetic demand.

In equine saliva, amino acids like tyrosine and phenylalanine rose almost 2-fold after exercise. Tyrosine results from hydroxylation of phenylalanine, an essential amino acid [37], and it is metabolized to acetoacetate and fumarate or used as a precursor of catecholamines that play an important role in athletic performance during exercise [38]. We can speculate that the rise in salivary tyrosine after exercise could reflect an increased production to encounter the need for catecholamines during exercise, but the lack of similar changes in serum needs to be clarified.

Sarcosine (N-methylglycine) is a non-proteinogenic amino acid derivative and occurs in the body as a product of the metabolism of glycine and creatine [39]. In addition to its multiple functions in the body and its use as a potential marker in various diseases, it has been an ingredient in toothpaste for decades as it prevents tooth decay and causes foaming [39]. The reason for the slight rising of sarcosine and creatine in equine saliva after exercise is not clear, but its physiological function could be protective for teeth health.

Ornithine levels increased up to 2.1-fold in saliva but not in serum, differently from what has been observed in human athletes where it is down regulated 60–70 min after exercise as a consequence of the accelerated urea cycle due to the higher production of ammonia [8,30]. However, the differences in the sampling times of our study compared with others might explain this discrepancy. Betaine, a small molecule acting as a methyl donor in minor pathways [37], decreased in saliva after exercise, but this finding needs further investigations.

The marked elevation in serum glycerol in our horses following exercise resulted from an extensive lipolysis and is consistent with the study by Lewis et al. [30], who found higher serum glycerol up-regulation after exercise in fitter athletes. Considering that all the horses included in our study were well trained, we could consider glycerol a marker of adaptation to training also in equine species.

Salivary methanol was reduced 1.8 times by exercise. The same result was recently found in the saliva of soccer players, and the change was ascribed to the higher evaporation of volatile compounds in the mouth during exercise [21]. In a recent study, methanol was considered a marker of inflammation in horses affected by equine asthma when detected in exhaled breath condensate (EBC) [40].

Uracil nucleotide, together with cytosine, are the major pyrimidine components of RNA. Uracil can be utilized in glycogen synthesis [41] and its rise in saliva could be related to the higher energetic demand during exercise.

2-hydroxyisobutyrate was up-regulated in the serum of standardbred horses following exercise. This metabolite is a normal constituent of human serum and saliva [42] that has been proposed as a biomarker for glycogen storage disease type 1a in juveniles and acute coronary syndrome [43,44]. It was also found decreased in plasma of Alzheimer's disease patients vs. controls [45], however no information is available about its role during sport activity.

3-hydroxybutyrate decreased in serum after exercise. An elevation of this ketone was documented in marathon runners due to ketone production [30]. It was also found to be increased in runners after a three-day intensified exercise as a result of fatty acid oxidation [29]. On the other hand, Berton et al. [8] did not find any change in 3-hydroxybutyrate after a leg press resistance exercise. Our result might suggest that this metabolic pathway is of negligible importance in trained standardbred horses performing acute exercise of short duration.

4-aminobutyrate (γ-Aminobutiryc acid or GABA) is the most inhibitory neurotransmitter in the central nervous system [46] and its presence has been already demonstrated in human saliva [47] as well as in salivary glands of men and rats [48,49], being associated with its biosynthetic and metabolic enzymes. However, this is the first report showing the presence of GABA in equine saliva and the down-regulatory effect of exercise in this species. GABA and its receptors can be found in other non-neuronal organs but its role in peripheral tissue remains to be established, although it is known to be involved in cellular proliferation [48]. Authors suggested a role of GABA(A)-R in the suppression of salivary secretion in rat salivary glands [50]. Similarly, the down-regulation of salivary GABA during exercise in horses could be associated with this regulatory function of saliva secretion, possibly linked to thermoregulation.

Significantly lower levels of myo-inositol have been found both in the serum and saliva of horses after exercise. Myo-inositol promotes the maturation of pulmonary surfactants and supports respiratory function [51], modulates cytoskeleton dynamics, thus allowing alveolar cells to counteract collapsing forces and promoting mechanical stabilization of cell shape [52], and it recruits water and organic compounds in the alveolar space, decreasing surface tension through the formation of a biofilm layer at the interface [53]. Myo-inositol is also the most effective allosteric effector identified to date, being able to increase the tissue delivering of oxygen bound to hemoglobin [54]. Since its administration can improve sport performance in laboratory mice, its analogues have been suspected to be abused in the horse racing industry [55]. Up to now, no study about its effect in equine species has been performed. Recently, reduced concentrations of myo-inositol have been found in the bronchoalveolar lavage fluid of horses affected by equine asthma in comparison with healthy horses [56]. The exact role of this metabolite in physical exercise is not known, however, the decrease observed in both serum and saliva could indicate an impairment in the normal respiratory function during exercise.

4. Materials and Methods

4.1. Animals

Twelve clinically healthy standardbred horses (7 males and 5 females), mean age 6.7 years old, mean body condition score 3 out of 5 [57], were included in the study with informed owner consent.

All horses were stabled in individual boxes and were fed the same polyphyte hay at a ratio of 2 kg/100 kg BW, and commercial horse feed in the amount of 0.8 kg/100 kg of BW (protein 11%; fat 4%; fiber 8.56%; Ca/P 2:1) twice a day (at 10 am and at 5 pm). Clean potable drinking water was offered ad libitum.

The sampling herein described was part of a normal procedure for monitoring the training status of the animals. All procedures were approved by the Animal Care Committee of Camerino University (Registration number E81AC.10/Ac) and were in accordance with the standard recommended by the EU Directive 2010/63/EU for experimental animals.

4.2. Experimental Procedure

All horses were trained 6 days per week and the training program consisted of 7 runs of 1000 m each at different speed. Briefly, the first warm-up run clockwise at light trot (mean speed 5 m/s) was followed by 3 counter clockwise runs (mean speed 10 m/s), then 1 recovery run at walking and 2 runs at full speed trot (mean speed 12 m/min), and finally 2 runs at light trot (mean speed 5 m/s). All animals were sampled the day before the rest-day.

Saliva and blood samples were collected before (T_0) and immediately after full speed exercise (T_1). Saliva samples were collected by using cotton swabs and Salivette® tubes (Sarstedt AG & Co., Nümbrecht, Germany). The cotton swab was grasped with a surgical clamp, inserted at the angle of the lips into the mouth of the horse and placed gently on the tongue surface for 5 min and then inserted into the Salivette® tube.

Venous blood samples were collected at T_0 and T_1 by jugular venepuncture into 4 mL vacutainer sterile tubes containing EDTA (Vacuette® Greiner Bio-One, Cassina de Pecchi, Italy) and 9 mL vacutainer sterile tubes containing clot activators (Vacuette® Greiner Bio-One, Cassina de Pecchi, Italy). A complete blood count (CBC) and biochemical profile was performed on blood and serum samples to exclude systemic disorders.

Whole blood was directly tested in the field for blood lactate concentration by using a portable device (Accutrend Plus®, Roche, Mannheim, Germany) at T_0 and T_1 as a part of the fitness evaluation routine.

All the sampling procedures were performed before feeding, at the same time of the day (06:00–8:00 a.m.) and, immediately after collection, all the samples were stored at 5 °C and delivered to the lab within 2 h.

4.3. Laboratory Analysis.

Saliva samples were obtained by centrifugation of Salivette® tubes (10 min at 1000 g) (Universal 32, Hettich Zentrifugen, Tuttlingen, Germany) and 2 mL aliquots of supernatant were stored at −20 °C until metabolomic analysis.

Blood samples with clot activators were centrifuged for 10 min at 1000 g (Universal 32, Hettich Zentrifugen, Tuttlingen, Germany) and the obtained sera were stored at −20 °C until analysis.

For ^{1}H-NMR analysis, we created a stock solution composed of 3-(trimethylsilyl)-propionic-2,2,3,3-d4 acid sodium salt (TSP) 10 mmol/L and NaN_3 2 mmol/L in D_2O. The former served as the NMR spectra chemical-shift reference, while the latter avoided bacteria proliferation. The solution was set to pH at 7.00 ± 0.02 by phosphate buffer (1 M). Both serum and saliva samples were prepared for ^{1}H-NMR by thawing and centrifuging 1 mL of each sample at 4 °C for 15 min at 18,630 g. The supernatant (700 μL) was added to 100 μL of the NMR analysis solution and centrifuged again.

The spectra were recorded with an AVANCE III spectrometer (Bruker, Milan, Italy), controlled by the Topspin software (Ver. 3.5), at a frequency of 600.13 MHz and a temperature of 298 K. The residual signal from the water was suppressed by pre-saturation, while broad signals from large molecules were reduced by a CPMG-filter, set as outlined by Zhu et al. [58]. Each spectrum was acquired by summing up 256 transients registering 32 K data points over a 7184 Hz spectral window, with an acquisition time of 2.28 s and relaxation delay of 5 s.

In Topspin, a manual correction phase was applied to each spectrum, together with a line-broadening of 0.3 Hz. The subsequent steps were performed in R computational language by means of scripts developed in-house. The spectra were aligned toward the right peak of the alanine doublet, set to 1.473 ppm. Baseline was then corrected, after having removed the residual water signal, by isolating irregularities of the baseline by peak detection, according to the "rolling ball" principle [59].

For signals' assignment, chemical shift and multiplicity were compared with the Chenomx software library (Chenomx Inc., Edmonton, AB, Canada, ver 8.3). The added TSP was employed as internal standard in the first sample analyzed. Differences in water content among samples were then taken into consideration by probabilistic quotient normalization [27]. Rectangular integration was employed to quantify each molecule, by focusing on one signal per molecule free from superimpositions.

4.4. Statistical Analysis

Differences between the two groups were looked for by means of a paired *t*-test applied to the concentrations of each molecule, transformed by the Box-Cox algorithm for normal distribution [60].

We highlighted any trend characterizing the samples with robust principal component analysis (rPCA) models [61], based on the molecules accepted by the univariate analysis. To this purpose, we employed the PcaHubert algorithm implemented in the rrcov package. The main features of each rPCA model are summarized by a score plot and by a Pearson correlation plot. The former is the projection of the samples in the PC space and highlights the underlying structure of the data. The latter relates the concentration of each variable to the components of the model.

Metabolic pathway analysis was conducted by relying on the Reactome pathway knowledgebase [62], with overrepresentation analysis (ORA) based on a hypergeometric test [63].

5. Conclusions

In the present study, data related to the metabolomics of saliva and serum in trained standardbred horses have been provided for the first time. Several molecules are common to metabolic pathways already described in human athletes (e.g., energetic metabolism). However, the findings of 2-hydroxyisobutyrate and 3- hydroxybutyrate in serum and GABA in equine saliva, as well as their modifications following exercise, provide new insights about the physiology of exercise in athletic horses. Furthermore, the use of metabolomic analysis allowed to identify metabolites like glycerol that might represent novel biomarkers for fitness evaluation in sport horses. However, further investigations are needed to validate the metabolites herein found as indices of fitness levels in equine species.

Author Contributions: Conceptualization, F.L; L.L and M.B; methodology, F.L., C.Z. and L.L.; formal analysis, L.L. and C.Z.; investigation, E.L. and S.S.; resources, S.S and B.T.; writing—original draft preparation, F.L., L.L. and M.B.; writing—review and editing, F.L., L.L., M.B. and B.T.; supervision, F.L. All authors have read and agreed to the published version of the manuscript.

Acknowledgments: Chenglin Zhu gratefully acknowledges financial support from Chinese Scholarship Council (grant number 201606910076).

References

1. Dunn, W.B.; Broadhurst, D.I.; Atherton, H.J.; Goodacre, R.; Griffin, J.L. Systems level studies of mammalian metabolomes: The roles of mass spectrometry and nuclear magnetic resonance spectroscopy. *Chem. Soc. Rev.* **2011**, *40*, 387–426. [CrossRef] [PubMed]
2. Gorostiaga, E.M.; Navarro-Amezqueta, I.; Calbet, J.A.; Hellsten, Y.; Cusso, R.; Guerrero, M.; Granados, C.; González-Izal, M.; Ibañez, J.; Izquierdo, M. Energy metabolism during repeated sets of leg press exercise leading to failure or not. *PLoS ONE* **2012**, *7*, e40621. [CrossRef]
3. Schoenfeld, B.J. Potential mechanisms for a role of metabolic stress in hypertrophic adaptations to resistance training. *Sports Med.* **2013**, *43*, 179–194. [CrossRef] [PubMed]
4. Tesch, P.A.; Colliander, E.B.; Kaiser, P. Muscle metabolism during intense, heavy-resistance exercise. *Eur. J. Appl. Physiol. Occup. Physiol.* **1986**, *55*, 362–366. [CrossRef] [PubMed]
5. Mendez-Villanueva, A.; Edge, J.; Suriano, R.; Hamer, P.; Bishop, D. The recovery of repeated-sprint exercise is associated with PCr resynthesis, while muscle pH and EMG amplitude remain depressed. *PLoS ONE* **2012**, *7*, e51977. [CrossRef] [PubMed]
6. Psychogios, N.; Hau, D.D.; Peng, J.; Guo, A.C.; Mandal, R.; Bouatra, S.; Young, N. The human serum metabolome. *PLoS ONE* **2011**, *6*, 16957. [CrossRef]
7. Ryan, D.; Robards, K. Metabolomics: The greatest omics of them all? *Anal. Chem.* **2006**, *78*, 7954–7958. [CrossRef]
8. Berton, R.; Conceição, M.S.; Libardi, C.A.; Canevarolo, R.R.; Gáspari, A.F.; Chacon-Mikahil, M.P.; Zeri, A.C.; Cavaglieri, C.R. Metabolic time-course response after resistance exercise: A metabolomics approach. *J. Sports Sci.* **2017**, *35*, 1211–1218. [CrossRef]
9. Contreras-Aguilar, M.D.; Escribano, D.; Martínez-Subiela, S.; Martín-Cuervo, M.; Lamy, E.; Tecles, F.; Cerón, J.J. Changes in saliva analytes in equine acute abdominal disease: A sialochemistry approach. *BMC Vet. Res.* **2019**, *15*, 187. [CrossRef]
10. Ghizoni, J.S.; Nichele, R.; de Oliveira, M.T.; Pamato, S.; Pereira, J.R. The utilization of saliva as an early diagnostic tool for oral cancer: microRNA as a biomarker. *Clin. Transl. Oncol.* **2020**, *22*, 804–812. [CrossRef]
11. Greabu, M.; Battino, M.; Mohora, M.; Totan, A.; Didilescu, A.; Spinu, T. Saliva—A diagnostic window to the body, both in health and in disease. *J. Med. Life* **2009**, *2*, 124–132. [PubMed]
12. Tecles, F.; Escribano, D.; Martínez-Miró, S.; Hernández, F.; Contreras, M.D.; Cerón, J.J. Cholinesterase in porcine saliva: Analytical characterization and behavior after experimental stress. *Res. Vet. Sci.* **2016**, *106*, 23–28. [CrossRef] [PubMed]
13. Tecles, F.; Contreras-Aguilar, M.D.; Martínez-Miró, S.; Tvarijonaviciute, A.; Martínez-Subiela, S.; Escribano, D.; Cerón, J.J. Total esterase measurement in saliva of pigs: Validation of an automated assay, characterization and changes in stress and disease conditions. *Res. Vet. Sci.* **2017**, *114*, 170–176. [CrossRef] [PubMed]
14. Tecles, F.; Rubio, C.P.; Contreras-Aguilar, M.D.; Lopez-Arjona, M.; Martinez-Miro, S.; Martinez-Subiela, S.; Cerón, J.J. Adenosine deaminase activity in pig saliva: Analytical validation of two spectrophotometric assays. *J. Vet. Diagn. Investig.* **2018**, *30*, 175–179. [CrossRef] [PubMed]
15. Contreras-Aguilar, M.D.; Escribano, D.; Quiles, A.; López-Arjona, M.; Cerón, J.J.; Martínez-Subiela, S.; Hevia, M.L.; Tecles, F. Evaluation of new biomarkers of stress in saliva of sheep. *Animal* **2019**, *13*, 1278–1286. [CrossRef]
16. Contreras-Aguilar, M.D.; Escribano, D.; Martín-Cuervo, M.; Tecles, F.; Cerón, J.J. Salivary alpha-amylase

activity and cortisol in horses with acute abdominal disease: A pilot study. *BMC Vet. Res.* **2018**, *14*, 156. [CrossRef]

17. Barranco, T.; Tvarijonaviciute, A.; Tecles, F.; Carrillo, J.M.; Sanchez-Resalt, C.; Jimenez-Reyes, P.; Rubio, M.; García-Balletbó, M.; Cerón, J.; Ramon Cugat, R. Changes in creatine kinase, lactate dehydrogenase and aspartate aminotransferase in saliva samples after an intense exercise: A pilot study. *J. Sports Med. Phys.* **2018**, *58*, 910–916.
18. Jacobsen, S.; Top Adler, D.M.; Bundgaard, L.; Sørensen, M.A.; Andersen, P.H.; Bendixen, E. The use of liquid chromatography tandem mass spectrometry to detect proteins in saliva from horses with and without systemic inflammation. *Vet. J.* **2014**, *202*, 483–488.
19. Bilancio, G.; Cavallo, P.; Lombardi, C.; Guarino, E.; Cozza, V.; Giordano, F.; Palladino, G.; Cirillo, M. Salivary levels of phosphorus and urea as indices of their plasma levels in nephropathic patients. *J. Clin. Lab. Anal.* **2018**, *32*, 1–6. [CrossRef]
20. de Sousa-Pereira, P.; Cova, M.; Abrantes, J.; Ferreira, R.; Trindade, F.; Barros, A.; Gomes, P.; Colaço, B.; Amado, F.; Esteves, P.J.; et al. Cross-species comparison of mammalian saliva using an LC-MALDI based proteomic approach. *Proteomics* **2015**, *15*, 1598–1607. [CrossRef]
21. Pitti, E.; Petrella, G.; Di Marino, S.; Summa, V.; Perrone, M.; D'Ottavio, S.; Bernardini, A.; Cicero, D.O. Salivary Metabolome and Soccer Match: Challenges for Understanding Exercise induced Changes. *Metabolites* **2019**, *9*, 141. [CrossRef] [PubMed]
22. Navas de Solis, C. Cardiovascular Response to Exercise and Training, Exercise Testing in Horses. *Vet. Clin. N. Am. Equine Pract.* **2019**, *35*, 159–173. [CrossRef] [PubMed]
23. Bazzano, M.; Giudice, E.; Rizzo, M.; Congiu, F.; Zumbo, A.; Arfuso, F.; Di Pietro, S.; Bruschetta, D.; Piccione, G. Application of a combined global positioning and heart rate monitoring system in jumper horses during an official competition—A preliminary study. *Acta Vet. Hung.* **2016**, *64*, 189–200. [PubMed]
24. Munsters, C.C.; van Iwaarden, A.; van Weeren, R.; Sloet van Oldruitenborgh-Oosterbaan, M.M. Exercise testing in Warmblood sport horses under field conditions. *Vet. J.* **2014**, *202*, 11–19. [PubMed]
25. Luck, M.M.; Le Moyec, L.; Barrey, E.; Triba, M.N.; Bouchemal, N.; Savarin, P.; Robert, C. Energetics of endurance exercise in young horses determined by nuclear magnetic resonance metabolomics. *Front. Physiol.* **2015**, *6*, 198.
26. Mach, N.; Ramayo-Caldas, Y.; Clark, A.; Moroldo, M.; Robert, C.; Barrey, E.; Lòpez, J.M.; Le Moyec, L. Understanding the response to endurance exercise using a systems biology approach: Combining blood metabolomics, transcriptomics and miRNomics in horses. *BMC Genom.* **2017**, *18*, 187. [CrossRef]
27. Laghi, L.; Picone, G.; Capozzi, F. Nuclear magnetic resonance for foodomics beyond food anaysis. *Trend Anal. Chem.* **2014**, *59*, 93–102.
28. Wishart, D.S.; Jewison, T.; Guo, A.C.; Wilson, M.; Knox, C.; Liu, Y.; Scalbert, A. HMDB 3.0—The human metabolome database in 2013. *Nucleic Acids Res.* **2013**, *41*, D801–D807. [CrossRef]
29. Nieman, D.C.; Shanely, R.A.; Gillitt, N.D.; Pappan, K.L.; Lila, M.A. Serum metabolic signatures induced by a three-day intensified exercise period persist after 14 h of recovery in runners. *J. Proteome Res.* **2013**, *12*, 4577–4584.
30. Lewis, G.D.; Farrell, L.; Wood, M.J.; Martinovic, M.; Arany, Z.; Rowe, G.C.; Souza, A.; Cheng, S.; McCabe, E.L.; Yang, E.; et al. Metabolic signatures of exercise in human plasma. *Sci. Transl. Med.* **2010**, 2, 33–37.
31. Nelson, D.L.; Cox, M.M. *Lehninger Principles of Biochemistry*, 7th ed.; International ed.; W.H. Freeman & Company: New York, NY, USA, USA; 2016; p. 612.
32. Gorostiaga, E.M.; Navarro-Amezqueta, I.; Calbet, J.A.; Sanchez-Medina, L.; Cusso, R.; Guerrero, M.; Izquierdo, M. Blood ammonia and lactate as markers of muscle metabolites during leg press exercise. *J. Strength Cond. Res.* **2014**, *28*, 2775–2785. [CrossRef] [PubMed]
33. Harris, R.A. Carbohydrate Metabolism: Major Metabolic Pathways and their Control. In *Textbook of Biochemistry with Clinical Correlation*, 4th ed.; Devlin, T.M., Ed.; Wiley-Liss: New York, NY, USA, 1997; pp. 268–317.
34. Krog-Mikkelsen, I.; Hels, O.; Tetens, I.; Holst, J.J.; Andersen, J.R.; Bukhave, K. The effects of L-arabinose on intestinal sucrase activity: Dose-response studies in vitro and in humans. *Am. J. Clin. Nutr.* **2011**, *94*, 472–478. [CrossRef]
35. Décombaz, J.; Jentjens, R.; Ith, M.; Scheurer, E.; Buehler, T.; Jeukendrup, A.; Boesch, C. Fructose and galactose enhance postexercise human liver glycogen synthesis. *Med. Sci. Sports Exerc.* **2011**, *43*, 1964–1971. [PubMed]

36. Felig, P.W. Amino acid metabolism in exercising man. *J. Clin. Investig.* **1971**, *50*, 2703–2714. [CrossRef]
37. Coomes, M.W. Amino Acid Metabolism. In *Textbook of Biochemistry with Clinical Correlation*, 4th ed.; Devlin, T.M., Ed.; Wiley-Liss: New York, NY, USA, 1997; pp. 446–483.
38. Sutton, E.E.; Coill, M.R.; Deuster, P.A. Ingestion of tyrosine: Effects on endurance, muscle strength, and anaerobic performance. *Int. J. Sport Nutr. Exerc. Metab.* **2005**, *15*, 173–185. [CrossRef] [PubMed]
39. Pundir, C.S.; Deswal, R.; Kumar, P. Quantitative analysis of sarcosine with special emphasis on biosensors: A review. *Biomarkers* **2019**, *24*, 415–422.
40. Bazzano, M.; Laghi, L.; Zhu, C.; Magi, G.E.; Serri, E.; Spaterna, A.; Tesei, B.; Laus, F. Metabolomics of tracheal wash samples and exhaled breath condensates in healthy horses and horses affected by equine asthma. *J. Breath Res.* **2018**, *12*, 046015. [CrossRef]
41. Olson, M.S. Bioenergetics and Oxidative Metabolism. In *Textbook of Biochemistry with Clinical Correlation*, 4th ed.; Devlin, T.M., Ed.; Wiley-Liss: New York, NY, USA, 1997; pp. 218–261.
42. Dame, Z.T.; Aziat, F.; Mandal, R.; Krishnamurthy, R.; Bouatra, S.; Borzouie, S.; Guo, A.C.; Tanvir Sajed, T.; Deng, L.; Lin, H.; et al. The human saliva metabolome. *Metabolomics* **2015**, *11*, 1864–1883. [CrossRef]
43. Duarte, I.F.; Goodfellow, B.J.; Barros, A.; Jones, J.G.; Barosa, C.; Diogo, L.; Garcia, P.; Gil, A.M. Metabolic characterisation of plasma in juveniles with glycogen storage disease type 1a (GSD1a) by high-resolution (1)H NMR spectroscopy. *NMR Biomed.* **2007**, *20*, 401–412. [CrossRef]
44. Laborde, C.M.; Mourino-Alvarez, L.; Posada-Ayala, M.; Alvarez-Llamas, G.; Serranillos-Reus, M.G.; Moreu, J.; Vivanco, F.; Padial, L.R.; Barderas, M.G. Plasma metabolomics reveals a potential panel of biomarkers for early diagnosis in acute coronary syndrome. *Metabolomics* **2014**, *10*, 414–424. [CrossRef]
45. Trushina, E.; Dutta, T.; Persson, X.M.; Mielke, M.M.; Petersen, R.C. Identification of altered metabolic pathways in plasma and CSF in mild cognitive impairment and Alzheimer's disease using metabolomics. *PLoS ONE* **2013**, *8*, e63644. [CrossRef] [PubMed]
46. Koolman, J.; Roehm, K.H. *Color. Atlas of Biochemistry*, 2nd ed.; Thieme: Stuttgart, Germany, 2005; p. 352.
47. Marukawa, H.; Shimomura, T.; Takahashi, K. Salivary substance P, 5-hydroxytryptamine, and gamma-aminobutyric acid levels in migraine and tension-type headache. *Headache* **1996**, *36*, 100–104. [CrossRef] [PubMed]
48. Jezewska, E.; Scinska, A.; Kukwa, W.; Sobolewska, A.; Turzynska, D.; Samochowiec, J.; Bienkowski, P. Gamma-aminobutyric acid concentrations in benign parotid tumours and unstimulated parotid saliva. *J. Laryngol. Otol.* **2011**, *125*, 492–496. [CrossRef] [PubMed]
49. Sawaki, K.; Ouchi, K.; Sato, T.; Kawaguchi, M. Existence of gamma-aminobutyric acid and its biosynthetic and metabolic enzymes in rat salivary glands. *Jpn. J. Pharmacol.* **1995**, *67*, 359–363. [CrossRef] [PubMed]
50. Okubo, M.; Kawaguchi, M. Rat submandibular gland perfusion method for clarifying inhibitory regulation of GABAA receptor. *J. Pharmacol. Sci.* **2013**, *122*, 42–50. [CrossRef] [PubMed]
51. Bizzarri, M.; Fuso, A.; Dinicola, S.; Cucina, A.; Bevilacqua, A. Pharmacodynamics and pharmacokinetics of inositol(s) in health and disease. *Expert Opin. Drug Metab. Toxicol.* **2016**, *12*, 1181–1196. [CrossRef]
52. Gilmore, A.P.; Burridge, K. Regulation of vinculin binding to talin and actin by phosphatidylinositol-4-5-bisphosphate. *Nature* **1996**, *381*, 531–535. [CrossRef]
53. Hallman, M.; Spragg, R.; Harrell, J.H.; Moser, K.M.; Gluck, L. Evidence of lung surfactant abnormality in respiratory failure. Study of bronchoalveolar lavage phospholipids, surface activity, phospholipase activity, and plasma myoinositol. *J. Clin. Investig.* **1982**, *70*, 673–683. [CrossRef]
54. Benesch, R.; Benesch, R.E. The effect of organic phosphates from the human erythrocyte on the allosteric properties of hemoglobin. *Biochem. Biophys. Res. Commun.* **1967**, *26*, 162–167. [CrossRef]
55. Lam, G.; Zhao, S.; Sandhu, J.; Yi, R.; Loganathan, D.; Morrissey, B. Detection of myo-inositol tris pyrophosphate (ITPP) in equine following an administration of ITPP. *Drug Test. Anal.* **2014**, *6*, 268–276.
56. Bazzano, M.; Laghi, L.; Zhu, C.; Magi, G.E.; Tesei, B.; Laus, F. Respiratory metabolites in bronchoalveolar lavage fluid (BALF) and exhaled breath condensate (EBC) can differentiate horses affected by severe equine asthma from healthy horses. *BMC Vet. Res.* **2020**. under review. [CrossRef] [PubMed]
57. Carroll, C.L.; Huntington, P.J. Body condition scoring and weight estimation of horses. *Equine Vet. J.* **1988**, *20*, 41–45. [CrossRef] [PubMed]
58. Zhu, C.; Faillace, V.; Laus, F.; Bazzano, M.; Laghi, L. Characterization of trotter horses urine metabolome by means of proton nuclear magnetic resonance spectroscopy. *Metabolomics* **2018**, *14*, 106. [CrossRef] [PubMed]

59. Kneen, M.A.; Annegarn, H.J. Algorithm for fitting XRF, SEM and PIXE X-ray spectra backgrounds. *Nucl. Instrum. Methods Phys. Res. Sect. B Beam Interact. Mater. At.* **1996**, *109–110*, 209–213. [CrossRef]
60. Box, G.E.P.; Cox, D.R. An Analysis of Transformations. *J. R. Stat. Soc. Ser. B* **2018**, *26*, 211–243. [CrossRef]
61. Hubert, M.; Rousseeuw, P.J.; Vanden Branden, K. ROBPCA: A new approach to robust principal component analysis. *Technometrics* **2005**, *47*, 64–79. [CrossRef]
62. Croft, D.; Mundo, A.F.; Haw, R.; Milacic, M.; Weiser, J.; Wu, G.; Jassal, B. The Reactome pathway knowledgebase. *Nucleic Acids Res.* **2014**, *42*, D472–D477. [CrossRef]
63. Johnson, N.L.; Kotz, S.; Kemp, A.W. *Univariate Discrete Distributions*, 2nd ed.; Wiley: New York, NY, USA, 2005.

9

Profiling of Metabolomic Changes in Plasma and Urine of Pigs Caused by Illegal Administration of Testosterone Esters

Kamil Stastny [1,*], Kristina Putecova [1], Lenka Leva [1], Milan Franek [1], Petr Dvorak [2] and Martin Faldyna [1]

1 Veterinary Research Institute, Hudcova 296/70, 62100 Brno, Czech Republic; putecova@vri.cz (K.P.); leva@vri.cz (L.L.); franek@vri.cz (M.F.); faldyna@vri.cz (M.F.)
2 Faculty of Veterinary Hygiene and Ecology, University of Veterinary and Pharmaceutical Sciences Brno, Palackeho tr. 1946/1, 61242 Brno, Czech Republic; dvorakp@vfu.cz
* Correspondence: stastny@vri.cz

Abstract: The use of anabolic steroid hormones as growth promoters in feed for farm animals has been banned in the European Union since 1988 on the basis of Council Directive 96/22/EC. However, there is still ongoing monitoring and reporting of positive findings of these banned substances in EU countries. The aim of this work was to investigate the efficacy and discriminatory ability of metabolic fingerprinting after the administration of 17β-testosterone esters to pigs. Plasma and urine samples were chromatographically separated on a Hypersil Gold C18 column. High resolution mass spectrometry metabolomic fingerprints were analysed on a hybrid mass spectrometer Q-Exactive. Three independent multivariate statistical methods, namely principal component analysis, clustre analysis, and orthogonal partial least squares discriminant analysis showed significant differences between the treated and control groups of pigs even 14 days after the administration of the hormonal drug. Plasma samples were also analysed by a conventional quantitative analysis using liquid chromatography with tandem mass spectrometry and a pharmacokinetic curve was constructed based on the results. In this case, no testosterone residue was detected 14 days after the administration. The results clearly showed that a metabolomics approach can be a useful and effective tool for the detection and monitoring of banned anabolic steroids used illegally in pig fattening.

Keywords: metabolomic; anabolic practices; testosterone; plasma; urine; pigs

1. Introduction

The use of hormones as growth promoters for fattening purposes in livestock has been banned in the European Union since 1988 by Council Directive 96/22/EC. However, the banned substances are still reported as positive in the European residue monitoring plans [1]. One of these banned substances is the anabolic and androgenic steroid testosterone, which naturally occurs in animal organisms. Testosterone is endogenously secreted by Leydig cells (testes) and is able to accelerate muscle growth (anabolic effect) and improve the development of male characteristics (androgenic effect). Testosterone is then secreted into the bloodstream where it primarily (98%) binds to a specific protein beta-globulin termed sex hormone binding globulin (SHBG) and to a lesser extent to albumin. By this binding, testosterone is biologically protected from inactivation in the liver, and is subsequently transported to the target tissues via the bloodstream. A small amount of circulating testosterone is converted to estradiol, but the greater part of free testosterone is converted to 17-ketosteroids, particularly androsterone and its isomer etio-cholanolone (androsterone metabolites) [2]. In some target tissues, testosterone is reduced to 5α-dihydrotestosterone (DHT) by the cytochrome P_{450} enzyme 5α-reductase, an enzyme highly expressed in male sex organs, skin, and hair follicles. The inactivation and degradation of testosterone and its metabolites in cattle and pigs occurs mainly in the liver and, to a lesser extent, in the kidneys. These mechanisms of inactivation and degradation of testosterone occur with the participation of specific enzymes involved in the catalytic action of the partially transformed steroid molecule. Inactivation and degradation include the following: addition of two hydrogens (reduction) to a double bond or ketone group; removal of two hydrogens (oxidation) from a hydroxyl group; addition of a hydroxyl group (hydroxylation) to a carbon in the steroid molecule; and conjugation of testosterone by reaction of sulfuric acid or glucuronic acid with a hydroxyl group on the steroid molecule, forming testosterone sulphates and glucuronides, respectively. The sulfated or glucuronide conjugated form of testosterone is then excreted in the urine [3].

Testosterone (a natural steroid) is illegally administered to animals in the form of synthetic steroid esters, but these are rapidly hydrolysed to a natural steroid in vivo. For example, after oral administration of testosterone undecanoate, unchanged ester was found in athletes' plasma for only 6 h [4]. In analytical practice, it is difficult to distinguish between metabolites of natural endogenous testosterone, which is always present in body fluids (plasma, urine), and metabolites of identical exogenous testosterone derived from hydrolysed synthetically prepared esters [5]. In human doping control, this problem is usually solved by determining the urinary ratio of 17β-testosterone/17α-testosterone levels (T/EpT ratio) or by using gas chromatography with isotopic mass spectrometry (GC-IRMS) and application of the $^{13}C/^{12}C$ isotope ratio [6]. The World Anti-Doping Agency (WADA) has established a decision limit if a T/E ratio is equal to or greater than 4, or an epitestosterone (17α-testosterone) concentration is greater than 200 ng mL^{-1} which would require a testing procedure to confirm doping [7,8]. Although important in humans, these analytical parameters have failed in animals because of differences in their metabolism [9]. In food safety practices, relatively high or low levels of 17β-testosterone and 17α-testosterone in urine are often ignored due to a lack of statistically valid reference data on naturally occurring endogenous background levels in animals. However, the EU Community Reference Laboratories (CRLs) for analytical methods recommended in the national monitoring control plans to limit concentrations for plasma (CC_α for confirmatory methods) to 0.5 μg L^{-1} for heifers 18 months old, 10 μg L^{-1} for bullocks six months old, and 30 μg L^{-1} for bulls 6–18 months old [10]. For other animals, no such recommendations exist for 17β-testosterone in plasma or urine.

Over time, a number of targeted analytical methods for the determination of testosterone in various biological samples (plasma, urine, muscles and hair) have been developed and described in the literature. In the 1990s, testosterone measurements were often performed by radioimmunoassay (RIA) [11] and immunoassays (ELISA) [12]. Immunological methods are fast, easy-to-perform, cheap, and have a short time to result for a large number of samples, so today they are preferably used in many laboratories primarily for screening. However, cross-reactivity and sensitivity in these assays is a common problem, so these methods are no longer good enough for the detection and quantitative determination of testosterone.

Gas chromatography (GC) and liquid chromatography (LC) are other alternatives for the targeted analysis of testosterone and its esters. GC methods coupled with mass spectrometry (MS) are usually applied for the determination of anabolic steroid levels ranging from micrograms to nanograms in biological samples [13]. The detection of testosterone esters at 1 ng mL^{-1} in human plasma by GC/MS has been reported [14]. However, GC-MS methods require a complicated, time-consuming, and expensive step of sample derivatization for steroid analysis. In general, these derivatives are unstable and are susceptible to thermal degradation during analysis, which, in particular, significantly affects the reproducibility of the method [15]. In contrast, LC-MS is a good solution for quantitative analysis of steroids because the included sample preparation step is easy, fast, economical and requires no further derivatization step. The high performance liquid chromatography with mass spectrometry (HPLC-MS) used for the analysis of steroid esters in plasma showed greater sensitivity than GC-MS [16]. HPLC is also a commonly used separation technique for the determination of testosterone and its esters in body fluids due to its sufficient sensitivity, good resolution, robustness and short analysis time [17]. Ultra-high performance liquid chromatography (UHPLC) coupled with tandem mass spectrometry (MS/MS) or today with high-resolution (HR) mass spectrometry is another powerful approach to significantly improve peak resolution, selectivity, sensitivity and speed of the analysis [18–20]. It should be noted that an interesting alternative to inconclusive urine analyses (endogenous testosterone vs. synthetic testosterone) at veterinary inspection may be the analysis of intact natural steroid esters in the hair by UHPLC-MS/MS [21,22] or DESI-MS (desorption ionizing mass spectrometry) [5].

Recently, new synthetic xenobiotic growth promoters have been designed and new ways of application employed, such as the administration of low dose cocktails. However, metabolomics approaches to non-targeted screening for the detection of anabolic practices with natural steroid hormones might change this situation in the future [9,23]. Indeed, several scientific studies have demonstrated the efficiency of mass spectrometry with high resolution based on urinary fingerprinting to discriminate anabolised animals from control ones. Rijk et al. [24] in their work showed the use of a novel untargeted metabolomics based strategy for the measurement of the anabolic steroid DHEA (dehydroepiandrosterone) and pregnenolone in bovine urine with liquid chromatography coupled with time-of-flight mass spectrometry (LCT Premier). In the same year, Kieken et al. [25] presented a metabolomics strategy involving the characterization of global metabolomic fingerprints in urine samples of non-treated and reGH (recombinant equine growth hormone)-treated horses by LC-HRMS (LTQ-Orbitrap) as a new screening tool for growth hormone abuse in horseracing.

Anizan et al. [26] presented in their study a metabolomics approach to 4-androstenedione (AED) detection after its administration to heifers. Using untargeted profiling by GC-MS, they identified 5α-androst-2-en-17-one in urine as a new biomarker of anabolic AED abuse. From 2011 to the present, several studies have been conducted in cattle in relation to the administration of banned substances for fattening, which have confirmed the correctness of the research focus on non-targeted analyses based on metabolomic approaches [27–32]. However, all these studies were in all cases carried out only in cattle, although in many European countries, for example, pork was consumed significantly more than beef. The only metabolomic study published so far for another animal species was conducted in 2017 in pigs to which a banned beta-agonist substance, ractopamine, was administered [33].

The present study aimed to investigate the efficacy of metabolomic profiling of pig plasma and urine samples by high resolution mass spectrometry (HRMS) to discriminate between the testosterone ester group and the control group. The experiment was performed in two independent groups of pigs, where individual animals were assigned to groups based on randomization. Plasma and urine samples were continuously collected at specified time intervals, prepared and subsequently measured on a high-resolution hybrid tandem mass spectrometer (QExactive). The obtained metabolomic fingerprints were processed and statistically analyzed using principal component analysis (PCA) and orthogonal partial least squares discriminant analysis (OPLS-DA) multivariate methods. Furthermore, the results of the non-targeted metabolomic analysis obtained in this way were compared with the results of the targeted determination of 17β-testosterone in the same plasma and urine samples. All pigs in the experiment were weighed at weekly intervals and the anabolic effect of testosterone was studied based on the body weight gain.

2. Results

2.1. *Anabolic Effect of 17β-Testosterone (Esters)*

All animals from both groups were weighed at regular weekly intervals during the treatment experiment, and the body weight gain (BW) in kg is shown in Tables 1 and 2. The anabolic effect was expressed in a graphical form of the dependence of the body weight of experimental pigs on the time interval for both treatment and control groups. Two linear regression models were used to highlight the growth trends of both groups of pigs and the statistical assessment of the anabolic effect of 17β-testosterone (Figure 1). The average weekly body weight gains were calculated from the detected

BW data for each group of pigs and are shown in Supplementary Materials Table S1 and Figure S1. All animals were clinically monitored during the experiment and were in good health until slaughter at the end of the experiment. The treatment experiment was performed without any problems.

Table 1. Body weight of the individual pigs treated with a hormone preparation containing a testosterone ester combination.

N_i	Ear Number	Sex	BW (kg)/Week						
			1	2	3	4	5	6	7
1	1	♂	28.1	35.0	41.3	45.0	52.1	58.0	69.0
2	2	♂	29.4	33.4	36.9	43.8	52.1	53.8	63.6
3	4	♂	34.1	40.5	44.2	51.6	58.8	64.2	74.0
4	5	♂	26.0	30.8	36.6	45.5	52.0	60.5	66.5
5	7	♂	23.5	27.0	29.5	33.7	39.2	41.8	50.2
6	11	♀	19.9	23.2	24.7	30.0	35.6	38.6	47.0
7	13	♀	29.6	33.9	37.3	45.0	50.2	54.6	63.5
8	14	♀	37.4	43.4	45.5	55.0	59.2	65.8	75.0
9	16	♀	30.4	35.3	37.4	44.8	51.2	56.7	66.0
10	25	♀	32.7	32.0	38.0	40.0	45.0	52.0	56.0
11	26	♀	22.5	46.0	54.0	59.0	67.5	73.0	81.0
12	27	♂	19.9	34.0	38.0	40.5	47.0	54.0	60.0
13	28	♂	26.7	36.5	42.0	46.0	52.5	60.0	66.0
	Female								
	Average		26.8	33.9	38.4	43.7	50.5	56.1	64.2
	CV		20.1	18.2	23.3	30.5	36.7	53.0	57.2
	SD		4.5	4.3	4.8	5.5	6.1	7.3	7.5
	CI (95%) lower		22.7	30.0	33.9	38.6	44.9	49.3	57.2
	CI (95%) upper		31.0	37.8	42.8	48.8	56.1	62.8	71.2
	Median		26.7	34.0	38.0	45.0	52.1	58.0	66.0
	SD_m		3.6	3.4	3.7	4.6	5.0	5.7	6.1
	Male								
	Average		28.8	35.6	39.5	45.6	51.5	56.8	64.7
	CV		42.3	67.7	95.4	108.7	122.4	140.5	152.8
	SD		6.5	8.2	9.8	10.4	11.1	11.9	12.4
	CI (95%) lower		21.9	27.0	29.2	34.7	39.8	44.3	51.8
	CI (95%) upper		35.6	44.3	49.7	56.6	63.1	69.2	77.7
	Median		30	34.6	37.7	44.9	50.7	55.7	64.8
	SD_m		4.5	5.8	7.5	7.4	8.1	8.8	8.7

Table 2. Body weight of the individual pigs in the control group.

N_i	Ear Number	Sex	BW (kg)/Week						
			1	2	3	4	5	6	7
1	8	♂	26.1	31.1	36.0	43.0	49.0	51.9	57.0
2	9	♂	26.3	30.2	34.6	38.6	44.2	48.1	54.5
3	12	♀	28.8	33.9	37.5	43.5	50.4	52.6	58.9
4	15	♀	29.2	33.8	36.1	43.5	50.0	54.2	60.5
5	21	♀	29.4	33.1	38.0	42.5	48.0	52.1	58.6
6	22	♀	30.4	36.0	43.0	45.5	51.0	55.0	60.0
7	23	♂	30.9	34.5	40.5	44.0	48.0	52.5	60.5
8	24	♂	24.9	35.0	39.5	42.0	44.5	51.5	58.1
	Average		28.3	33.5	38.2	42.8	48.1	52.2	58.5
	CV		4.8	3.8	7.5	4.0	6.6	4.2	4.1
	SD		2.2	1.9	2.7	2.0	2.6	2.1	2.0
	CI (95%) lower		26.4	31.8	35.9	41.1	46.0	50.5	56.8
	CI (95%) upper		30.1	35.1	40.4	44.5	50.3	54.0	60.2
	Median		29.0	33.9	37.8	43.3	48.5	52.3	58.8
	SD_m		1.5	1.5	2.1	1.8	1.7	1.8	1.5

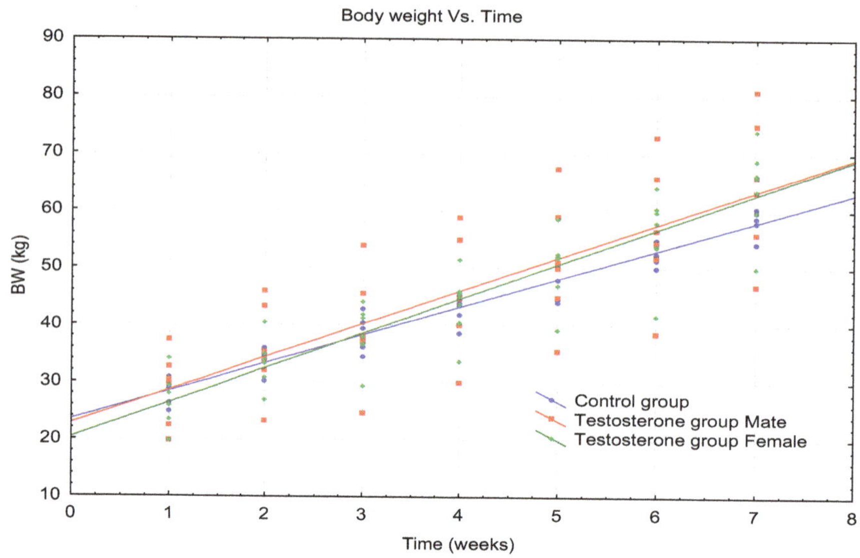

Figure 1. A combined diagram of two regression models for testosterone treated pigs (male and female) and the control group of pigs.

2.2. *Targeted Analysis of 17 β-Testosterone in Plasma*

2.2.1. Identification of Analytes

Standards of 17β-testosterone, testosterone propionate, testosterone isocaproate, testosterone decanoate, and 17β-testosterone-D_2 internal standard were always identified on the basis of the retention time (see Figure 2) obtained from the chromatogram and mass accuracy (MA) parameters calculated from mass spectra for precursor and product ions of each analyte by comparing the theoretical mass *m*/*z* with the measured experimental mass *m*/*z*. The obtained results and the calculated MA values determined by the standards are presented in Supplementary Materials Table S3 and in the diagrams showing the detected experimental mass spectra of the analyte standards, always in comparison with the theoretical mass spectra (Supplementary Materials Figures S2 and S3).

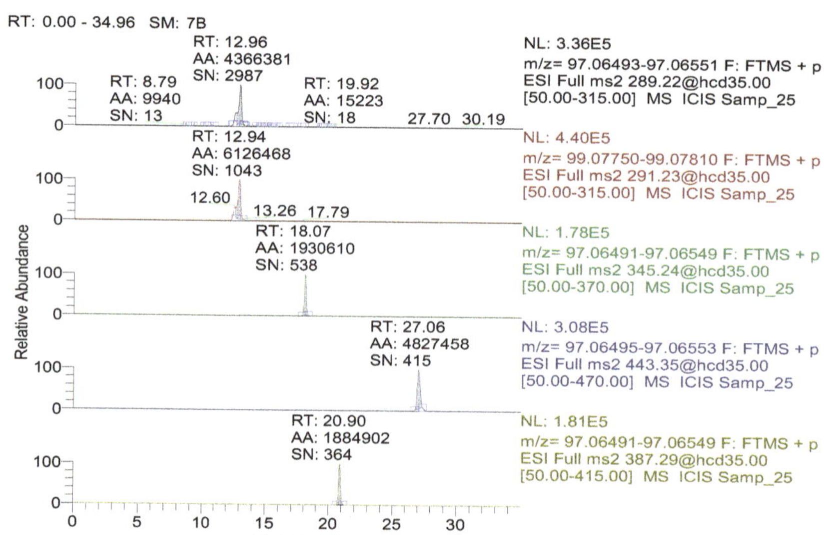

Figure 2. Chromatogram of the analysed plasma samples fortified with standards at 10 μg L^{-1} (ppb), RT indicates retention time, AA indicates the peak area, SN indicates signal to noise ratio, black chromatogram represents 17β-testosterone, red chromatogram represents IS 17β-testosterone-D2, green chromatogram represents 17β-testosterone propionate, blue chromatogram represents 17β-testosterone decanoate and the yellow chromatogram represents 17β-testosterone isocaproate.

2.2.2. Study Validation

The targeted quantitative method was in-house validated complies with model-dependent performance characteristic covering specificity, selectivity, precision, repeatability, within-laboratory reproducibility, the calibration curve, detection limit (LOD), limit of quantification (LOQ), decision limit (CCα), detection capability (CCβ), and ruggedness according to the recommendation defined in Commission Decision 2002/657 / EC [34] and the reference guidelines in VICH GL49 [35].

The linearity of the quantitative method was determined for testosterone and testosterone ester analytes that were fortified into real pig plasma samples with increasing concentrations. The model samples were prepared according to the procedure described in Section 4.5. A matrix calibration curve was constructed based on the measured peak area ratios (Std. area/IS area) and the corresponding concentration levels. The parameters of the linear regression models were calculated by the least squares method (with a weight coefficient w = 1/2) based on ISO 11843: 2 [36]. Correlation coefficients (r), linear regression model parameters (y = a + bx) and critical curve limits (LOD, LOQ) were calculated and reported in Table 3. The calibration curve for 17β-testosterone, which was used to back-estimate the results of real plasma samples obtained during the experiment, was shown graphically (Supplementary Materials Figure S4). The complete standard area and internal standard area data that was used to calculate the 17β-testosterone calibration curve are presented in Supplementary Materials Table S4.

To determine the precision and repeatability (within-laboratory reproducibility) of the targeted analysis method, the standard deviation (SD) and variation coefficient (CV, %) were determined and calculated by repeated measurement of fortified plasma samples at two concentration levels. The calculated CV (n = 12) was less than 3.09% for a concentration level of 5 ng mL^{-1} 17β-testosterone in plasma, demonstrating the good precision and repeatability required for confirmatory residual analyses by the Commission Decision 2002/657/EC. The results of the validation study for precision, repeatability, and other calculated statistics are shown in Supplementary Materials Table S5.

Table 3. Regression parameters of matrix calibration curves in the concentration range 0 to 80 ng mL^{-1}.

Analyte	Intercept (a)	Slope (b)	SD of the Slope	Correlation Coefficient r	LOD (ng mL^{-1})	LOQ (ng mL^{-1})
17β-testosterone	0.0518	0.0690	0.00096	0.9982	0.32	0.63
17β-testosterone propionate	0.0291	0.0482	0.00086	0.9991	0.19	0.52
17β-testosterone decanoate	0.0115	0.0713	0.00089	0.9979	0.21	0.54
17β-testosterone isocaproate	0.0248	0.0448	0.000073	0.9996	0.17	0.43

Note: LOD and LOQ were estimated according to IUPAC (Direct Signal Method) methodology.

2.2.3. Pharmacokinetic Profile of 17β-Testosterone

The experiment included targeted analysis of the primary testosterone metabolite in porcine plasma after a single i.m. administration and subsequent determination of the pharmacokinetic curve. Plasma concentrations of free 17β-testosterone for individual pigs were determined based on an estimation from the matrix calibration curve (see Supplementary Materials Table S6). The resulting plasma concentrations of 17β-testosterone were used to construct a pharmacokinetic curve, and a graph of concentration versus time is shown in Figure 3.

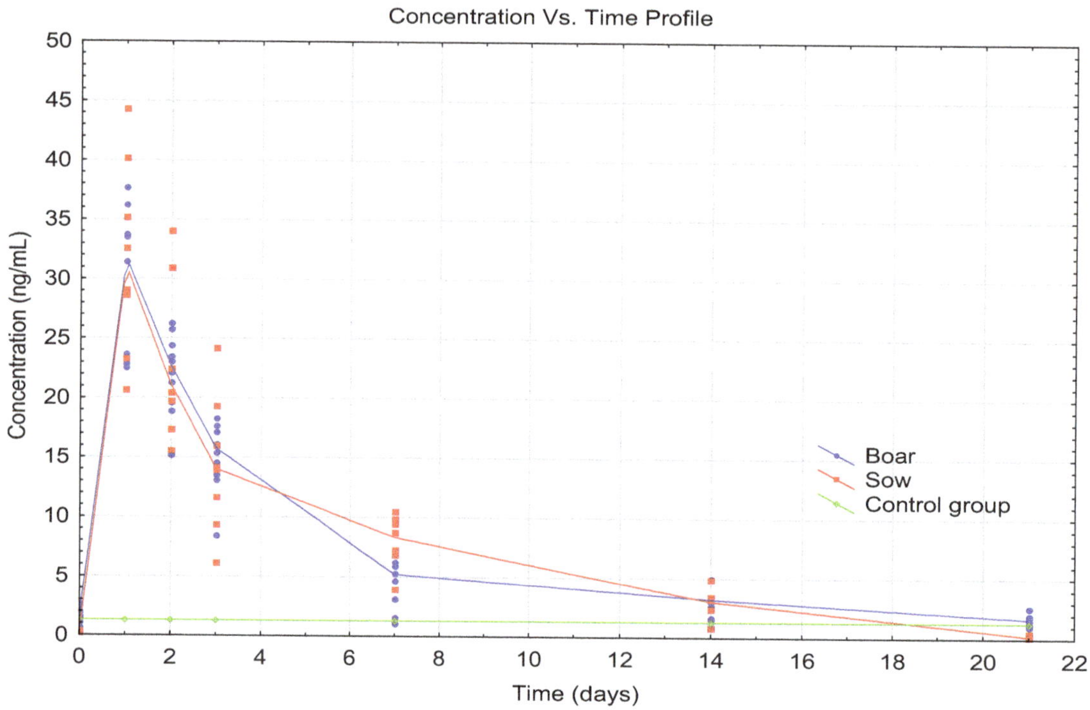

Figure 3. Plasma concentrations in pigs—time profile of testosterone after a single i.m. administration of 0.6 mL Sustanon 250 mg/mL inj.; the points on the curve represent the detected 17β-testosterone plasma concentrations in individual pigs.

2.3. Metabolomic Study of Blood Plasma and Urine

The obtained plasma and urine samples were processed in the laboratory as described in Section 4.2. The metabolomic profiles of the individual samples on day 14 after the administration of the hormonal preparation SUSTANON were measured as described in Section 4.4. Both groups of plasma and urine metabolomic profiles were processed for a comparison in XCMS software and, alternatively, using the SIEVE company software. Both variants of data processing identified the approximately corresponding number of ions (m/z) of peaks or metabolites: 2500 ions were found in plasma and 1400 ions were found in urine. In both cases, the original number of ions in these data sets was further reduced based on a p-volume ≤ 0.05 for further statistical processing. The source data of sets X (n × m) after reduction each contained $n = 21$ rows (animal objects) and m = 254 columns of statistically significant identified peak areas or metabolites for plasma and m = 213 columns for urine, respectively. Datasets were transformed using two different methods, i.e., column centering [37] and probabilistic quotient normalization (PQN) [38], and a natural logarithm was applied for their scaling before the subsequent multivariate statistical analysis. The hotelling T^2 test criterion did not identify any outlier in both data sets (Supplementary Materials Figures S5 and S6).

Multidimensional statistical methods such as principal component analysis (PCA), clustering analysis (CA) and orthogonal partial squares discriminant analysis (OPLS-DA) were applied for finding relationships between metabolomics datasets of plasma and urine. PCA score plots for plasma and urine samples (Figure 4) and a dendrogram from CA (Figure 5) visibly differentiated between the control group and the treated group of pigs after 17 beta-testosterone administration. The main graphical results from the OPLS-DA analysis of data matrix of X mass spectra of plasma and urine samples versus data matrix Y for binary variables (1 = group of treated pigs, 2 = control group) were generated by the proposed statistical model and are shown in Figure 6. Furthermore, the coefficients $R^2(X) = 0.616$, $R^2(Y) = 0.987$ for the fit and $Q^2(Y) = 0.898$ for prediction of the model (according to cross validation) were calculated by OPLS-DA analysis for plasma and the coefficients $R^2(X) = 0.469$, $R^2(Y) = 0.997$ and $Q^2(Y) = 0.879$ for urine data. The OPLS-DA permutation tests further confirmed that the proposed statistical models are correct and robust (Supplementary Materials Figures S7 and S8).

A volcano plot and variable importance in the projection (VIP) plot and S-plot from OPLS-DA were employed to determine the most discriminating metabolites between the treatment group and the control group (Figure 7).

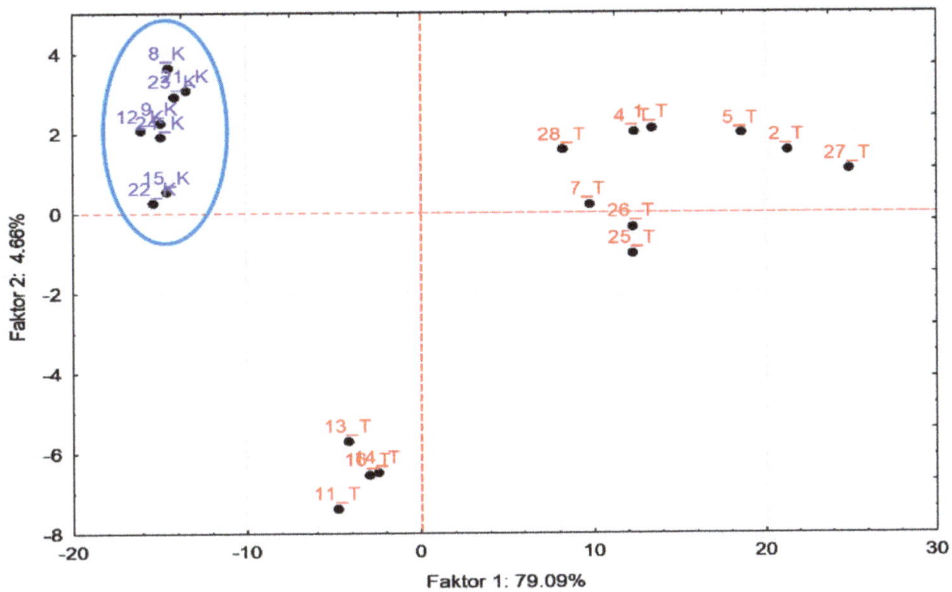

(A)

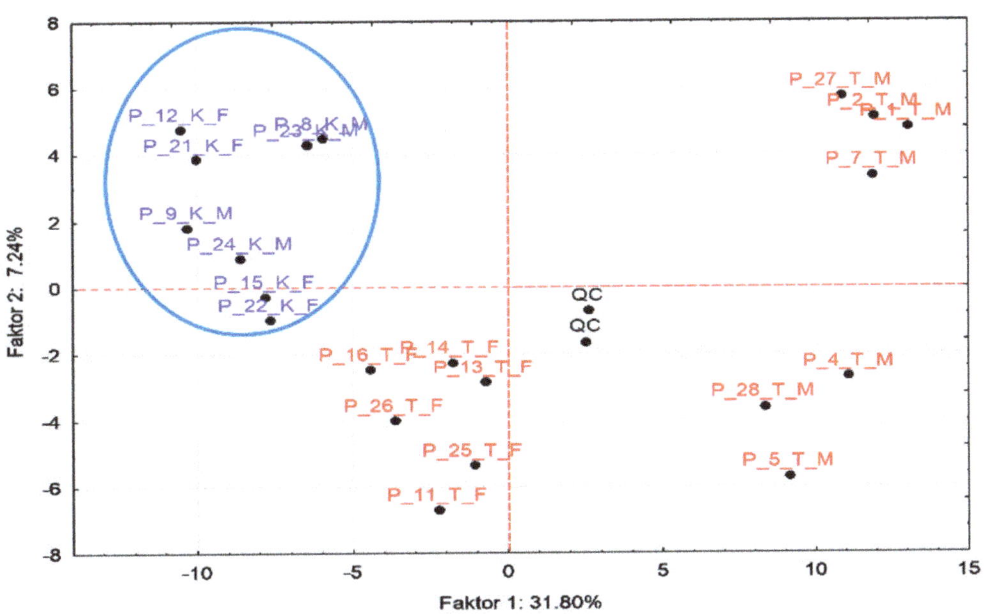

(B)

Figure 4. PCA Score plots for plasma (**A**) and urine (**B**) data matrix, blue ellipse, and blue point descriptions (K) represent statistically significantly different samples from the control group of pigs versus the treated (T) group of pigs; added urine labelling: M—male and F—female (Centering, STATISTICA).

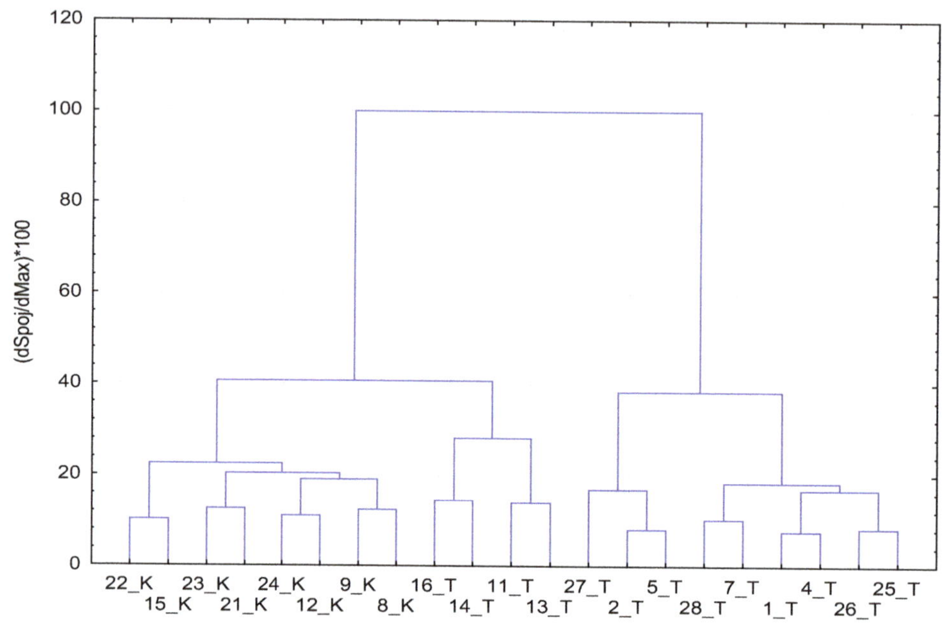

(A)

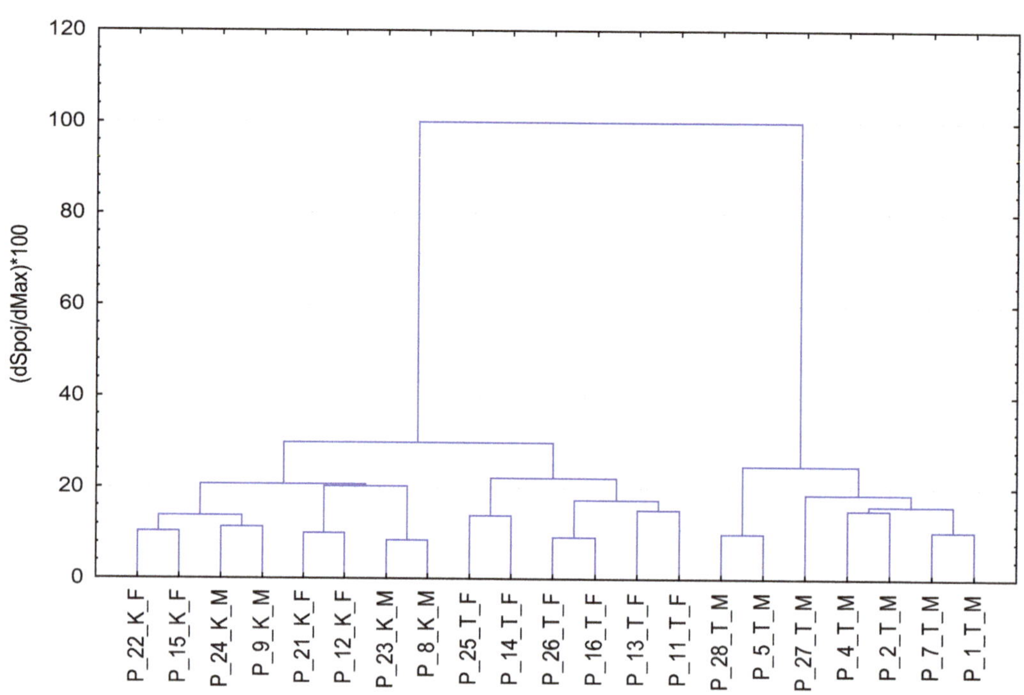

(B)

Figure 5. CA dendrogram of matrix data objects for plasma (**A**) and urine (**B**), labels: K—control group, T—treated group and M—male, F—female (by Euclidean distance method, STATISTICA).

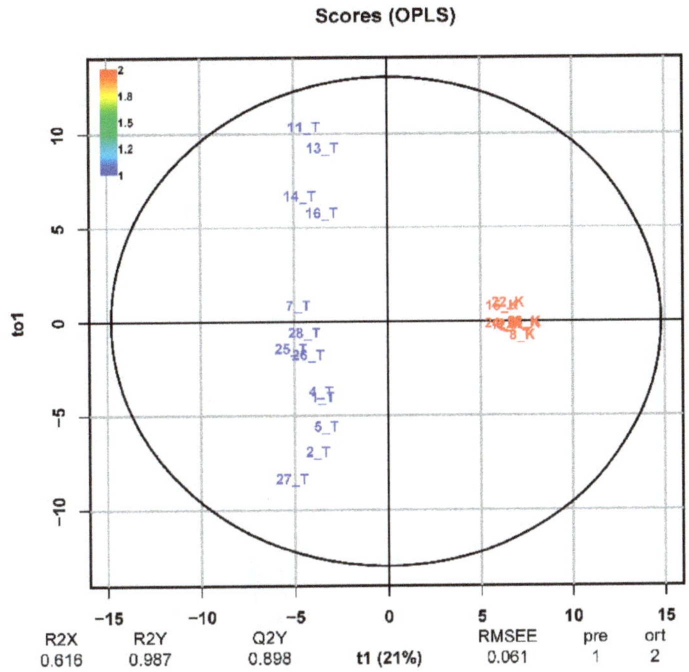

(A)

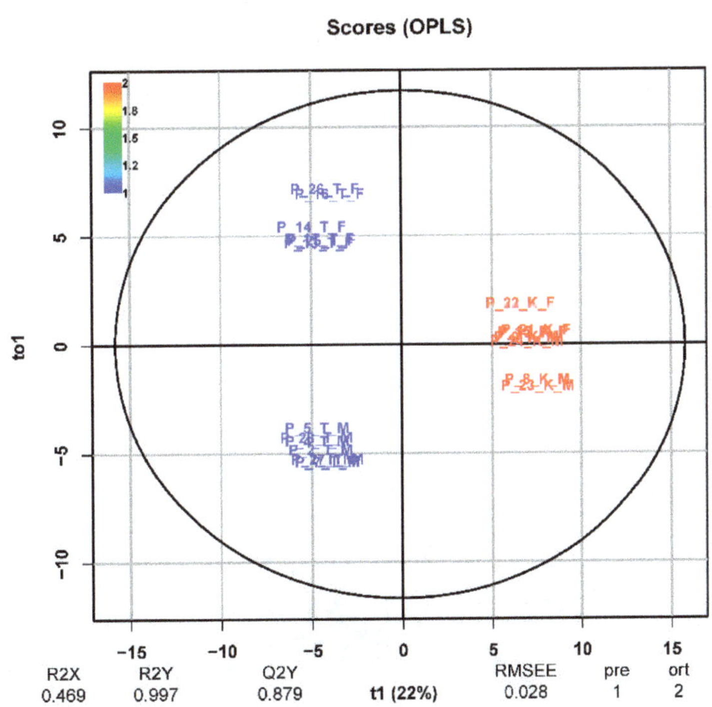

(B)

Figure 6. The OPLS-DA score plots for plasma (**A**) and urine (**B**) data matrix demonstrate robust discrimination between the control group of pigs marked with red colour and the group of treated pigs marked with blue colour (PQN scaling, R package). The control group indicated by red Pig number 8, 9, 12, 15, 21, 22, 23, and 24. The treated group indicated by blue Pig number 1, 2, 4, 5, 7, 11, 13, 14, 16, 25, 26, 27, and 28.

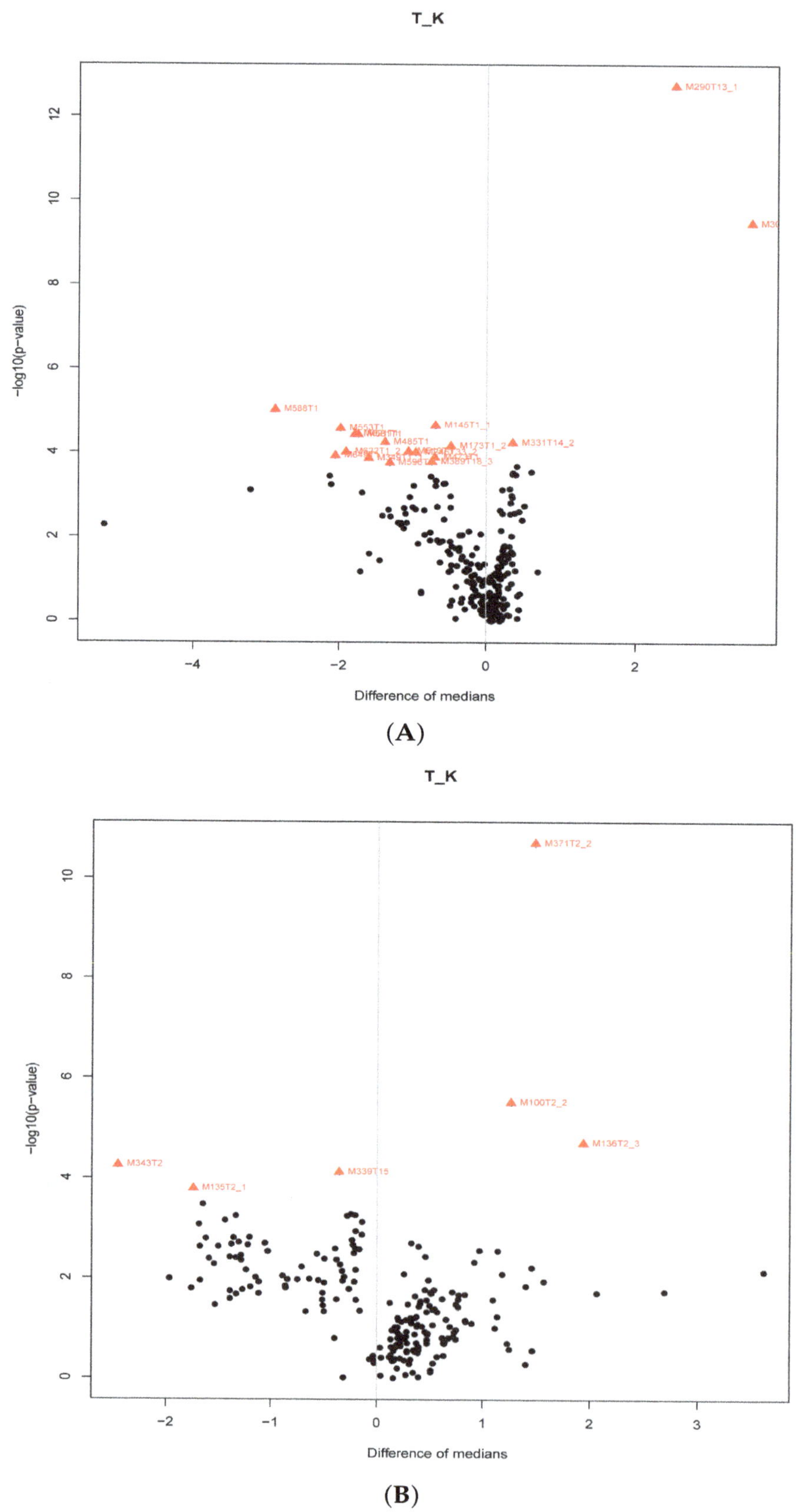

Figure 7. The OPLS-DA Vulcano plots for plasma (**A**) and urine (**B**) data matrix, only metabolites with the VIP scores above 2 were considered significant. A list of specific numbers of metabolites is given in Table 4.

Table 4. An overview of the most discriminating compounds (metabolites) found based on Vulcano-plots, VIP's and S-plots from OPLS-DA; x—statisticaly significant.

No.	Name	*p*-Value	*m/z*	RT	Putative Metabolite	Vulcano	VIP	S-Plot	Name	*p*-Value	*m/z*	RT	Putative Metabolite	Vulcano	VIP	S-Plot
	Plasma samples								Urine samples							
1	M145T1_1	1.26×10^{-8}	144.96248	0.75		x		x	M127T5	3.60×10^{-3}	127.01611	4.93			x	x
2	M288T13_3	8.58×10^{-5}	288.34193	13.10			x	x	M135T2_1	7.30×10^{-3}	135.00340	1.65		x	x	
3	M290T13_1	1.95×10^{-5}	290.21876	12.96	$C_{19}H_{30}O_2$, DHT	x	x	x	M136T2_3	2.70×10^{-3}	136.07603	1.65		x	x	x
4	M303T3_1	4.81×10^{-5}	303.12451	3.41	$C_{17}H_{19}O_5$, myco-phenolic acide	x	x	x	M228T12	4.63×10^{-2}	228.19606	11.98			x	x
5	M349T1_1	1.88×10^{-9}	348.92403	0.74		x	x		M228T6	3.51×10^{-2}	228.15024	5.57			x	x
6	M417T1	5.91×10^{-8}	416.91155	0.74			x	x	M271T20	2.42×10^{-2}	271.20530	20.10	$C_{19}H_{28}O_2$-H_2O, Androstane-dione, DHA, DHEA	x	x	
7	M485T1	8.63×10^{-11}	484.89886	0.74		x	x	x	M339T15	0.90×10^{-4}	339.17783	14.66				x
8	M553T1	2.50×10^{-8}	552.88600	0.75		x	x	x	M343T2	4.03×10^{-2}	342.85174	1.68		x	x	x
9	M581T1	1.25×10^{-9}	580.88112	0.74		x	x	x	M371T2_2	1.80×10^{-3}	371.22826	1.66		x	x	x
10	M588T1	3.50×10^{-9}	587.88214	0.74		x	x	x	M399T3	3.06×10^{-2}	399.11460	2.55			x	
11	M596T1	5.38×10^{-6}	595.86801	0.75		x	x	x	M441T14	3.73×10^{-2}	441.21007	13.68			x	
12	M601T1_2	6.98×10^{-5}	601.38312	0.75			x	x	M476T2	2.27×10^{-2}	476.30732	2.18			x	
13	M621T1	5.28×10^{-7}	620.87317	0.74		x	x	x	M520T3	2.65×10^{-2}	520.33401	2.59			x	
14	M622T1_2	5.39×10^{-8}	621.87566	0.75		x	x	x	M667T31	1.43×10^{-2}	666.61850	30.81			x	x
15	M632T30	6.82×10^{-5}	632.17150	30.35			x	x								
16	M633T30	3.76×10^{-6}	633.16927	30.36			x	x								
17	M649T1	8.55×10^{-9}	648.86805	0.74		x	x									

Note: DHA—Dehydroandrosterone, DHEA—Dehydroepiandrosterone, DHT—5α-dihydrotestosterone.

2.4. QC Samples

To control the quality of the metabolic profile (fingerprint) measurement, a constant amount of the internal 17β-testosterone-D2 standard was added to each plasma and urine sample (Figure 8), and pooled QC samples were included in each measured acquisition. The RT (deviation up to 10%), peak area (deviation up to 10%) and MA (Δppm < 3) were checked in each measured metabolomic profile.

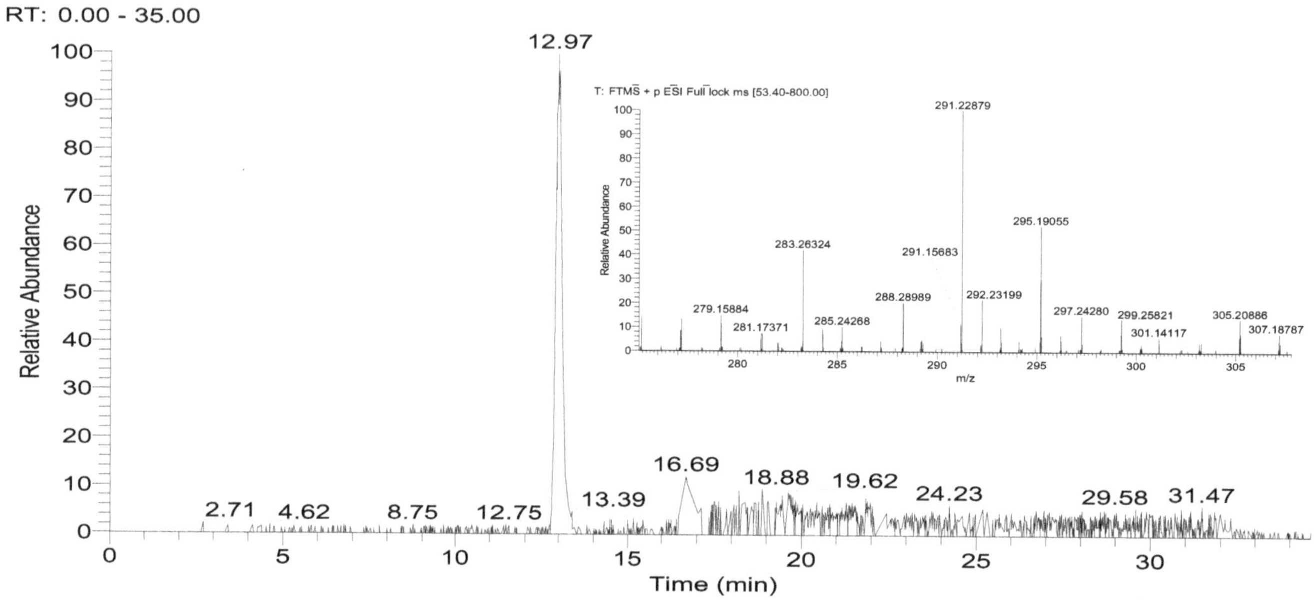

Figure 8. Pooled urine QC sample, control chromatogram for internal standard and corresponding mass spectrum (experimental m/z = 291.22879, Δppm = 0.1).

3. Discussion

The experimental study was carried out in pigs with the main aim of demonstrating the suitability or unsuitability of metabolomic approaches and techniques for detecting the use of banned androgenic anabolic steroids in animal feed and food of animal origin in European countries. All experimental data obtained from the performed study were statistically evaluated using interactive computer-oriented approaches and specialized statistical software with the main emphasis on the correct interpretation of results and endeavour for obtaining a more comprehensive view of the analyzed and assessed topics of our study. The results were presented primarily in a graphical form, as opposed to the mathematical evaluation of the performed statistical analyses, because they generally have a higher predictive ability for the overall evaluation of the achieved effect.

3.1. Anabolic Effect of 17β-Testosterone (Esters)

From the obtained experimental data of BW measurements in a weekly time interval, estimates of standard univariate statistics (mean, variance, standard deviation, median, median standard deviation and 95% confidence interval) were calculated for each time period, separately for the treated pigs and the control group (Tables 1 and 2). The mean BW of pigs from the treated group was lower (mean = 27.7 kg, CI = [24.5; 30.9]) 1 week after 17β-testosterone administration than the mean BW of pigs from the control group (mean = 28.3 kg, CI [26.4; 30.1]). At the end of the weighing at week 7 after the administration, this situation was reversed. The mean BW of pigs from the treated group (mean = 64.4 kg, CI = [58.6; 70.2]) was 9% higher than the mean BW of pigs from the control group (mean = 58.5 kg, CI = [56.8; 60.2]) and averages differed significantly (t-test, $p < 0.05$).

Two linear regression models of BW growth versus time were designed to test the anabolic effect, each model especially for the group of treated pigs and the data of the control group (Figure 1). Both models were tested by regression diagnostics. Testing of the regression triplet (data + model

+ method) showed that the proposed linear models were significant and correct. Estimates of the regression parameters of both linear models are given in Supplementary Materials Table S2. Statistical testing of a comprehensive comparison of both models over the whole-time interval was performed. The Chow test [39] was used to test equality between sets of coefficients in two linear models, using the Fisher-Snedecor distribution with m and r degrees of freedom as statistics. The calculated statistic $F_{Ch} = 13.583$ was greater than the critical value of $F_{0.95(2.62)} = 3.150$. It could be concluded that the hypothesis H_0 was rejected at the significance level $\alpha = 0.05$ and both models were not equal. Applying the conclusion of the Chow test, it can be concluded that the results of BW growth detected in pigs in relation to the time interval can be considered different for both groups of pigs, e.g., the effect of the anabolic effect of testosterone was demonstrated. Since the initial weights of the two groups of pigs did not differ significantly, it can be stated that in this case the slope of the two linear models compared is statistically significantly different. This partial conclusion of the study is also evident from the graphical comparison of the linear curves of both groups of pigs (see Figure 1). The anabolic effect, but in this case of nandrolone (19-nor 17β-testosterone), a synthetic analogue of testosterone, has recently been similarly demonstrated in treated barrows vs. control barrows [40].

3.2. *Targeted Determination of 17β-Testosterone in Plasma*

An analytical method for the targeted determination of 17β-testosterone in pig plasma based on LC-MS/(HR)MS was developed as part of this study to estimate the pharmacokinetic curve. The quantitative method of analysis was developed and validated as a confirmatory method as required by the European Directive for residues [34]. The correct identification of targeted analytes using the mass accuracy (MA) criterion [41,42] and quantification based on a matrix calibration curve with parameter estimates for precision and repeatability were part of the validation of the confirmation method; the results are given in Section 2.2. A detailed description of the methodology used for the identification and validation of targeted analytes has been previously described by Stastny et al. [43].

All calculated MA values (Supplementary Materials Table S2) ranged from 0.1 to 1.9 (Δppm) for the individual analytes determined, and these calculated values were lower than the allowed instrumental tolerance ≤ 3 ppm for the QExactive mass spectrometer used. The tolerance of retention times (RT) was below ± 10% in all cases. Furthermore, a very good chromatographic separation of the analytical method from the chromatogram shown in Figure 2 is evident, where the signal-to-noise ratio (SN) value shown is significantly higher than the generally recommended value (SN > 3), e.g., for 17β-testosterone, the ratio was SN = 2987:1. Correlation coefficients (r) estimated for the target analytes were ≥0.9991 (17β-testosterone), ≥0.9979 (17β-testosterone propionate) and 0.9996 (17β-testosterone decanoate) in plasma (Table 4). The obtained regression models showed good linearity. The sensitivity of the method was determined by the critical values LOD = 0.32 ng mL^{-1} and LOQ = 0.63 ng mL^{-1} for 17β-testosterone based on estimated from the matrix calibration curve model. The precision of the method as a simple repeatability, expressed as the value of the relative standard deviation (RSD), was 3.09% for 17β-testosterone to a concentration level of 5 ng mL^{-1}. Based on the results obtained from the validation study, it was possible to conclude that the developed analytical method is suitable for the identification and quantification of free 17β-testosterone in pig plasma in the range of matrix calibration 0.5 to 80 ng mL^{-1}.

The pharmacokinetic curve of free 17β-testosterone in pig plasma after the administration of 0.6 mL/pig of the hormonal preparation SUSTANON 250 was constructed on the basis of the results of detected concentrations in selected time intervals (Figure 3). The application dosage of the hormonal drug used was calculated on the basis of the recommended dosage for this drug in human medicine. The pharmacokinetic curves for castrated boars and sows were identical. The maximum C_{max} concentration (29.31 ng mL^{-1} ≈ 102 nmol L^{-1} for castrated pigs, 31.73 ng mL^{-1} ≈ 110 nmol L^{-1} for sows) was reached t_{max} 24 h after the administration. Plasma testosterone levels returned to the mean endogenous testosterone levels in castrated boars and sows in approximately 21 days. The PK results for pigs corresponded to the pharmacokinetic properties of i.m. administration of testosterone in men listed in the summary of product characteristics (SPC) of the hormonal drug used (C_{max} = 70 nmol L^{-1}, t_{max} = 24–48 h, elimination time approximately 21 days). Although there are a number of published results on PK testosterone levels in human medicine, the authors of this study of testosterone pharmacokinetics could not compare their results with other authors because such results in animals (pigs) have not been yet published in veterinary medicine.

The determined concentrations of endogenous free 17β-testosterone in the plasma of castrated boars of the hybrid Lage White/Czech White breed (50/50) ranged from 0.55 to 2.58 ng mL^{-1} (mean = 1.56 ng mL^{-1}). These determined concentrations for endogenous testosterone corresponded to the values reported in the literature: White composite breed 4.0 ng mL^{-1} [44], Yorkshire 2.5–5.1 ng mL^{-1}, and Duroc 0.7–3.1 ng mL^{-1} [45]. The determined concentrations of endogenous free 17β-testosterone in sow plasma were below the limit of detection (<LOD) of the analytical method.

The individual testosterone esters contained in SUSTANON 250 hydrolyzed very rapidly in the bloodstream of pigs and were no longer detecable 24 h after application, the concentration was <LOD of the targeted quantitative method. Authors Rejtharová et al. [46] describes the methodology of targeted analysis of testosterone esters in model samples of bovine and porcine serum by LC-MS/MS. On the contrary, as the results of our study show, the targeted determination of testosterone esters in pig serum samples to directly demonstrate the illegal use of banned testosterone is not very suitable for a very short hydrolysis time after application in a real biological system.

3.3. Metabolomic Profiling of Changes in Plasma and Urine

The matrices of metabolomics data X obtained from an experiment performed in pigs and measured on a high-resolution mass spectrometer contain data of the same type *m/z*, i.e., they are homogeneous matrices (low molecular weight compounds, metabolites). To search for latent structure and reveal interrelationships in characters (X-variables) and mainly in objects, cases (pigs) in metric scale, two generally used multivariate statistical methods for character reduction to latent variables were used: principal component analysis (PCA) and cluster analysis (CA). To investigate the dependences between the independent matrix X of metabolomics data and the second dependent variable matrix Y (single-column matrix with binary data, treated group = 1 and control group = 2), another powerful statistical method, partial least squares projection to latent structures PLS was used in the form of differential PLS-DA analysis and in the orthogonal variant O-PLS-DA. Unsupervised PCA and supervised OPLS-DA are today the most widely used multidimensional statistical methods in metabolomics for non-targeted monitoring of changes in biochemical pathways in various biological samples, for their ability to reduce data dimensions or reduce large numbers of variables without much loss of information contained in their first few principal components (most often 2–3).

The metabolomics study was performed in a total of 21 pigs (objects), which were allocated into two groups: a treated group (n = 13) and a control group (n = 8). The problem of the first metabolomics studies in the field of food safety and illegal use of prohibited substances in livestock fattening performed and published between 2009 and 2010 was mainly the small number of experimental animals [24–26]. The authors of these first metabolomic studies were aware of this problem and subsequent studies carried out and published since 2011 have already been performed in an adequate number of animals (n > 10), for example [28–31].

The main results of the PCA method were score plots for plasma and urine data matrix (Figure 4), which showed significant discrimination of all objects (pigs) into two large clusters for both cases of biological matrices. A compact cluster of control group pigs (indicated by a blue ellipse) against two clearly separated clusters (point descriptions marked in red) of pigs from the treated group was found. In the treated group of pigs after testosterone administration, there was another incomplete discrimination according to the characteristics of the sex. However, in the case of plasma, two sows (objects 25 and 25) remained assigned to a cluster of castrated boars, which means that they correlated more with this group. The above results lead to a significant partial conclusion that the built PCA models were able to clearly differentiate the group of tested pigs from the control group.

A CA dendrogram of matrix data objects for plasma and urine (Figure 5) were constructed based on the average Euclidean distance and also showed a reliable distinction between the group of tested pigs and the control group, so CA analysis confirmed the previous partial conclusion from PCA. The following graphical outputs from the third supervised OPLS-DA method (Figure 6) scatter plots also confirmed the previous conclusions and were able to significantly differentiate the group of control pigs (marked in blue) from the group of tested pigs (descriptions marked in red). The R^2 (X), R^2 (Y) and Q^2 (Y) statistics for the OPLS-DA models were calculated. Multiple correlation R^2 and cross-validated coefficient Q^2 for control vs. treated group R^2 (X) = 0.616, R^2 (Y) = 0.987 and Q^2 (Y) = 0.898 for plasma and R^2 (X) = 0.469, R^2 (Y) = 0.997 and Q^2 (Y) = 0.879 for urine, confirmed good class separation and a high predictive ability. Coefficients Q^2 (Y) expressing the predictive abilities of the proposed model were calculated for 75% cross-validation.

The group of tested pigs, analogously as in the PCA and CA models, was further divided in the plane t1 and to1 into two clusters according to their sex. Therefore, the OPLS-DA model for urine was further designed and tested, where in matrix Y (two-column matrix with binary data; treated group = 1, control group = 2 and male group = 1, female group = 2) there was a differentiation according to gender (Figure 9). Statistics R^2 (X) = 0.417, R^2 (Y) = 0.973, and Q^2 (Y) = 0.85 also showed good separation and high prediction. The observed differentiation of groups of pigs by sex in all three models used after testosterone administration brings a whole new dimension to the whole issue of using metabolomics profiling to prove illegal use of banned substances. This genetic factor will have to be taken into account and further investigated in additional metabolomics studies in pigs.

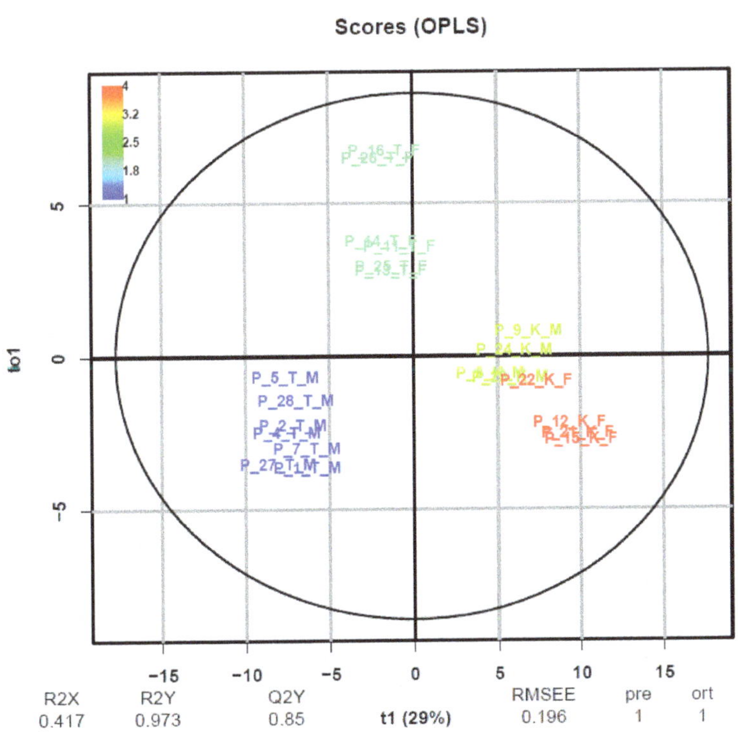

Figure 9. The OPLS-DA score plots for urine data matrix demonstrate discrimination between the control group of pigs and secondary discrimination by sex, objects (pigs) marked with: K—control group, T—treated group, M—male and F—female (PQN scaling, R package). A list of specific numbers of pigs is given in Tables 1 and 2.

All three used multivariable statistical methods, PCA, CA, and OPLS-DA, mathematically independent, were able to significantly differentiate the use of synthetic exogenous testosterone from naturally occurring (endogenous) testosterone in pigs of the same breed. This conclusion is in contrast to the findings published in the only study performed to date on ractopamine (a group of banned β-agonists) in 2017 [33]. Here, the authors of this study state that no significant difference between the samples from the control group and the ractopamine treated group was observed in the PCA analysis when all features were used. The use of metabolomics approaches and techniques also seems to be very promising from the point of view of the time of detection of banned anabolic steroids. Synthetic exogenous 17β-testosterone was demonstrably detected in plasma and urine in the treated group of pigs 14 days after administration. A similar conclusion was reached by the authors of a metabolomics study performed in cattle with β-agonists [31], when urine samples taken on days 27 and 48 after the administration could no longer be distinguished from the control group using PCA and OPLS-DA statistical models.

Supervised OPLS-DA models for urine and plasma, as one of the important practical results of this work, will be used for further testing of real samples to verify their predictive abilities. The models will be supplemented with other banned anabolic steroids and, especially, with increased numbers of test (training) data. Subsequently, the models will be verified by screening real plasma and urine samples taken as part of the monitoring of foreign substances in the Czech Republic. This is in line with the findings of other metabolomics studies [28,30–32], which also suggest increasing the number of samples used in statistical models and testing the proposed models on real samples obtained from national monitoring of contaminants in other EU countries (e.g., France, The Netherlands, Spain).

3.4. Metabolomic Profiling for Identification of Biomarkers

In the sequence of metabolomics multivariate statistical analysis, the last and often the most time-consuming step is to identify the most discriminating metabolites which are essential from the point of view of elucidating metabolomics pathways. A volcano plot with variable importance in projection plot (VIP) and S-plot from OPLS-DA were used to determine the most discriminating metabolites between the treatment group and control group (Figure 7 and Supplementary Materials Figure S9). The most discriminating compounds found were compared with the METLIN database, and their list and characteristics are shown in Table 4. Some very discriminating compounds have been identified with known human testostrone metabolites and confirmed against standards based on RT and MA criteria, such as M290T13_1 which corresponds to 5α-dihydrotestosterone. Nevertheless, further work needs to be done to identify the compounds and gain detailed explanation of their chemical structure.

4. Materials and Methods

4.1. Animal Experiment and Urine/Plasma Sampling

The animal experiments were performed at the Veterinary Research Institute in Brno, Czech Republic. Twenty clinically healthy 90-day-old male and female pigs (approximately 28 kg body weight) were randomly assigned to test (13 animals) and control (8 animals) groups. Animals from the test group were treated with an i.m. injection (0.6 mL/pig) of the hormonal preparation (30 mg mL^{-1} 17β-testosterone propionate, 60 mg mL^{-1} 17β-testosterone phenylpropionate, 60 mg mL^{-1} 17β-testosterone isocaproate, 100 mg mL^{-1} 17β-testosterone decanoate; Sustanon 250, N.V. Organon, CZ Reg.56/357/91-C). Experimental animals were grower pigs (hybrids of Large White × Landrace (sow) × Duroc (boar)) which were fed twice a day with a standard commercial diet according to the weight category. Pigs were housed in two separate pens (one pen/treatment and one pen/control) of 2.80 × 2.00 m.

The animals were injected on day 12 after the start of the experiment and were euthanized on day 90 of the experiment. Urine samples were collected from both groups 14, 28, 42 and 90 days after treatment and all samples were kept frozen until analysis at −20 °C. Plasma samples were collected from day 1, 2, 3, 4, 5, 7, 14, and 28 after treatment to day 90. After blood clotting and 10 min of centrifugation at 6000× *g* of the samples, serum was removed and kept frozen until analysis at −20 °C. All pigs were weighted every week within the experiment. All pigs were slaughtered at a body weight of 90–110 kg and the treated animal carcasses were destroyed. The study was performed in compliance with Act No. 246/1992 Coll. of the Czech National Council for the protection of animals against cruelty and with the agreement of the Branch Commission for Animal Welfare of the Ministry of Agriculture of the Czech Republic (permission no. MZe 17214).

4.2. *Reagents and Materials*

Reference analyte standards (17β-testosterone, 17β-testosterone-D2 as isotopically labeled internal standards) were purchased from Sigma-Aldrich, Prague, Czech Republic. The standards were dissolved in methanol, diluted to a low concentration (mg mL^{-1}) and used as working solutions. The organic solvents used were obtained from Merck (Darmstadt, Germany) and were in the SupraSolv® class. Used water prepared in an ultrapure water system of Golgman's water (Prague, Czech Republic). Centrifugal membrane filters Vivacon 500, cut off at 10 kDa, were obtained from Sartorius, Prague, Czech Republic.

4.3. *Sample Preparation*

4.3.1. Plasma Samples for Targeted Analysis

Each defrosted plasma sample (500 μL) was transferred to a 15 mL centrifuge tube. After adding a methanol solution of 50 μL of 17β-testosterone-D2 internal standard and 5 mL of ethyl acetate, the samples were shaken vigorously for 3 min on a vortex and then centrifuged at 4000× *g* for 10 min. The supernatant (ca. 4 mL) was subsequently transferred to evaporator tubes. The samples were evaporated to dryness at 25 °C using a gentle stream of nitrogen (99.99% N_2). The samples were reconstituted with 200 μL in mobile phase solution (methanol: water, 70:30, *v/v*) and filtered through a 0.45 μm (Hydrophilic PTFE) membrane centrifuge filter. The resulting samples were transferred to the insert in chromatography vials (250 μL). Then, 10 μL samples were injected directly the LC-MS/MS system.

4.3.2. Samples for Metabolomics Profiling

Plasma samples were defrosted at room temperature, homogenized and normalized by the creatinine concentration (Chemistry Analyzer BS-200, MINDRAY, Nanshen, China). Subsequently, 200 μL of each sample was filtered through centrifugal devices (Vivacon 500, cut-off at 10 kDa, 14,000× *g*, 4 °C, 30 min) to remove high molecular weight proteins. Filtrates (120 μL) were mixed with 30 μL of internal standard (testosterone-D_2 in methanol at the concentration of 1 ng mL^{-1}). After thorough shaking, 5 μL of each sample was injected into the chromatographic system.

Urine samples were defrosted at room temperature and normalized by specific density adjustment [47]. The specific density of the samples was measured using a refractometer (digital refractometer 30GS, Mettler-Toledo, Prague, Czech Republic) and, if necessary, adjusted to approximately 1.010–1.030 kg m^{-3} by dilution with deionized water. The urine samples (500 μL) were then centrifuged through a centrifugal membrane filter (Vivacon 500, cut-off 10 kDa) at 14,000× *g*, 4 °C for 30 min to remove high molecular weight proteins. To the sample filtrate (ca. 450 μL), 50 μL of 17β-testosterone-D2 internal standard was added. The urine samples were transferred to chromatographic vials and subsequently analysed.

QC samples for quality control of metabolomics profiling were prepared by the pool method so that they have the same or very similar (bio)-chemical varieties in the same range as individual samples of the study. The QC sample was prepared as a pool of all individual plasma or urine samples that were included in the study. Each QC sample was generated by mixing 20 μL of the filtrate (cut-off) of each individual sample and was analysed at the same time as the study samples as part of the overall measurement sequence.

4.4. *Targeted Quantitative Analytical Method*

4.4.1. LC Condition

Plasma samples were injected directly into a Thermo Fisher Scientific, (Waltham, MA, USA) LC system Accela 1200 equipped with an autosampler with a temperature controlled tray and column. Chromatographic separation was performed on Waters C18 XTerra MS analytical columns (150 × 2.1 mm, size 3.5 μm) with a Waters C18 XTerra MS guard column (10 × 2.1 mm, size 3.5 μm). The column and autosampler tray temperatures were set at 35 °C. The mobile phase consists of 0.1% formic acid in water: methanol (95:5, *v*/*v*) A and 0.1% formic acid in water: methanol (5:95, *v*/*v*) B, the flow rate was constant 300 μL min^{-1}. Gradient elution of 0–2 min with mobile phase 95% A and 5% B was started, 2.1–20 min linear gradient from 5% to 90% B, 20.1–25 min 10% A and 90% B, 25.1–30 min linear gradient from 5% to 95% A and 30.1–35 min 95% A and 5% B. The runtime of the method was 35 min.

4.4.2. MS/MS Parameters

The tandem hybrid mass spectrometer Q Exactive (Thermo Fisher Scientific, (Massachusetts, USA) equipped with a heated electrospray ionisation probe measured in a positive mode (H-ESI+). For targeted quantification analysis, the mass spectrometer worked in the parallel reaction monitoring mode PRM (corresponding to the selected reaction monitoring mode SRM) with high resolution RP = 17,500 (FWHM) at 200 *m*/*z*. Before the start of each acquisition series, the mass spectrometer was externally calibrated to the mass accuracy with the positive ion calibration solution and the negative ion calibration solution (both Thermo Fisher Scientific). The "lock-mass" calibration was set to the molecular mass of $[M+Na]^+$ = 64.01577 g mol^{-1} and $[M_2+H]^+$ = 83.06037 g mol^{-1} for the acetonitrile ion, and was run continuously during the acquisition. Instrument and collision cell (HCD) parameters were optimised by direct syringe infusion of working solutions of 50 ng mL^{-1} of each targeted compound with a 5 μL min^{-1} flow-rate. The mass spectrometer setting was as follows: sheath gas flow rate 30 (unit), aux gas flow rate 5 (unit), spray voltage 4.0 kV, capillary temperature 320 °C, heater temperature 220 °C, S-lens RF level 50, AGC target of 5×10^6, collision energy 35 eV, and a maximum injection time of 100 ms.

The precursor ions and the four most intense product ions for each analyte were measured for quantification and identification (confirmation), respectively. The whole LC-(HR) MS system was controlled and the acquired data were stored and processed using Xcalibur 3.1 software, and then evaluated using Mass Frontier v. 7.0 for the identification.

4.4.3. Method Validation

The targeted quantitative method for the determination of 17β-testosterone and its esters in plasma has been validated to the extent required by European Directive 657/2002/EC [34] used for the determination of residues of foreign substances in biological matrices and according to the VICH GL49 [35] reference guide for validation method. To determine the validation characteristics of the matrix calibration curve, critical values detection limit (LOD), limit of quantification (LOQ), decision limit (CC_α), detection capability (CC_β) and calibration range 0.5–80 ng mL^{-1} was used for 17β-testosterone and its esters. Samples of pig plasma (blank) were supplemented with the standards of 17β-testosterone and its esters at concentration corresponding to 0.5, 1, 5, 10, 20, 40, and 80 ng mL^{-1}. The concentrations of the internal standards were constantly 10 ng mL^{-1}. For each concentration level, two model samples were prepared and each sample was measured two times. Linear regression was carried out by plotting the peak area ratios of the analyte against the internal standard (dependent variable Y) versus the analyte concentration (independent variable X). To evaluate the precision of the method, repeatability and within-laboratory reproducibility, standard deviation (SD), and the coefficient of variation (CV, %) were determined. Six model samples of pig plasma (n = 6) were prepared at concentrations of 5 and 80 ng mL^{-1} for all standards and measurements which were repeated two times for each sample on three different days (3 × 6, $n = 18$).

4.5. Metabolomic Profiling

4.5.1. LC Separation

Plasma and urine samples prepared for non-targeted analysis (fingerprinting) were injected directly into the Accela 1200 LC system with a mass spectrometer. Chromatographic separation was performed on a C18 Hypersil GOLD (50 × 2.1 mm, 1.9 μm size) analytical column equipped with a C18 Hypersil GOLD (10 × 2.1 mm, 1.9 μm size) guard column, both from Thermo Fisher Scientific. The temperatures of the column and the autosampler tray were set at 35 °C and 20 °C, respectively. The flow rate through the column was constantly set at 200 μL min^{-1}. The injection volume of the sample was 5 μL. The mobile phase consists of 0.1% formic acid in water: methanol (95:5, *v*/*v*) A and 0.1% formic acid in water: methanol (5:95, *v*/*v*) B, the flow rate was constant 300 $\mu Lmin^{-1}$. Gradient elution of 0–2 min with mobile phase 95% A and 5% B was started, 2.1–20 min linear gradient from 5% to 90% B, 20.1–25 min 10% A and 90% B, 25.1–30 min linear gradient from 5% to 95% A and 30.1–35 min 95% A and 5% B. The runtime of the method was 35 min.

4.5.2. Non-Targeted Mass Spectrometry

For non-targeted metabolomic analysis, the tandem hybrid mass spectrometer Q Exactive worked in the positive full scan mode with resolving power (PR) = 70,000 (FWHM) at 200 *m*/*z* in the range 50 to 750 *m*/*z* and in the centroid mode. Before the start of each acquisition series, the mass spectrometer was externally calibrated to the mass accuracy with a positive ion calibration solution and a negative ion calibration solution (both Thermo Fisher Scientific (Massachusetts, USA). The "lock-mass" calibration was set to the molecular mass of $[M+Na]^+$ = 64.01577 g mol^{-1} and $[M_2+H]^+$ = 83.06037 g mol^{-1} for the acetonitrile ion, and was run continuously during the acquisition. The mass spectrometer setting was as follows: sheath gas flow rate 30 (unit), aux gas flow rate 5 (unit), spray voltage 4.5 kV, capillary temperature 320 °C, heater temperature 180 °C, S-lens RF level 50, AGC target of 6×10^6 and a maximum injection time of 200 ms. The whole LC-(HR)MS system was controlled and the obtained data were stored and processed using Xcalibur 3.1 software.

4.6. Data Processing

The generated metabolomics profiling data sets were processed by the control software of the Xcalibur® mass spectrometer and saved in a specific data format (*.raw). The first step was to convert data from Excalibur-specific raw files to open format files (*.mzXML) using MS Convertor software (ProteoWizard) [48]. Subsequently, metabolomics data were processed using the XCMS Online web version platform [49]. All results and images for processing were downloaded as zip files for offline analysis, including putative METLIN identities for each metabolite.

4.7. Statistical Data Analysis

Univariate and multivariate statistical analysis was performed in an interactive manner using statistical software Statistica (Version 13.3, TIBCO Software, Palo Alto, CA, USA) and R-statistic software in the Metabol package [50]. The data for processing was exported in the form of data matrix X (n × m) from an Excel file (output from data processing) to individual statistical programs. Before the application of multivariate statistical analysis, exploratory data analysis was conducted [51]. This included the assessment of primarily found outlier objects (or features thereof), assuming linear relationship and verifying the data provided (normality, non-correlation, homogeneity). Subsequently, the data matrices were standardized by two different methods. In the first case, the data were transformed by mean centering [37], and in the second case, they were transformed by probabilistic quotient normalization PQN [38,52] and a natural logarithm was applied for their scaling before subsequent statistical analysis. Multidimensional statistical methods, such as principal component analysis (PCA), cluster analysis (CA), and orthogonal partial least squares discriminant analysis (OPLS-DA), were used to statistically evaluate data obtained from non-targeted metabolomics analyses.

5. Conclusions

The present study confirmed the ability of metabolomics approaches and techniques to significantly differentiate pigs administered an androgenic anabolic steroid which is on the list of banned substances (17β-testosterone) in EU countries from control pigs. Using metabolomics profiling for plasma and urine samples, it was possible to differentiate the used synthetic exogenous testosterone from naturally occurring (endogenous) testosterone based on the results from three statistical models PCA, CA and OPLS-DA. The metabolomics workflow was designed with widely used multivariate statistical methods, analytical techniques, and equipment so that, based on the results of our study, it will be possible to develop validated methodologies for routine screening to prove the illegal use of prohibited substances in pig fattening. Furthermore, the anabolic effect of testosterone in pigs was demonstrated by comparing BW gains during the fattening period, and the targeted analysis of plasma testosterone levels provided data for the pharmacokinetic curves.

Supplementary Materials:

Figure S1: Graph of average weekly weight gains in kg, Figure S2: The measured experimental MS1 of the 17β-testosterone standards (up), comparison with the theoretical MS1 (down), Figure S3: The measured experimental MS1 of the 17β-testosterone-D2 standards (up), comparison with the theoretical MS1 (down), Figure S4: Calibration curve of 17β-testosterone in plasma, Figure S5: Hotelling plot for identification of outliers objects in data source matrix X with calculated critical value T2 for plasma, (K—control grup vs. T—treated grup), Figure S6: Hotelling plot for identification of outliers objects in data source matrix X with calculated critical value T2 for urine, Figure S7: The PLS-DA score plots for plasma (A) and urine (B) data matrix demonstrates robust discrimination between the control group of pigs marked with blue color and the group of teated pigs marked with red color (Centering, Statistica), Figure S8: The OPLS-DA permutation tests further confirmed that the proposed statistical models are correct and robust; A—plasma, B—urine, Figure S9: Variable importance in projection (VIP) and S-plots from OPLS-DA were used to determine the most discriminating metabolites between treatments and controls, A—plasma and B—urine, Table S1: The average weekly body weight gains, Table S2: The regression parameters of both linear models of BW growth versus time, Table S3: Identification and confirmation of 17β-testosterone and testosterone esters by mass accuracy (MA) for MS1, MS2 data, Table S4: Linearity of 17β-testosterone in plasma, Table S5: Precision, repeatability and within-laboratory reproducibility, Table S6: Pharmacokinetic of testosterone in plasma.

Author Contributions: Conceptualization, K.S., K.P., L.L., M.F. (Milan Franek), P.D. and M.F. (Martin Faldyna); methodology, K.S., K.P., L.L., M.F. (Milan Franek), P.D. and M.F. (Martin Faldyna); validation, K.S., K.P., L.L., M.F. (Milan Franek), P.D. and M.F. (Martin Faldyna); formal Analysis, K.S., K.P., L.L.; investigation, K.S., K.P., L.L., M.F. (Milan Franek), P.D. and M.F. (Martin Faldyna); resources, K.S., K.P., M.F. (Milan Franek), P.D. and M.F. (Martin Faldyna); writing—original draft preparation, K.S., K.P. and M.F. (Milan Franek); All authors have read and agreed to the published version of the manuscript.

References

1. European Food Safety Authority. Report for 2018 on the results from the monitoring of veterinary medicinal product residues and other substances in live animals and animal products. *EFSA Support. Publ.* **2020**, *17*, 1775E.
2. Rosner, W. Plasma steroid-binding proteins. *Endocrinol. Metab. Clin. N. Am.* **1991**, *20*, 697–720. [CrossRef]
3. Westphal, U. Steroid-Protein Interactions Revisited. In *Steroid-Protein Interactions II*; Springer: Berlin/Heidelberg, Germany, 1986; pp. 1–7.
4. Peng, S.H.; Segura, J.; Farré, M.; González, J.C.; De La Torre, X. Plasma and urinary markers of oral testosterone undecanoate misuse. *Steroids* **2002**, *67*, 39–50. [CrossRef]

5. Nielen, M.W.F.; Nijrolder, A.W.J.M.; Hooijerink, H.; Stolker, A.A.M. Feasibility of desorption electrospray ionization mass spectrometry for rapid screening of anabolic steroid esters in hair. *Anal. Chim. Acta* **2011**, *700*, 63–69. [CrossRef]
6. Van De Kerkhof, D.H.; De Boer, D.; Thijssen, J.H.H.; Maes, R.A.A. Evaluation of testosterone/epitestosterone ratio influential factors as determined in doping analysis. *J. Anal. Toxicol.* **2000**, *24*, 102–115. [CrossRef]
7. Wada Laboratory Expert Group. *WADA Technical Document TD2014EAAS—Endogenous Anabolic Androgenic Steroids Measurement and Reporting*; World Anti-Doping Agency: Montreal, QC, Canada, 2014; pp. 1–8.
8. Le Bizec, B.; Monteau, F.; Gaudin, I.; André, F. Evidence for the presence of endogenous 19-norandrosterone in human urine. *J. Chromatogr. B Biomed. Sci. Appl.* **1999**, *723*, 157–172. [CrossRef]
9. Pinel, G.; Weigel, S.; Antignac, J.P.; Mooney, M.H.; Elliott, C.; Nielen, M.W.F.; Le Bizec, B. Targeted and untargeted profiling of biological fluids to screen for anabolic practices in cattle. *Trends Anal. Chem.* **2010**, *29*, 1269–1280. [CrossRef]
10. Community Reference Laboratories. *CRL Guidance Paper (7 December 2007)—CRL's View on State of the Art Analytical Methods for National Residue Control Plans*; RIVM—National Institute for Public Health and the Environment: Bilthoven, The Netherlands, 2007; pp. 1–8.
11. Boschi, S.; De Iasio, R.; Mesini, P.; Bolelli, G.F.; Sciajno, R.; Pasquali, R.; Capelli, M. Measurement of steroid hormones in plasma by isocratic high performance liquid chromatography coupled to radioimmunoassay. *Clin. Chim. Acta* **1994**, *231*, 107–113. [CrossRef]
12. Di Benedetto, L.T.; Dimitrakopoulos, T.; Davy, R.M.; Iles, P.J. Testosterone determination using rapid heterogeneous competitive-binding for enzyme-linked immunosorbent assay in flow injection. *Anal. Lett.* **1996**, *29*, 2125–2139. [CrossRef]
13. Zhang, Z.; Duan, H.; Zhang, L.; Chen, X.; Liu, W.; Chen, G. Direct determination of anabolic steroids in pig urine by a new SPME-GC-MS method. *Talanta* **2009**, *78*, 1083–1089. [CrossRef]
14. De la Torre, X.; Segura, J.; Polettini, A.; Montagna, M. Detection of testosterone esters in human plasma. *J. Mass Spectrom.* **1995**, *30*, 1393–1404. [CrossRef]
15. He, C.; Li, S.; Liu, H.; Li, K.; Liu, F. Extraction of testosterone and epitestosterone in human urine using aqueous two-phase systems of ionic liquid and salt. *J. Chromatogr. A* **2005**, *1082*, 143–149. [CrossRef] [PubMed]
16. Shackleton, C.H.L.; Chuang, H.; Kim, J.; De La Torre, X.; Segura, J. Electrospray mass spectrometry of testosterone esters: Potential for use in doping control. *Steroids* **1997**, *62*, 523–529. [CrossRef]
17. Konieczna, L.; Plenis, A.; Oldzka, I.; Kowalski, P.; Bczek, T. Optimization of LC method for the determination of testosterone and epitestosterone in urine samples in view of biomedical studies and anti-doping research studies. *Talanta* **2011**, *83*, 804–814. [CrossRef]
18. Wang, G.; Hsieh, Y.; Cui, X.; Cheng, K.-C.; Korfmacher, W.A. Ultra-performance liquid chromatography/tandem mass spectrometric determination of testosterone and its metabolites in vitro samples. *Rapid Commun. Mass Spectrom.* **2006**, *20*, 2215–2221. [CrossRef]
19. You, Y.; Uboh, C.E.; Soma, L.R.; Guan, F.; Li, X.; Liu, Y.; Rudy, J.A.; Chen, J.; Tsang, D. Simultaneous separation and determination of 16 testosterone and nandrolone esters in equine plasma using ultra high performance liquid chromatography-tandem mass spectrometry for doping control. *J. Chromatogr. A* **2011**, *1218*, 3982–3993. [CrossRef]
20. Ponzetto, F.; Boccard, J.; Baume, N.; Kuuranne, T.; Rudaz, S.; Saugy, M.; Nicoli, R. High-resolution mass spectrometry as an alternative detection method to tandem mass spectrometry for the analysis of endogenous steroids in serum. *J. Chromatogr. B Anal. Technol. Biomed. Life Sci.* **2017**, *1052*, 34–42. [CrossRef]
21. Aqai, P.; Stolker, A.A.M.; Lasaroms, J.J.P. Effect of sample pre-treatment on the determination of steroid esters in hair of bovine calves. *J. Chromatogr. A* **2009**, *1216*, 8233–8239. [CrossRef]
22. Matraszek-Żuchowska, I.; Woźniak, B.; Sielska, K.; Posyniak, A. Determination of steroid esters in hair of slaughter animals by liquid chromatography with tandem mass spectrometry. *J. Vet. Res.* **2019**, *63*, 561–572. [CrossRef]
23. Dervilly-Pinel, G.; Courant, F.; Chéreau, S.; Royer, A.L.; Boyard-Kieken, F.; Antignac, J.P.; Monteau, F.; Le Bizec, B. Metabolomics in food analysis: Application to the control of forbidden substances. *Drug Test. Anal.* **2012**, *4*, 59–69. [CrossRef]

24. Rijk, J.C.W.; Lommen, A.; Essers, M.L.; Groot, M.J.; Van Hende, J.M.; Doeswijk, T.G.; Nielen, M.W.F. Metabolomics approach to anabolic steroid urine profiling of bovines treated with prohormones. *Anal. Chem.* **2009**, *81*, 6879–6888. [CrossRef] [PubMed]
25. Kieken, F.; Pinel, G.; Antignac, J.P.; Monteau, F.; Paris, A.C.; Popot, M.A.; Bonnaire, Y.; Le Bizec, B. Development of a metabonomic approach based on LC-ESI-HRMS measurements for profiling of metabolic changes induced by recombinant equine growth hormone in horse urine. *Anal. Bioanal. Chem.* **2009**, *394*, 2119–2128. [CrossRef] [PubMed]
26. Anizan, S.; Bichon, E.; Duval, T.; Monteau, F.; Cesbron, N.; Antignac, J.-P.; Le Bizec, B. Gas chromatography coupled to mass spectrometry-based metabolomic to screen for anabolic practices in cattle: Identification of 5α-androst-2-en-17-one as new biomarker of 4-androstenedione misuse. *J. Mass Spectrom.* **2012**, *47*, 131–140. [CrossRef] [PubMed]
27. Pinel, G.; Rambaud, L.; Monteau, F.; Elliot, C.; Le Bizec, B. Estranediols profiling in calves' urine after 17β-nandrolone laureate ester administration. *J. Steroid Biochem. Mol. Biol.* **2010**, *121*, 626–632. [CrossRef]
28. Dervilly-Pinel, G.; Weigel, S.; Lommen, A.; Chereau, S.; Rambaud, L.; Essers, M.; Antignac, J.P.; Nielen, M.W.F.; Le Bizec, B. Assessment of two complementary liquid chromatography coupled to high resolution mass spectrometry metabolomics strategies for the screening of anabolic steroid treatment in calves. *Anal. Chim. Acta* **2011**, *700*, 144–154. [CrossRef]
29. Regal, P.; Anizan, S.; Antignac, J.P.; Le Bizec, B.; Cepeda, A.; Fente, C. Metabolomic approach based on liquid chromatography coupled to high resolution mass spectrometry to screen for the illegal use of estradiol and progesterone in cattle. *Anal. Chim. Acta* **2011**, *700*, 16–25. [CrossRef]
30. Blokland, M.H.; Van Tricht, E.F.; Van Rossum, H.J.; Sterk, S.S.; Nielen, M.W.F. Endogenous steroid profiling by gas chromatography-tandem mass spectrometry and multivariate statistics for the detection of natural hormone abuse in cattle. *Food Addit. Contam. Part A Chem. Anal. Control Expo. Risk Assess.* **2012**, *29*, 1030–1045. [CrossRef]
31. Dervilly-Pinel, G.; Chereau, S.; Cesbron, N.; Monteau, F.; Le Bizec, B. LC-HRMS based metabolomics screening model to detect various β-agonists treatments in bovines. *Metabolomics* **2015**, *11*, 403–411. [CrossRef]
32. Blokland, M.H.; van Tricht, E.F.; van Ginkel, L.A.; Sterk, S.S. Applicability of an innovative steroid-profiling method to determine synthetic growth promoter abuse in cattle. *J. Steroid Biochem. Mol. Biol.* **2017**, *174*, 265–275. [CrossRef]
33. Peng, T.; Royer, A.L.; Guitton, Y.; Le Bizec, B.; Dervilly-Pinel, G. Serum-based metabolomics characterization of pigs treated with ractopamine. *Metabolomics* **2017**, *13*, 77. [CrossRef]
34. European Commission. 2002/657/EC: Commission Decision of 12 August 2002 Implementing Council Directive 96/23EC Concerning the Performance of Analytical Methods and Interpretation of Results. *Off. J. Eur. Communities* **2002**, *45*, 8–36.
35. European Medicines Agency. *VICH GL49 Studies to Evaluate the Metabolism and Residue Kinetics of Veterinary Drugsin Food-Producing Animals: Validation of Analytical Methods Used in Residue Depletion Studies*; European Medicines Agency: Amsterdam, The Netherlands, 2015.
36. International Organization for Standardization. *ISO 11843-1:1997. Capability of Detection—Part 1: Terms and Definitions*; ISO: Geneva, Switzerland, 1997.
37. Worley, B.; Powers, R. Multivariate Analysis in Metabolomics. *Curr. Metab.* **2013**, *1*, 92–107.
38. Filzmoser, P.; Walczak, B. What can go wrong at the data normalization step for identification of biomarkers? *J. Chromatogr. A* **2014**, *1362*, 194–205. [CrossRef] [PubMed]
39. Chow, G.C. Tests of Equality Between Sets of Coefficients in Two Linear Regressions. *Econometrica* **1960**, *28*, 591–605. [CrossRef]
40. Groot, M.J.; Lasaroms, J.J.P.; van Bennekom, E.O.; Meijer, T.; Vinyeta, E.; van der Klis, J.D.; Nielen, M.W.F. Illegal treatment of barrows with nandrolone ester: Effect on growth, histology and residue levels in urine and hair. *Food Addit. Contam. Part A Chem. Anal. Control Expo. Risk Assess.* **2012**, *29*, 727–735. [CrossRef] [PubMed]
41. Holčapek, M.; Jirásko, R.; Lísa, M. Recent developments in liquid chromatography–mass spectrometry and related techniques. *J. Chromatogr. A* **2012**, *1259*, 3–15. [CrossRef]
42. Rochat, B.; Kottelat, E.; McMullen, J. The future key role of LC–high-resolution-MS analyses in clinical laboratories: A focus on quantification. *Bioanalysis* **2012**, *4*, 2939–2958. [CrossRef]

43. Stastny, K.; Stepanova, H.; Hlavova, K.; Faldyna, M. Identification and determination of deoxynivalenol (DON) and deepoxy-deoxynivalenol (DOM-1) in pig colostrum and serum using liquid chromatography in combination with high resolution mass spectrometry (LC-MS/MS (HR)). *J. Chromatogr. B Anal. Technol. Biomed. Life Sci.* **2019**, *1126–1127*. [CrossRef]
44. McCoard, S.; Wise, T.; Ford, J. Endocrine and molecular influences on testicular development in Meishan and White Composite boars. *J. Endocrinol.* **2003**, *178*, 405–416. [CrossRef]
45. Park, C.S.; Yi, Y.J. Comparison of semen characteristics, sperm freezability and testosterone concentration between Duroc and Yorkshire boars during seasons. *Anim. Reprod. Sci.* **2002**, *73*, 53–61. [CrossRef]
46. Rejtharová, M.; Rejthar, L.; Čačková, K. Determination of testosterone esters and nortestosterone esters in animal blood serum by LC-MS/MS. *Food Addit. Contam. Part A Chem. Anal. Control Expo. Risk Assess.* **2018**, *35*, 233–240. [CrossRef]
47. Jacob, C.C.; Dervilly-Pinel, G.; Biancotto, G.; Le Bizec, B. Evaluation of specific gravity as normalization strategy for cattle urinary metabolome analysis. *Metabolomics* **2014**, *10*, 627–637. [CrossRef]
48. Adusumilli, R.; Mallick, P. Data conversion with proteoWizard msConvert. In *Methods in Molecular Biology*; Humana Press Inc.: Totowa, NJ, USA, 2017; Volume 1550, pp. 339–368.
49. Smith, C.A.; Want, E.J.; O'Maille, G.; Abagyan, R.; Siuzdak, G. XCMS: Processing mass spectrometry data for metabolite profiling using nonlinear peak alignment, matching, and identification. *Anal. Chem.* **2006**, *78*, 779–787. [CrossRef] [PubMed]
50. Gardlo, A.; Friedecký, D.; Najdekr, L.; Karlíková, R.; Adam, T. *Metabol: The Statistical Analysis of Metabolomic Data*; Laboratory for Inherited Metabolic Disorders, Faculty of Medicine and Dentistry, Palacky University in Olomouc, University Hospital Olomouc: Olomouc, Czech Republic, 2019.
51. Gelman, A. Exploratory data analysis for complex models. *J. Comput. Graph. Stat.* **2004**, *13*, 755–779. [CrossRef]
52. Cook, T.; Ma, Y.; Gamagedara, S. Evaluation of statistical techniques to normalize mass spectrometry-based urinary metabolomics data. *J. Pharm. Biomed. Anal.* **2020**, *177*, 112854. [CrossRef] [PubMed]

10

Characterization of Lipid Profiles after Dietary Intake of Polyunsaturated Fatty Acids using Integrated Untargeted and Targeted Lipidomics

Satoko Naoe [1], Hiroshi Tsugawa [1,2,*], Mikiko Takahashi [2], Kazutaka Ikeda [1,3] and Makoto Arita [1,3,4,*]

1. Laboratory for Metabolomics, RIKEN Center for Integrative Medical Sciences, Yokohama 230-0045, Japan; satoko.naoe@mochida.co.jp (S.N.); kazutaka.ikeda@riken.jp (K.I.)
2. Metabolome informatics research team, RIKEN Center for Sustainable Resource Science, Yokohama 230-0045, Japan; mikiko.takahashi@riken.jp
3. Cellular and Molecular Epigenetics Laboratory, Graduate School of Medical Life Science, Yokohama City University, Tsurumi, Yokohama 230-0045, Japan
4. Division of Physiological Chemistry and Metabolism, Graduate School of Pharmaceutical Sciences, Keio University, Minato-ku, Tokyo 105-8512, Japan

* Correspondence: hiroshi.tsugawa@riken.jp (H.T.); makoto.arita@riken.jp (M.A.).

Abstract: Illuminating the comprehensive lipid profiles after dietary supplementation of polyunsaturated fatty acids (PUFAs) is crucial to revealing the tissue distribution of PUFAs in living organisms, as well as to providing novel insights into lipid metabolism. Here, we performed lipidomic analyses on mouse plasma and nine tissues, including the liver, kidney, brain, white adipose, heart, lung, small intestine, skeletal muscle, and spleen, with the dietary intake conditions of arachidonic acid (ARA), eicosapentaenoic acid (EPA), and docosahexaenoic acid (DHA) as the ethyl ester form. We incorporated targeted and untargeted approaches for profiling oxylipins and complex lipids such as glycerol (phospho) lipids, sphingolipids, and sterols, respectively, which led to the characterization of 1026 lipid molecules from the mouse tissues. The lipidomic analysis indicated that the intake of PUFAs strongly impacted the lipid profiles of metabolic organs such as the liver and kidney, while causing less impact on the brain. Moreover, we revealed a unique lipid modulation in most tissues, where phospholipids containing linoleic acid were significantly decreased in mice on the ARA-supplemented diet, and bis(monoacylglycero)phosphate (BMP) selectively incorporated DHA over ARA and EPA. We comprehensively studied the lipid profiles after dietary intake of PUFAs, which gives insight into lipid metabolism and nutrition research on PUFA supplementation.

Keywords: arachidonic acid; omega-3 fatty acids; lipidomics; mass spectrometry; dietary fat; fatty acid metabolism

---

1. Introduction

Polyunsaturated fatty acids (PUFAs) are essential nutrients that have a range of biological effects such as on brain function, cardiovascular disease, obesity, cancer, and bone health in humans [1,2], and also affect the reproduction quality of livestock animals [3]. Among PUFAs, arachidonic acid (ARA), eicosapentaenoic acid (EPA), and docosahexaenoic acid (DHA) are known as long chain PUFAs, and their physicochemical properties and metabolisms provide various functions in mammalian cells. ARA and EPA/DHA are also known as $\omega6$ and $\omega3$ polyunsaturated fatty acids, respectively, and their biological functions and the importance of $\omega3/\omega6$ fatty acid balance in maintaining homeostasis have attracted increasing attentions [1,2]. Mammals cannot synthesize ARA, EPA, and DHA from their precursors in sufficient quantities, i.e., linoleic acid (LA) and alpha-linoleic acid (ALA), and they must rely on dietary intake. Therefore, the dietary intake of ARA, EPA, and DHA is essential to maintain human health, and it is well known that livestock animals and marine foods are the primary nutrient sources for ARA and EPA/DHA, respectively.

The lipid profiles in tissues are affected by different balances of dietary PUFAs [4–9]. In fact, the intake of ω3 PUFAs increases tissue levels of ω3 PUFAs and their oxidized forms, namely oxylipins, while the amount of ω6 PUFAs and their oxidized forms is decreased [4,5]. Moreover, a previous study revealed that dietary intake of ω3 PUFA inhibits ARA biosynthesis from LA by the suppression of fatty acid desaturase 2 (FADS2) expression [6], and competes with ARA for membrane phospholipid remodeling. In addition, several ω3 PUFA intervention studies have shown that the supplementation of ω3 PUFA increases ω3-PUFA-derived oxylipins while decreasing ARA-derived oxylipins in human peripheral blood [7]. Although oxylipin profiles under different dietary conditions have been reported [4,5,8,9], there have been few studies that comprehensively investigated the lipid profile after the dietary intake of ARA, EPA, or DHA side-by-side in various tissues, which would be necessary to grasp the effects of different PUFA-containing diets.

After dietary intake, PUFAs are actively incorporated into cells as acyl chains of membrane phospholipids and other lipid classes, such as triacylglycerol and cholesteryl esters. Intake of ARA, EPA, and DHA affects several physiological functions, such as membrane scaffold formation to harmonize biomolecule interactions [10], energy storage and related functions to maintain metabolism [11], and signal transduction via bioactive lipid mediators produced by cyclooxygenases (COX), lipoxygenases (LOX), and cytochrome P450 (CYP) [12]. Since these functions are accomplished by a molecular diversity of lipid species, the comprehensive profiling of lipids in cells and tissues is important to understanding the mechanisms underlying the effect of different PUFA supplementations.

Liquid chromatography coupled with high resolution tandem mass spectrometry (LC-HR-MS/MS) and triple quadruple mass spectrometry (LC-QqQ/MS) are popular techniques for untargeted and targeted lipidomic analysis, respectively [13,14]. LC-QqQ/MS-based targeted lipidomics has high sensitivity, and has often been used for the profiling of oxylipins, where the concentrations in plasma range from 10 pM to 100 nM [14]. In contrast, LC-HR-MS/MS, which has high resolution and scanning speed, has often been used for the profiling of glycerol (phospho) lipids, sphingolipids, and sterols, where concentrations in plasma range from 10 nM to 100 μM [14]. The current informatics technique for untargeted lipidomics enabled us to characterize more than 90 lipid classes by untangling the MS/MS spectrum [15]. Mass-spectrometry-based lipidomics, therefore, has the potential to comprehensively examine tissue lipid profiles in detail, and this lipidomic profiling could give us new insights into dietary PUFA distribution and metabolism in the body.

In this study, we investigated the lipid profiles in plasma and tissues including the liver, kidney, white adipose, skeletal muscle, heart, small intestine, lung, brain, and spleen after dietary intake of the ethyl ester form of ARA (ARA-E), EPA (EPA-E) or DHA (DHA-E) in mice. In total, 1026 molecular species of lipid were identified, of which 915 and 111 were annotated in untargeted and targeted analyses, respectively.

2. Materials and Methods

2.1. Standard Chemicals

ARA-d8, 15-HETE-d8, LTB_4-d4 and PGE_2-d4 were obtained from Cayman Chemical (Ann Arbor, MI, USA). Highly purified EPA ethyl ester (EPA-E) (> 98%), DHA ethyl ester (DHA-E) (> 97%), and ethyl arachidonate (ARA-E) (> 99%) were obtained from Nippon Suisan Kaisha, Ltd. (Tokyo, Japan), Harima foods, Inc. (Osaka, Japan) and NuChek Prep, Inc. (Elysian, MN, USA), respectively. All solvents of LC/MS grade were obtained from Wako (Tokyo, Japan).

2.2. Animals

Male C57BL/6J mice (Japan SLC, Inc., Shizuoka, Japan) were purchased at 10 weeks of age, and housed under controlled temperature and lighting (12 h light/dark cycle) with free access to water and a controlled diet (fish-meal-free F1, Funabashi Farm, Chiba, Japan). The fatty acid composition and other nutritional factors for fish-meal-free F1 are described in Supplementary Table S1. In this study, the fish-meal-free F1 was used as an ARA-, EPA-, and DHA-free diet, and each PUFA was supplemented into the feed in its ethyl ester form. After 1 week of acclimation, mice were assigned to four groups (n = 5) and fed control diet (control group), control diet supplemented with 1% EPA-E (*w/w*) (EPA-fed group), 1% DHA-E (*w/w*) (DHA-fed group), or 1% ARA-E (*w/w*) (ARA-fed group) for 2 weeks. Animals were dissected under isoflurane anesthesia, after which organs, such as the liver, kidney, white adipose, skeletal muscle, heart, small intestine, lung, brain, and spleen, were collected and plasma was isolated. For the lipidome analyses, four biological replicates (n = 4) were used for liver tissue, and five biological replicates (n = 5) were used for the other tissues and plasma. Two of the white adipose samples from the EPA-fed group and one of the skeletal muscle samples from the ARA-fed group in were excluded from the targeted lipidomics results owing to LC-MS analysis failure. All animal experiments were carried out in accordance with the guidelines for the use and care of laboratory animals of Mochida Pharmaceutical, and were approved by the Institutional Animal Care and Use Committee (identification code: PMS14-041).

2.3. Determination of Plasma Triglyceride, Cholesterol, and Fatty Acid Composition

Standard enzymatic methods were used to determine plasma total cholesterol (TC) and triglyceride (TAG) with commercially available kits purchased from Wako Pure Chemical Industries Ltd. Total fatty acids in plasma were analyzed by gas chromatography (Japan SLC, Inc., Shizuoka, Japan). C16:0, C16:1, C18:0, C18:1, C18:2, C20:3, C20:4, C20:5, C22:5, and C22:6 levels were determined using the standard substances: palmitic acid, palmitoleic acid, stearic acid, oleic acid, linoleic acid, dihomo-γ-linoleic acid, arachidonic acid, eicosapentaenoic acid, docosapentaenoic acid, and docosahexaenoic acid, respectively.

2.4. Untargeted LC-MS/MS-Based Lipidomics

Total lipids from organs were extracted as previously described [16] using internal standards: 1 μM TAG (8:0/8:0/8:0)-^{13}C3 (Larodan, Inc., Monroe, MI, USA), 0.5 μM phosphatidyl glycerol (17:0/14:1) (Avanti Polar Lipids, Inc., Alabaster, AL, USA), 0.5 μM lyso phosphatidylcholine (17:1) (Avanti Polar Lipids, Inc.), 5 μM acylcarnitine (18:0)-d3 (Larodan, Inc., Monroe, MI, USA), 25 μM palmitic acid-d3 (Olbracht Serdary Research Laboratories, Toronto, ON, Canada), 25 μM stearic acid-d3 (Olbracht Serdary Research Laboratories, Toronto, ON, Canada), 5 μM cholic acid-d4 (Cambridge Isotope Laboratories, Inc., Tewksbury, MA, USA), and 0.5 μM Cer/Sph mixture I (Avanti Polar Lipids, Inc., Alabaster, AL, USA) (Supplementary Table S2). Untargeted analysis was performed using an ACQUITY UPLC system (Waters, Milford, MA, USA) coupled with a quadruple time-of-flight/MS (TripleTOF 5600+, SCIEX, Framingham, MA, USA). LC separation was performed using a reverse-phase column (Acquity UPLC BEH peptide C18; 2.1 × 50 mm, 1.7 μm particle size; Waters, Milford, MA, USA) with a gradient elution of mobile phase A (methanol: acetonitrile: water = 1:1:3, *v/v/v* for volume ratio containing 5 mM ammonium acetate and 10 nM EDTA) and mobile phase B (100% isopropanol containing 5 mM ammonium acetate and 10 nM EDTA), and the composition was produced by mixing those solvents. LC gradient consisted of holding solvent (A/B: 100/0) for 1 min, then linearly converting to solvent (A/B: 60/40) for 4 min, linearly converting to solvent (A/B:36/64) for 2.5 min and holding for 4.5 min, then linearly converting to solvent (A/B: 17.5/82.5) for 0.5 min, linearly converting to solvent (A/B: 15/85) for 8.5 min, and linearly converting to solvent (A/B: 5/95) for 1 min followed by returning to solvent (A/B: 100/0) and holding for 5 min for re-equilibration. The flow rate and column temperature were set to 0.3 mL/min and 45 °C, respectively. Data dependent MS/MS acquisition mode was applied as previously described [16]. Briefly, the temperature and ion spray voltage floating were set to 300 °C and −5.5 kV, respectively. The accumulation times for MS1 and MS/MS were set to 100 ms and 50 ms, respectively, for scanning a mass range from *m/z* 75 to *m/z* 1250. The collision energy (CE) was set to 35 eV with a CE spread of 15 eV in high-resolution mode, and other settings of DDA mode were as follows: 10 most intense ions, 100 cps intensity threshold, and 100 sec exclusion time.

2.5. Targeted LC-MS/MS-Based Lipidomics

The LC-MS/MS analysis was performed as described previously [12]. Samples were extracted by solid-phase extraction using Sep-Pak Vac 3cc C18 cartridges (Waters) with deuterium-labeled internal standards at a final concentration of 10 pg/μL ARA-d8, 10 pg/μL 15-HETE-d8, 10 pg/μL LTB_4-d4, 10 pg/μL PGE_2-d4 per samples. The targeted analysis was performed using a UPLC system (Waters UPLC, Waters, Milford, MA, USA) with a triple quadruple linear ion trap mass spectrometer (QTRAP 5500; SCIEX, Framingham, MA, USA), equipped with Acquity UPLC BEH C18 column (1.0 × 150 mm, 1.7 μm particle size; Waters, Milford, MA, USA). Samples were eluted with a mobile phase composed of water/acetate (100:0.1, *v/v*) and acetonitrile/methanol (4:1, *v/v*) (73:27) for 5 min, and ramped to 30:70 over 15 min, to 20:80 over 25 min and held for 8 min, ramped to 0:100 over 35 min, and held for 10 min with flow rates of 70 μL/min (0–30 min), 80 μL/min (30–33 min), and 100 μL/min (33–45 min). The MS/MS analyses were performed in negative ion mode, and oxylipins were identified and quantified by multiple reaction monitoring (Supplementary Table S3). Compounds were quantified by using stable internal standards. The calibration curves of all compounds were acquired in triplicate (Supplementary Figure S1). The extraction and matrix recovery were calculated with internal standards (IS) and the compound quantification was conducted with the calibration curves. The lipid compounds of (group 1) free fatty acids, (group 2) monohydrides and epoxides, (group 3) diols, leukotrienes, thromboxanes, lipoxines, and resolvins, and (group 4) prostaglandins and others were corrected with ARA-d8, 15-HETE-d8, LTB_4-d4, and PGE_2-d4, respectively. The compound concentrations were calculated as:

$$\text{Compound concentration} = \frac{(\text{target compound peak area})\,/\,(\text{slope of calibration curve})}{(\text{IS peak area})\,/\,(\text{IS slope of calibration curve})}$$

2.6. Data Analysis

The targeted analysis data was analyzed using MultiQuant software (SCIEX, Framingham, MA, USA). The chromatogram peaks were manually curated. The untargeted analysis data was analyzed using MS-DIAL software [17,18] version 2.90 (http://prime.psc.riken.jp/) with the following parameters: retention time start: 0 min; retention time end: 100 min; mass range start: 0 Da; mass range end: 5000 Da; accurate mass tolerance (MS1) tolerance: 0.01 Da; MS2 tolerance: 0.025 Da; maximum charge number: two; smoothing method: linear weighted moving average; smoothing level: 3; minimum peak width: five scans; minimum peak height: 1000; mass slice width: 0.1 Da; sigma window value: 0.5; MS2Dec amplitude cut off: 0; exclude after precursor: true; keep isotope until: 0.5 Da; keep original precursor isotopes: false; exclude after precursor: true; retention time tolerance for identification: 10 min; MS1 for identification: 0.01 Da; accurate mass tolerance (MS2) for identification: 0.05 Da; identification score cut off: 80%; using retention time for scoring: true; relative abundance cut off: 0; top candidate report: true; retention time tolerance for alignment: 0.1 min; MS1 tolerance for alignment: 0.015 Da; peak count filter: 0; adduct ion setting: $[M+H]^+$, $[M+NH_4]^+$, $[M+Na]^+$, $[M+C_2H_3N+H]^+$, $[M+H-H_2O]^+$, $[M+C_3H_8O+H]^+$, $[M+C_2H_3N+Na]^+$, $[M-C_6H_{10}O_5+H]^+$, $[2M+H]^+$, $[2M+NH_4]^+$, $[2M+Na]^+$ in positive ion mode and $[M-H]^-$, $[M-H_2O-H]^-$, $[M+CH_3COO]^-$, $[M-C_6H_{10}O_5-H]^-$, $[2M-H]^-$, and $[2M+CH_3COO]^-$ in negative ion mode. All lipid classes supported in the MS-DIAL version were used for the targeted lipids. The annotation was performed using the in silico spectral library of lipids, and the false positive annotations were manually curated by mass spectrometry experts. The principal component analysis and circus plots were executed in the R language with the packages of "prcomp" and "OmicCircos". Our untargeted lipidomic data were discussed in terms of relative quantification, defined by lipidomic standard initiative (https://lipidomics-standards-initiative.org/), among dietary conditions in each tissue. All of the mass spectrometry data are available at RIKEN DROP Met (http://prime.psc.riken.jp/menta.cgi/prime/drop_index, ID= DM0030).

3. Results

3.1. Lipid Profiles after Dietary Intake of ARA, EPA, or DHA

We first examined the profiles of total cholesterol, triacylglycerol (TAG), and total fatty acids in mouse plasma. Dietary intake of 1% (*w/w*) ARA-EE, EPA-EE, or DHA-EE for 2 weeks significantly decreased plasma cholesterol and decreased triacylglycerol levels as compared to the levels of the control group (Figure 1a,b). There were no body weight loss, suggesting that PUFA supplementation did not affect the amount of dietary consumption (Figure 1c). Total fatty acid levels, including C16:0 (palmitic acid), C18:1, and C18:2, were decreased after dietary intake of PUFAs, while C20:4 (ARA), C20:5 (EPA), and C22:6 (DHA) were increased in the respective dietary conditions (Figure 1d). Here, abbreviations of C18:1 and C18:2 were used because of the unclear chromatogram separation from several isomers including vaccenic acid (18:1 n-7), oleic acid (18:1 n-6), linoleic acid (n-6), and conjugated linoleic acids. In this study, we fed PUFAs in synthetic ethyl ester form (PUFA-EE) because this method of producing ethyl esters can achieve a high degree of fatty acid purity. However, the bioavailability of the ethyl ester forms has been reported to be lower than the TAG form, which is the major form of PUFAs in natural fish oil [19,20]. Dietary PUFA-EE and TAG are both hydrolyzed to free fatty acid by pancreatic lipase and passively transported into enterocytes. In the enterocytes, free fatty acids are re-esterified to TAG and then transferred into lymph and blood circulation. Among those processes, the digestion of the ester bond and re-esterification may cause the difference of bioavailability [21]. However, the plasma PUFA levels were increased as expected after dietary intake of PUFA-EEs (Figure 1d), supporting the validity of this study design.

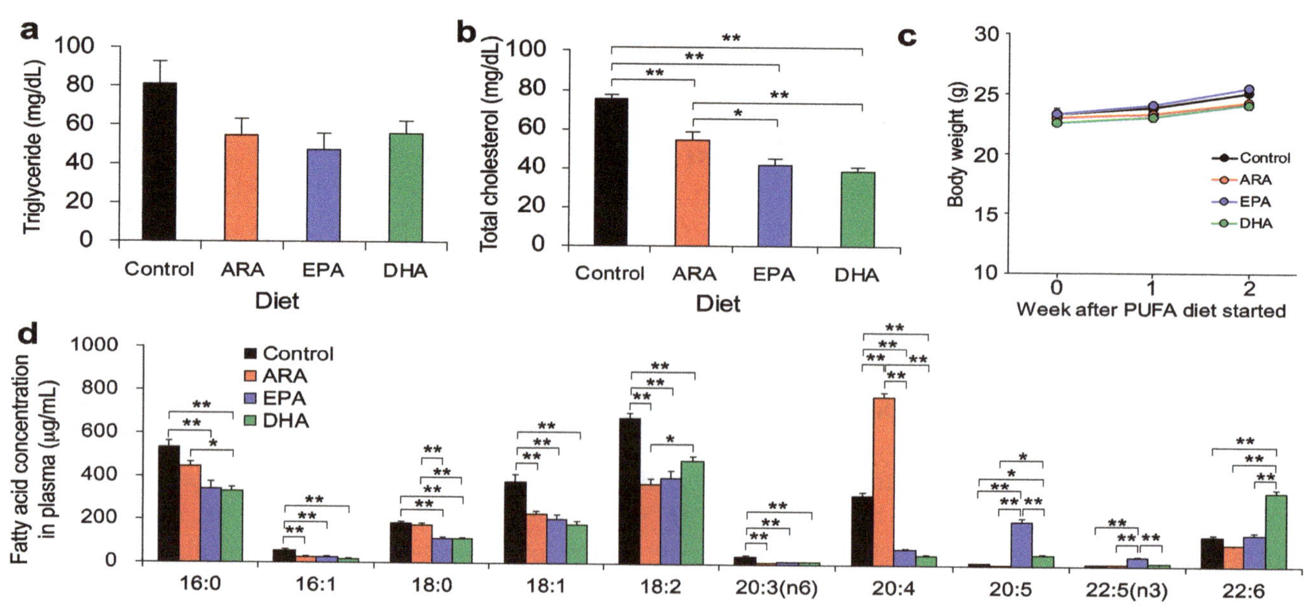

Figure 1. Effects on mouse plasma and body tissues following dietary supplementation with polyunsaturated fatty acids (PUFAs). (**a**) The total triglyceride and (**b**) total cholesterol in plasma were analyzed after two weeks of each PUFA diet. (**c**) The body weight in each PUFA diet condition was traced for two weeks, where the intake was started at 11 weeks of age. (**d**) The concentrations of total fatty acids in plasma were analyzed. The abbreviations ARA, EPA, and DHA, refer to arachidonic acid, eicosapentaenoic acid, and docosahexaenoic acid, respectively. The statistical significance was evaluated by Tukey's testing (* $P < 0.05$ and ** $P < 0.01$) where one-way ANOVA (analysis of variance) showed the statistical significance among groups ($P < 0.05$).

Next, we performed untargeted and targeted lipidomic analysis on the plasma and the nine tissues. A total of 915 molecules from 33 lipid classes were characterized in the untargeted analysis, while 111 oxylipins derived from LA, ALA, ARA, EPA, and DHA were identified in the targeted analysis (Table 1) The score plots of the principal component analysis (PCA) from untargeted lipidomics were clearly clustered by dietary-fed group in the metabolic organs, such as the liver and kidney, and the factor loadings showed that the profiles of glycerolipids and glycerophospholipids contributed to the principal components (Figure 2a and Supplementary Figure S2). On the other hand, the lipidome in the brain and muscle tissues, including the heart and skeletal muscle, was not significantly changed between different dietary conditions. Interestingly, the score plots of DHA dietary intake were classified independently from other conditions in the skeletal muscle, adipose, and heart. According to the factor loadings, the profile of DHA-containing lipids substantially contributed to these score plot dimensions. In fact, the muscle tissues were enriched in DHA present in membrane phospholipids [22], and our results suggested that the efficiency of DHA incorporation into membrane phospholipids was higher in muscle tissues compared to other tissues, and that the PC, PE, and PI lipid classes highly incorporated DHA as an acyl chain component in both the skeletal muscle and heart. In contrast, the membrane phospholipids incorporating DHA and the score plots of dietary DHA intake were not significantly changed in the brain.

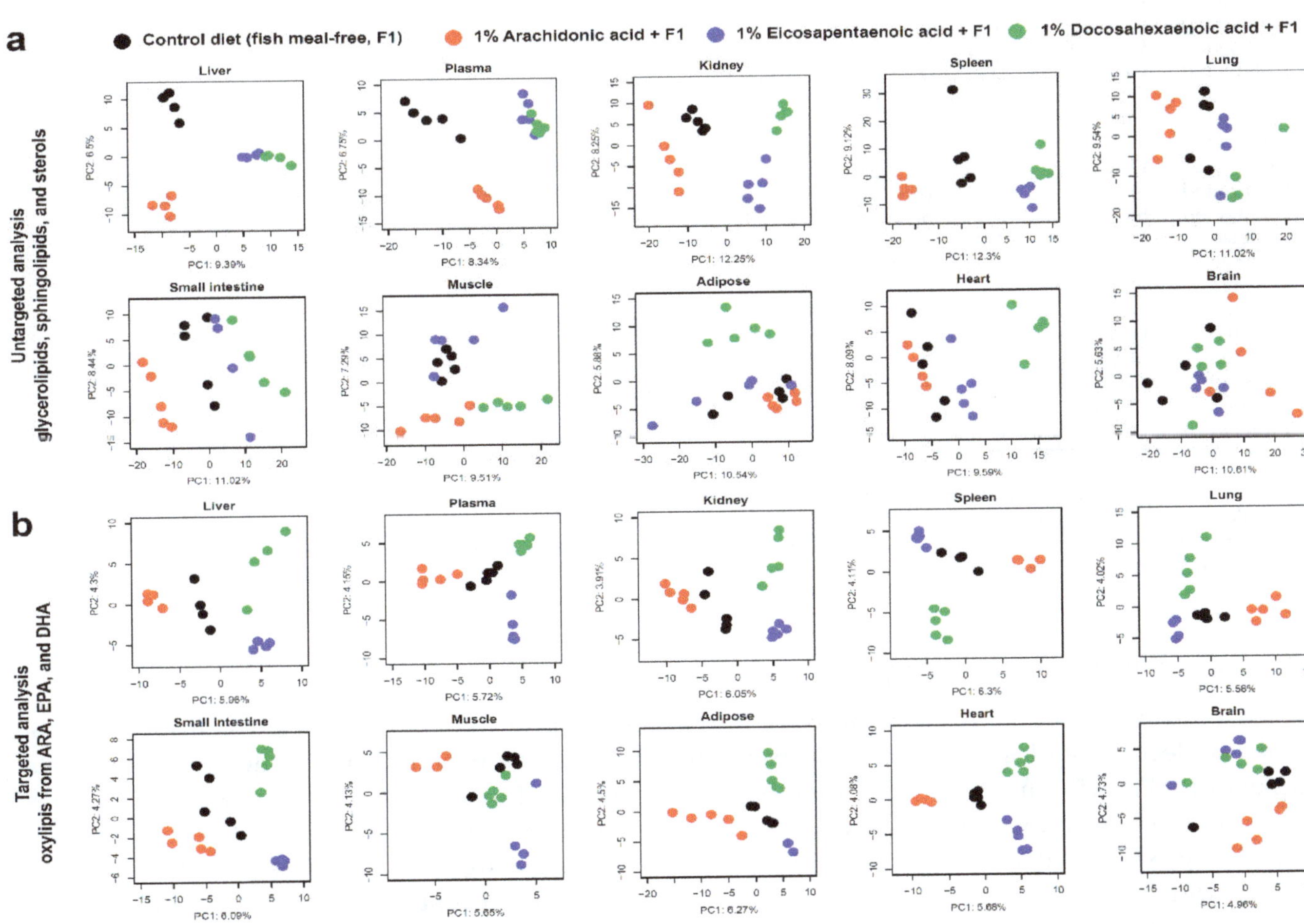

Figure 2. Principal component analyses (PCA) in untargeted and targeted lipidomics data. (**a**) The first and second principal components of PCA were described for plasma and all tissue metabolites obtained in the untargeted analysis. (**b**) The first and second components were also shown by the targeted lipidomics data. F1: fish-meal-free diet not containing ARA, EPA, or DHA.

Table 1. Summary of characterized lipids in untargeted and targeted lipidomics.

Method	Class	Liver	Plasma	Kidney	Spleen	Lung	Small Intestine	Muscle	Adipose	Heart	Brain
Untargeted	FA	12	16	18	17	15	22	15	14	13	14
Untargeted	ACar	2	2	9	7	10	7	14	0	12	6
Untargeted	MAG	1	2	2	2	3	4	2	1	0	3
Untargeted	DAG	14	8	14	15	18	29	13	28	6	5
Untargeted	TAG	70	80	73	88	116	96	80	156	116	6
Untargeted	CE	7	10	4	6	2	5	3	0	3	1
Untargeted	LPC	7	13	8	7	8	7	7	1	7	6
Untargeted	LPE	4	6	10	9	8	8	6	0	4	7
Untargeted	LPG	0	0	0	1	1	0	0	0	0	0
Untargeted	LPS	0	0	1	0	0	1	0	0	0	0
Untargeted	LPI	2	0	3	1	0	2	0	0	1	4
Untargeted	PC	33	30	25	43	45	38	47	6	33	22
Untargeted	EtherPC	0	1	18	17	9	3	4	0	7	2
Untargeted	PE	27	9	34	29	32	28	31	4	28	25
Untargeted	EtherPE	8	3	24	40	35	26	16	2	21	35
Untargeted	PG	5	1	13	23	22	9	7	1	8	11
Untargeted	BMP	1	0	3	6	7	7	0	0	1	0
Untargeted	PS	2	0	6	20	11	2	1	0	1	10
Untargeted	PI	15	14	18	25	18	16	15	1	12	10
Untargeted	OxPC	2	0	9	0	0	5	5	0	8	0
Untargeted	OxPE	13	0	29	11	2	16	10	0	18	2
Untargeted	OxPG	0	0	0	0	0	0	0	0	0	0
Untargeted	OxPI	2	0	5	0	0	2	1	0	0	0
Untargeted	OxPS	0	0	7	2	2	1	0	0	1	0
Untargeted	Cer-NS	10	6	19	16	24	12	12	6	13	12
Untargeted	Cer-NDS	0	0	1	2	1	4	1	0	0	1
Untargeted	Cer-NP	0	0	4	0	4	8	0	0	0	1
Untargeted	Cer-AS	0	0	2	0	6	0	2	0	0	5
Untargeted	Cer-AP	0	0	0	0	1	5	0	0	0	0
Untargeted	HexCer-NS	4	4	8	11	12	5	4	2	2	13
Untargeted	HexCer-AP	0	0	4	0	0	15	0	0	0	0
Untargeted	SHexCer	0	0	5	0	2	0	2	0	0	14
Untargeted	SM	6	6	13	13	18	11	6	2	9	5
Targeted	FFA	17	17	17	16	17	17	17	17	17	16
Targeted	LA-O	7	7	5	6	6	7	7	7	7	7
Targeted	ALA-O	3	3	3	3	3	3	3	3	2	2
Targeted	GLA-O	1	0	0	0	0	1	1	0	0	0
Targeted	DGLA-O	1	2	1	2	2	4	4	3	1	1
Targeted	MA-O	1	2	1	2	0	2	2	1	1	1
Targeted	ARA-O	26	23	28	27	30	29	27	30	28	27
Targeted	EPA-O	16	13	13	15	14	19	13	19	14	13
Targeted	DHA-O	15	17	15	16	16	17	18	16	15	15

The nomenclature of lipid classes in untargeted analysis is described in RIKEN PRIMe website (http://prime.psc.riken.jp/). FFA: free fatty acid; LA-O, ALA-O, GLA-O, DGLA-O, MA-O, ARA-O, EPA-O, and DHA-O: oxylipins derived from linoleic acid, alpha-linolenic acid, gamma-linolenic acid, dihomo-gamma-linolenic acid, mead acid, arachidonic acid, eicosapentaenoic acid, and docosahexaenoic acid, respectively.

3.2. Oxylipin Profiles in Targeted Lipidomics

According to the PCA score plots of the targeted lipidomic analysis, the oxylipins and their related molecules of plasma and tissues, except for the brain, were significantly affected by the dietary intake of PUFAs (Figure 2b and Supplementary Figure S2). Further detail of the oxylipin profiles is given in Figure 3. In most organs, except for the small intestine, lung, and spleen, the concentrations of 11,12-epoxyeicosatrienoic acid (EET) and 14,15-EET were higher than those of other oxylipins in ARA-derived metabolites, including epoxides, hydroxides, prostaglandins (PGs), and others. In contrast, COX-derived PGs such as PGE_2, PGD_2, and 6-keto-$PGF_{1\alpha}$ were enriched in the small intestine, lung, and spleen, and these tissues contained higher amounts of hydroxyeicosatetraenoic acids (HETEs).

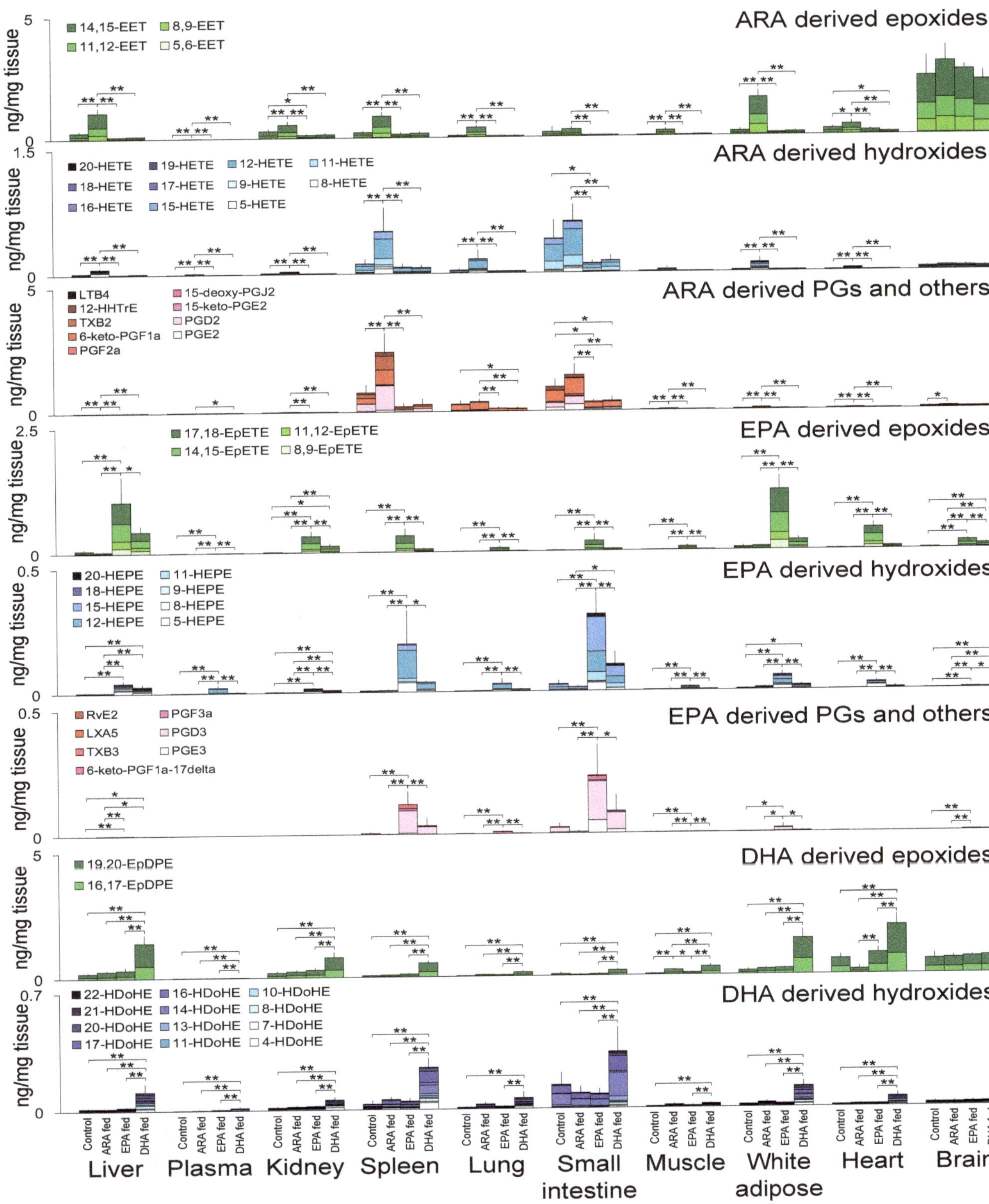

Figure 3. Oxylipin profiles in mouse tissues under each dietary condition. The total amounts of monoepoxides, hydroxides, and cyclized oxidized fatty acids including prostaglandins derived from ARA, EPA, and DHA are described. The results are given for each tissue in the four groups, i.e., the control, ARA-fed, EPE-fed, and DHA-fed groups. The ratio of oxidized fatty acids in each metabolite group is described by the stacked bar chart. The error bar indicates the standard deviation from the total amount values, and the statistical significance was evaluated by Tukey's testing (* $P < 0.05$ and ** $P < 0.01$) where one-way ANOVA showed the statistical significance among groups ($P < 0.05$). For plasma, the y-axis unit becomes ng/μL plasma instead of ng/mg tissue.

The levels of EPA and EPA-derived oxylipins were in fact lower than those of ARA and DHA in plasma and all tissues (Figure 1d and Supplementary Figure S3), while their amounts were substantially increased by dietary EPA intake as indicated by EPA-derived metabolites (Figure 3). Among the EPA-derived oxylipins, epoxyeicosatetraenoic acids (EpETEs) were the primary oxylipins present in the liver, kidney, white adipose tissue, skeletal muscle, heart, lung, and brain, while PGs and hydroxyeicosapentaenoic acids (HEPEs) were enriched in the small intestine and spleen. A similar pattern was observed in dietary DHA intake; epoxydocosapentaenoic acid (EpDPE) and hydroxydocosahexaenoic acid (HDoHE) were characteristically present in tissues.

The ARA-, EPA-, and DHA-derived epoxides, which are known as CYP metabolites, were enriched in metabolic tissues such as the liver, kidney, and white adipose tissue. ARA-derived epoxides activate various signaling pathways, and elicit functional responses such as vasorelaxation and anti-inflammation [23]. EPA-derived epoxides have anti-allergic and anti-inflammatory properties: 17,18-EpETE inhibits mast cell degranulation [24,25] and 12-hydroxy-17,18-EpETE suppresses neutrophil infiltration and eosinophilic inflammation [26,27]. These epoxides are converted into diols by soluble epoxide hydrolase, and the levels of diols and the diols/epoxide ratio were high in the small intestine (Supplementary Figure S4).

The $\omega 3$ oxylipins of EPA were increased following the dietary intake of DHA, while the amounts of ARA-derived oxylipins were decreased by $\omega 3$ PUFA dietary intake. Similar observations have also been reported in several other studies [5,7,28]. Interestingly, the increase in brain EPA levels following DHA intake was relatively high, and this result may represent the retro-conversion from DHA to EPA [29]. In contrast, ARA metabolism was competitively suppressed by a rich intake of EPA and DHA.

3.3. Lipidomic Signatures in Untargeted Analyses

We used circus plots to display the result of untargeted lipidomics (Figure 4) for the liver and brain, the metabolites of which were the most and least affected by the dietary intakes of PUFAs, respectively (Figure 2a). The heatmap layer and correlation linkages in the circus plot clearly showed that the PUFAs in liver were incorporated into glycerolipids, glycerophospholipids, and cholesterol esters, while the profile of sphingolipids was not affected by any dietary supplement. Moreover, the heatmap layer and the detail of fatty acid compositions indicating the existence of 18:2, 20:4/22:4, 20:5/22:5, and 22:6 as the acyl chain composition in a certain lipid class showed that glycerol (phospho) lipids containing 22:4 and 22:5 were increased in most tissues after the dietary intake of ARA and EPA, respectively. This result indicated that ARA and EPA were elongated to docosatetraenoic acid (DTA, 22:4) and docosapentaenoic acid (DPA, 22:5) by elongases such as elongation of very long chain fatty acids protein (ELOVL) 2 and 5 [30]. In contrast, an increase in lipids containing the elongated product of DHA, i.e., 24:6, was not observed following DHA dietary supplementation in the tissues examined.

A few differences in lipid profiles were observed in the mouse brain: free fatty acids (FA) of ARA, EPA, and DHA were affected by dietary intake (Figure 4). Moreover, several phospholipids, including phosphatidylcholine (PC), phosphatidylethanolamine (PE), phosphatidylserine (PS), and ether linked PE (EtherPE), partially reflected the changes in dietary conditions according to the heatmap layer. Interestingly, the amounts of most lipid classes, including sphingolipids, were increased in the ARA-supplemented mice. Moreover, untargeted lipidomics identified the mono-galactocyl/glycosyl-diacylglycerol (MGDG) lipids, the main component of plant membrane lipids, in the mouse brain, where the MGDG lipids were composed of saturated or monounsaturated fatty acids. A few studies have also shown the existence of MGDG in the brain, and its biosynthesis and physiological role have been examined in several in vitro studies, although their mechanisms are still unclear [31–33]. These results suggest that the annotation of untargeted lipidomics should be performed first in an unbiased manner.

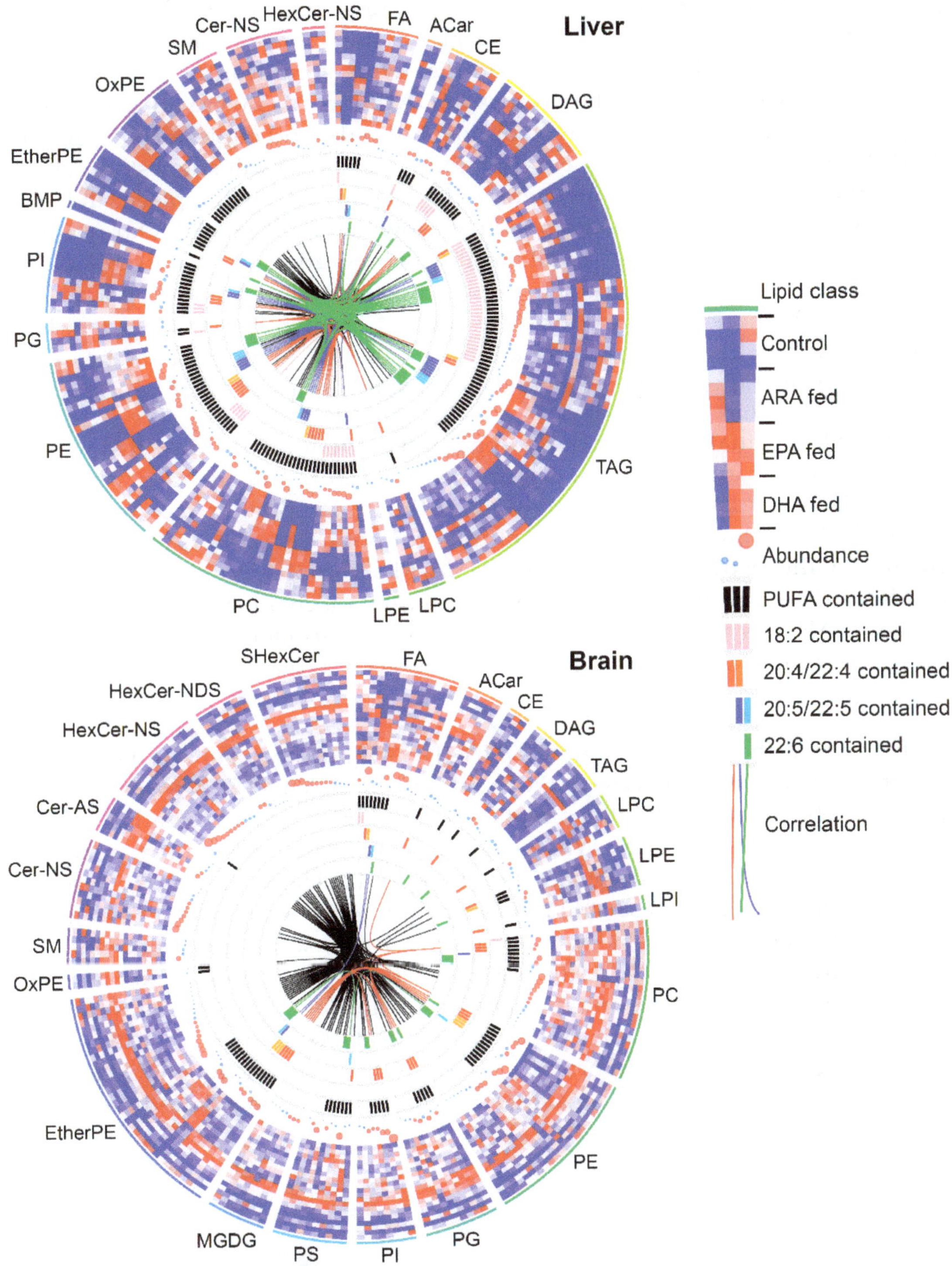

Figure 4. Lipid profiles in the liver and brain from the untargeted lipidomics data. In the circus plot, the lipid profiles are described separately by lipid class. The profile has been scaled in each metabolite from −1 to 1 in the heatmap. The ion abundance of each lipid is described by the circle size, where the average value in all conditions was utilized for the ion abundance size. Any lipid with more than two double bonds in an acyl chain was defined as a PUFA-containing metabolite. Lipids containing 18:2, 20:4, 22:4, 20:5, 22:5, and 22:6 are marked by pink, red, orange, blue, sky blue, and green colors, respectively. Metabolites were linked if the ion abundance correlation of two metabolites in biological samples was greater than 0.9.

Untargeted lipidomics also revealed unique lipid features in the other tissues. The diacylglycerol (DAG) and TAG lipids were the most abundant species in the white adipose tissue, as has previously been described [34], and the DAG and TAG profiles were significantly affected in most tissues. Interestingly, however, we found that the profiles of DAG and TAG were not affected in the spleen, although the profiles of free fatty acids, glycerophospholipids, and cholesteryl esters did reflect the intakes of ARA, EPA, and DHA (Supplementary Figure S5). This result suggested that the metabolic pathway has a limited ability to incorporate dietary PUFAs into TAG in the spleen. Indeed, diglyceride acyltransferase (DGAT) activity has been reported to be low in the spleen [35]. Our results also showed that the ether-linked PC and PE (EtherPC and EtherPE) incorporating *O*-alkenyl or *O*-alkyl chains were enriched in the kidney, spleen, small intestine, muscle, heart, and brain (Table 1), while the acyl chain profiles of EtherPC and EtherPE reflected the dietary PUFA in all tissues and plasma. In contrast, the low variety of *O*-alkenyl or *O*-alkyl chains containing PC and PE in the liver can be explained by the low expression of the alkyldihydroxyacetonephosphate synthase (AGPS) [36].

In addition, we found that the levels of phospholipids containing LA (LA-PLs) were substantially decreased after the dietary intake of ARA (Figure 5a), and profound changes were observed in PC and PE. This may have resulted from competition for the PL remodeling enzyme lysophospholipid acyltransferase 3 (LPCAT3), which has a high affinity for both ARA and LA [37]. Since free LA was decreased by dietary intake of ARA, in addition to LA-PLs (Supplementary Figure S5), the decrease of LA under ARA supplementation could reflect a potential mechanism whereby total $\omega 6$ fatty acid amount, i.e., LA + ARA, is maintained in mammalian cells: in fact, a previous study provided a complementary report in which an ARA decrease is observed in plasma following a higher intake of LA [38].

We also found that the BMP containing DHA (DHA-BMP) was substantially increased under DHA supplementation, while an increase in BMP containing ARA and EPA was not observed in ARA or EPA supplementation (Figure 5b): the annotation was performed by curating both positive and negative ion MS/MS spectra (Figure 5c). While DHA is known to be the predominant fatty acyl chain of BMP in the brain [39], our study suggested that there is a selective mechanism in which DHA is preferably incorporated into BMP in several tissues, including the small intestine, lung, heart, kidney, spleen, and liver. Although the BMP biosynthesis mechanism is still unclear, one study reported that lysocardiolipin acyltransferase 8 (AGPAT8), preferring unsaturated fatty acids as the acyl donor, catalyzed the acylation of BMP [40,41], and the gene expression level of AGPAT8 was high in the small intestine, heart, liver, and kidney in the BioGPS dataset (http://ds.biogps.org/?dataset=GSE10246&gene=225010). Importantly, these results indicate the importance of untargeted lipidomics, where unexpected and novel insights can be discovered in a data-driven manner.

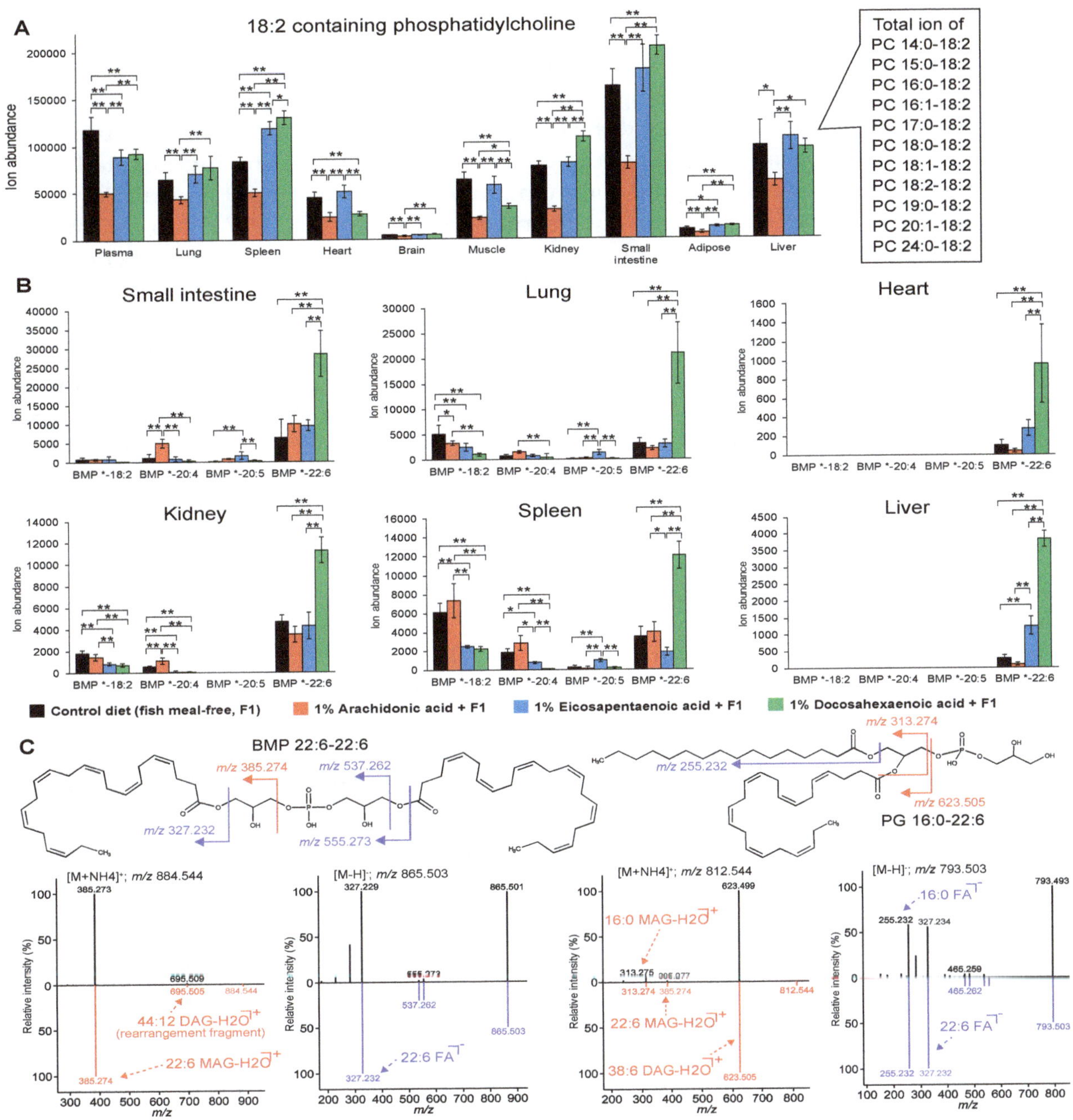

Figure 5. Unique lipid modulations in ARA or DHA dietary intakes. (**a**) The profile of phosphatidylcholines containing linoleic acid (18:2) as the longest PUFA chain and the ion abundances of each molecule. Note that a hyphen has been used for the description of acyl chains, e.g., 18:0-18:2, because the *sn1*/*sn2* acyl chain structural isomers were not distinguished in this study. (**b**) The profile of bis(monoacylglycero)phosphate (BMP): the asterisk (*) means any of the fatty acids and the ion abundances of lipids containing 18:2, 20:4, 20:5, or 22:6 were summed for quantification. Black, red, blue, and green represent the dietary conditions: control, ARA, EPA, and DHA supplementation, respectively. The statistical significance was evaluated by Tukey's testing (* $P < 0.05$ and ** $P < 0.01$) where one-way ANOVA showed the statistical significance among groups ($P < 0.05$). (**c**) The experimental positive and negative ion mode MS/MS spectra (black color) of BMP 22:6-22:6 and PG 16:0-22:6 with the in silico reference spectrum (red color for ESI(+)-MS/MS and blue color for ESI(−)-MS/MS). The cleaved places generating the major fragment ions are described by red color for ESI(+)-MS/MS and blue color for ESI(−)-MS/MS. The product ion of *m/z* 695.5 from BMP 22:6-22:6 in positive ion mode was interpreted as the result of structural rearrangement in the fragmentation process.

4. Discussion

In this study, we applied lipidomics techniques to monitor the comprehensive lipid profiles of mouse plasma and tissues, and their changes following the dietary intake of different PUFAs: ARA, EPA, and DHA. Importantly, we examined the effect of dietary supplementation of pure ARA, EPA, or DHA instead of dietary oils such as fish, linseed, and sunflower oils, and investigated the plasma and nine organs to better understand the common and/or differential lipid modulations in tissues. Mouse tissues, except for the brain, effectively incorporated the dietary PUFAs into glycerolipids and glycerophospholipids, and tissue levels of free fatty acids and oxylipins were well correlated with dietary PUFA intakes. It should be noted that our lipidomics data provided the lipidome result from "bulk" cells summing the heterogeneous nature in each tissue, and the results mainly reflect the lipid profiles of the major cell type or the major part of tissue. As for the brain, Arnold et al. reported similar results [42] when they examined cerebral cortex upon dietary EPA and DHA intake; the lipid changes in cerebral cortex were compared with other tissues including liver, kidney, heart, lung, and pancreas. Since cerebral cortex is the largest part of the brain, our whole brain data would have been significantly by the cerebral cortex. This was also the case for other tissues such as liver, white adipose, and skeletal muscle. Therefore, the lipidomes of different cell populations in each tissue should be investigated to further understand the mechanism of PUFA dietary intake in more detail.

ARA, EPA, and DHA supplementation induced various changes in lipid profiles. Dietary PUFA intake substantially increased the levels of free PUFAs and oxylipins, as well as the incorporation of PUFAs in phospholipids and triglycerides. In contrast, this side-by-side study revealed the unique features of ARA, EPA, and DHA metabolism. The basal amount of EPA was much lower than that of ARA and DHA in the plasma and tissues. This may have been because EPA easily undergoes degradation and/or conversion in tissues [43,44]. Dietary supplementation of EPA for 2 weeks elevated tissue EPA levels, and the profile of EPA-derived oxylipins was similar to that of ARA-derived oxylipins. However, the profile of DHA-derived oxylipins was a little bit different from those of ARA and EPA. These results may suggest that EPA and DHA have different metabolic compartments, which may be associated with their difference in biological effects.

In addition, dietary DHA was selectively incorporated into a unique phospholipid BMP. BMP is known as an acyl chain positional isomer of phosphatidylglycerol (PG), in which one acyl moiety is incorporated into the glycerol polar head, and the structures can be distinguished by positive ion mode MS/MS spectra, while the MS/MS spectra are the same in negative ion mode (Figure 5c). The biosynthetic and degradation pathways of BMP are still unclear; however, several enzymes, such as AGPAT8 and α/β hydrolase domain-containing 6 (ABHD6), have been reported as candidate enzymes involved in these pathways [40,41,45,46]. The subcellular localization of BMP is in the multivesicular membranes of late endosomes, where it enhances sphingolipid activator protein activity, resulting in (glyco) sphingolipid degradation [47–49]. In addition, BMP lipids contribute to membrane deformation, fusion, and transportation, as well as the incorporation of proteins and lipids [50], and one study reported that DHA-rich BMP is prone to peroxidation to prevent cholesterol oxidation [51]. However, the biological significance of DHA-BMP in endosome function and regulation is not well understood.

PCA analysis (Figure 2) and circus plots (Figure 4 and Supplementary Figure S5) displayed the metabolic fingerprints of different tissues. In the immune organs, the spleen, small intestine, and lung, high levels of PGs, TXs and fatty acid hydroxides were detected. In contrast, PUFA epoxides were the major metabolites in plasma and other tissues. CYPs are responsible for fatty acid epoxidation, and ARA-, EPA-, and DHA-derived epoxides have various actions, such as vasodilation, anti-inflammation, anticoagulation, and anti-allergy activity [23–27].

The comprehensive lipidomics approach revealed many findings about dietary PUFA metabolism. Our untargeted/targeted lipidomics data incorporating 1026 unique lipid molecules (Supplementary

Tables S2 and S3) are powerful when combined with other omics layers such as proteomics and transcriptomics. Recently, the lipidomics standards initiative (LSI) proposed three types of quantification in mass-spectrometry-based lipidomics: level 1, matching internal standards (IS) together with consideration of species-specific analytical response (essentially stating that stable-isotope-labeled lipids are preferred); level 2, matching IS where the lipid class between analyte and internal standard is identical; and level 3, non-matching IS where analytes are normalized with other lipid class molecules [52]. Although our untargeted lipidomics data have been displayed as the relative quantification in each tissue, our lipidome table can be normalized as level 3 and partially level 2 quantifications by using the internal standards. A consortium with the advances in analytical chemistry and informatics research would contribute to facilitating open data sciences through the reanalysis of deposited data [15].

This experimental design to investigate the PUFA metabolisms is expandable for further nutrient researches, and we believe that this study is a fundamental step towards clarification of the effects of dietary PUFA intakes in mice. Our lipidomics study offers a comprehensive picture of dietary PUFA metabolism in different tissues, and could provide an opportunity for data-driven hypotheses and biological insights into the molecular mechanisms of how different PUFA balances affect human and livestock health and disease.

Supplementary Materials:
Figure S1. Calibration curves of standard compounds including internal standards. All of calibration curves of 111 metabolites and four internal standards was described. Three technical replicates were obtained for each concentration. The R-square and coefficient value used for metabolite quantification were also described. The compound ID delimited by double underbars "__" is matched with MRM transition ID described in Supplementary Table S2. The 95% confidence interval was also described by gray color. Figure S2. Factor loadings of PCA described in Figure 2. The top 20 factor loadings from the maximum and minimum values for first- and second principal components were described for each PCA result. Figure S3. Concentration of free fatty acids in each tissue after dietary intakes. Black, red, blue, and green colors represent the dietary conditions; control, ARA, EPA, and DHA supplementation respectively. The statistical significance was evaluated by Tukey's testing (* $P < 0.05$ and ** $P < 0.01$) where one-way ANOVA showed the statistical significance among groups ($P < 0.05$). Figure S4. Detail of several oxylipin profiles under different dietary conditions. (a) The concentrations of 14,15-EET in control and ARA intake condition were described. (b) The concentrations of 17,18-EpETE in the control and following EPA intake were described. (c) The concentrations of 19,20-EpDPE in the control and following DHA intake were described. (d) The concentrations of 14,15-DHT in the control and following ARA intake were described. (e) The concentrations of 17,18-diHETE in the control and following EPA intake were described. (f) The concentrations of 19,20-diHDoPE in the control and following DHA intake were described. (g) The ratio of 14,15-DHT and 14,15-EET in ARA intake condition was described. (h) The ratio of 17,18-diHETE and 17,18-EpETE in EPA intake condition was described. (i) The ratio of 19,20-diHDoPE and 19,20-EpDPE in DHA intake condition was described. The significance was evaluated as * $P < 0.05$ and ** $P < 0.01$. Figure S5. Circus plots for plasma, adipose, kidney, small intestine, spleen, lung, muscle, and heart. Table S1. Details of fish-meal free F1 diet. Table S2. Details of untargeted lipidomics results. The false discovery rate (FDA) was corrected by Benjamini-Hochberg (BH) method, and the Q-value was also described. Table S3. Details of targeted lipidomics results. The false discovery rate (FDA) was corrected by Benjamini-Hochberg (BH) method, and the Q-value was also described.

Author Contributions: S.N. and M.A. designed the research. S.N. performed the experiments, and K.I. supported them. S.N. performed most of the data analysis for targeted lipidomics, and H.T. and M.T. performed the analysis for untargeted and targeted lipidomics. S.N., H.T., and M.A. primarily wrote the manuscript, and all authors approved the final manuscript.

Acknowledgments: We thank to Yuuya Senoo, Aya Hori, and Mie Honda (RIKEN) for LC-MS analyses, and Yasutaka Jinno and Hiroyuki Kawano (Mochida Pharmaceutical Co.) for study supports.

References

1. Saini, R.K.; Keum, Y.S. Omega-3 and omega-6 polyunsaturated fatty acids: Dietary sources, metabolism, and significance—A review. *Life Sci.* **2018**, *203*, 255–267. [CrossRef] [PubMed]
2. Zárate, R.; el Jaber-Vazdekis, N.; Tejera, N.; Pérez, J.A.; Rodríguez, C. Significance of long chain polyunsaturated fatty acids in human health. *Clin. Transl. Med.* **2017**, *6*, 25. [CrossRef] [PubMed]
3. Wathes, D.C.; Abayasekara, D.R.; Aitken, R.J. Polyunsaturated fatty acids in male and female reproduction. *Biol. Reprod.* **2007**, *77*, 190–201. [CrossRef] [PubMed]

4. Ostermann, A.I.; Schebb, N.H. Effects of omega-3 fatty acid supplementation on the pattern of oxylipins: A short review about the modulation of hydroxy-, dihydroxy-, and epoxy-fatty acids. *Food Funct.* **2017**, *8*, 2355–2367. [CrossRef] [PubMed]
5. Ostermann, A.I.; Waindok, P.; Schmidt, M.J.; Chiu, C.Y.; Smyl, C.; Rohwer, N.; Weylandt, K.H.; Schebb, N.H. Modulation of the endogenous omega-3 fatty acid and oxylipin profile in vivo—A comparison of the fat-1 transgenic mouse with C57BL/6 wildtype mice on an omega-3 fatty acid enriched diet. *PLoS ONE* **2017**, *12*, e0184470. [CrossRef] [PubMed]
6. Wijendran, V.; Downs, I.; Srigley, C.T.; Kothapalli, K.S.; Park, W.J.; Blank, B.S.; Zimmer, J.P.; Butt, C.M.; Salem, N.; Brenna, T. Dietary arachidonic acid and docosahexaenoic acid regulate liver fatty acid desaturase (FADS) alternative transcript expression in suckling piglets. *Prostaglandins Leukot. Essent. Fat. Acids* **2013**, *89*, 345–350. [CrossRef] [PubMed]
7. Arterburn, L.M.; Hall, E.B.; Oken, H. Distribution, interconversion, and dose response of n-3 fatty acids in humans. *Am. J. Clin. Nutr.* **2006**, *83*, 1467S–1476S. [CrossRef]
8. Balvers, M.G.J.; Verhoeckx, K.C.M.; Bijlsma, S.; Rubingh, C.M.; Meijerink, J.; Wortelboer, H.M.; Witkamp, R.F. Fish oil and inflammatory status alter the n-3 to n-6 balance of the endocannabinoid and oxylipin metabolomes in mouse plasma and tissues. *Metabolomics* **2012**, *8*, 1130–1147. [CrossRef]
9. Wood, J.T.; Williams, J.S.; Pandarinathan, L.; Janero, D.R.; Lammi-Keefe, C.J.; Makriyannis, A. Dietary docosahexaenoic acid supplementation alters select physiological endocannabinoid-system metabolites in brain and plasma. *J. Lipid Res.* **2010**, *51*, 1416–1423. [CrossRef]
10. Harayama, T.; Riezman, H. Understanding the diversity of membrane lipid composition. *Nat. Rev. Mol. Cell Biol.* **2018**, *19*, 281–296. [CrossRef]
11. Olzmann, J.A.; Carvalho, P. Dynamics and functions of lipid droplets. *Nat. Rev. Mol. Cell Biol.* **2019**, *20*, 137–155. [CrossRef] [PubMed]
12. Arita, M. Mediator lipidomics in acute inflammation and resolution. *J. Biochem.* **2012**, *152*, 313–319. [CrossRef] [PubMed]
13. Tsugawa, H.; Ikeda, K.; Arita, M. The importance of bioinformatics for connecting data-driven lipidomics and biological insights. *Biochim. Biophys. Acta Mol. Cell Biol. Lipids* **2017**, *1862*, 762–765. [CrossRef] [PubMed]
14. Burla, B.; Arita, M.; Arita, M.; Bendt, A.K.; Cazenave-Gassiot, A.; Dennis, E.A.; Ekroos, K.; Han, X.; Ikeda, K.; Liebisch, G.; et al. MS-based lipidomics of human blood plasma: A community-initiated position paper to develop accepted guidelines. *J. Lipid Res.* **2018**, *59*, 2001–2017. [CrossRef] [PubMed]
15. Tsugawa, H.; Satoh, A.; Uchino, H.; Cajka, T.; Arita, M.; Arita, M. Mass Spectrometry Data Repository Enhances Novel Metabolite Discoveries with Advances in Computational Metabolomics. *Metabolites* **2019**, *9*, 119. [CrossRef] [PubMed]
16. Tsugawa, H.; Ikeda, K.; Tanaka, W.; Senoo, Y.; Arita, M.; Arita, M. Comprehensive identification of sphingolipid species by in silico retention time and tandem mass spectral library. *J. Cheminformatics* **2017**, *9*, 1–12. [CrossRef]
17. Tsugawa, H.; Cajka, T.; Kind, T.; Ma, Y.; Higgins, B.; Ikeda, K.; Kanazawa, M.; VanderGheynst, J.; Fiehn, O.; Arita, M. MS-DIAL: Data-independent MS/MS deconvolution for comprehensive metabolome analysis. *Nat. Methods* **2015**, *12*, 523–526. [CrossRef]
18. Tsugawa, H.; Nakabayashi, R.; Mori, T.; Yamada, Y.; Takahashi, M.; Rai, A.; Sugiyama, R.; Yamamoto, H.; Nakaya, T.; Yamazaki, M.; et al. A cheminformatics approach to characterize metabolomes in stable-isotope-labeled organisms. *Nat. Methods* **2019**, *16*, 295–298. [CrossRef]
19. Hansen, J.B.; Olsen, J.O.; Wilsgard, L.; Lyngmo, V.; Svensson, B. Comparative effects of prolonged intake of highly purified fish oils as ethyl ester or triglyceride on lipids, haemostasis and platelet function in normolipaemic men. *Eur. J. Clin. Nutr.* **1993**, *47*, 497–507.
20. Neubronner, J.; Schuchardt, J.P.; Kressel, G.; Merkel, M.; von Schacky, C.; Hahn, A. Enhanced increase of omega-3 index in response to long-term n-3 fatty acid supplementation from triacylglycerides versus ethyl esters. *Eur. J. Clin. Nutr.* **2011**, *65*, 247–254. [CrossRef]
21. Schuchardt, J.P.; Hahn, A. Bioavailability of long-chain omega-3 fatty acids. *Prostaglandins Leukot. Essent. Fat. Acids* **2013**, *89*, 1–8. [CrossRef] [PubMed]
22. Hishikawa, D.; Valentine, W.J.; Iizuka-Hishikawa, Y.; Shindou, H.; Shimizu, T. Metabolism and functions of docosahexaenoic acid-containing membrane glycerophospholipids. *FEBS Lett.* **2017**, *591*, 2730–2744. [CrossRef] [PubMed]

23. Spector, A.A.; Norris, A.W. Action of epoxyeicosatrienoic acids on cellular function. *Am. J. Physiol. Physiol.* **2007**, *292*, C996–C1012. [CrossRef] [PubMed]
24. Nagatake, T.; Shiogama, Y.; Inoue, A.; Kikuta, J.; Honda, T.; Tiwari, P.; Kishi, T.; Yanagisawa, A.; Isobe, Y.; Matsumoto, N.; et al. The 17,18-epoxyeicosatetraenoic acid–G protein–coupled receptor 40 axis ameliorates contact hypersensitivity by inhibiting neutrophil mobility in mice and cynomolgus macaques. *J. Allergy Clin. Immunol.* **2018**, *142*, 470–484. [CrossRef]
25. Kunisawa, J.; Arita, M.; Hayasaka, T.; Harada, T.; Iwamoto, R.; Nagasawa, R.; Shikata, S.; Nagatake, T.; Suzuki, H.; Hashimoto, E.; et al. Dietary ω3 fatty acid exerts anti-allergic effect through the conversion to 17,18-epoxyeicosatetraenoic acid in the gut. *Sci. Rep.* **2015**, *5*, 9750. [CrossRef]
26. Mochimaru, T.; Fukunaga, K.; Miyata, J.; Matsusaka, M.; Kabata, H.; Ueda, S.; Suzuki, Y.; Goto, T.; Urabe, D.; Inoue, M.; et al. 12-OH-17,18-Epoxyeicosatetraenoic acid alleviates eosinophilic airway inflammation in murine lungs. *Allergy* **2018**, *73*, 369–378. [CrossRef]
27. Kubota, T.; Arita, M.; Isobe, Y.; Iwamoto, R.; Goto, T.; Yoshioka, T.; Urabe, D.; Inoue, M.; Arai, H. Eicosapentaenoic acid is converted via ω-3 epoxygenation to the anti-inflammatory metabolite 12-hydroxy-17,18-epoxyeicosatetraenoic acid. *FASEB J.* **2014**, *28*, 586–593. [CrossRef]
28. Dasilva, G.; Pazos, M.; García-Egido, E.; Pérez-Jiménez, J.; Torres, J.L.; Giralt, M.; Nogués, M.R.; Medina, I. Lipidomics to analyze the influence of diets with different EPA:DHA ratios in the progression of Metabolic Syndrome using SHROB rats as a model. *Food Chem.* **2016**, *205*, 196–203. [CrossRef]
29. Grønn, M.; Christensen, E.; Hagve, T.A.; Christophersen, B.O. Peroxisomal retroconversion of docosahexaenoic acid (22:6(n-3)) to eicosapentaenoic acid (20:5(n-3)) studied in isolated rat liver cells. *Biochim. Biophys. Acta (BBA) Lipids Lipid Metab.* **1991**, *1081*, 85–91. [CrossRef]
30. Zhang, J.Y.; Kothapalli, K.S.D.; Brenna, J.T. Desaturase and elongase limiting endogenous long chain polyunsaturated fatty acid biosynthesis. *Curr. Opin. Clin. Nutr. Metab. Care* **2016**, *19*, 103–110. [CrossRef]
31. Schmidt-Schultz, T.; Althaus, H.H. Monogalactosyl Diglyceride, a Marker for Myelination, Activates Oligodendroglial Protein Kinase C. *J. Neurochem.* **1994**, *62*, 1578–1585. [CrossRef] [PubMed]
32. Murray, R.K.; Narasimhan, R.; Levine, M.; Pinteric, L.; Shirley, M.; Lingwood, C.; Schacter, H. *Galactoglycerolipids of Mammalian Testis, Spermatozoa, and Nervous Tissue*; ACS Symposium Series; American Chemical Society: Washington, DC, USA, 1980; pp. 105–125.
33. Påhlsson, P.; Spitalnik, S.L.; Spitalnik, P.F.; Fantini, J.; Rakotonirainy, O.; Ghardashkhani, S.; Lindberg, J.; Konradsson, P.; Larson, G. Characterization of galactosyl glycerolipids in the HT29 human colon carcinoma cell line. *Arch. Biochem. Biophys.* **2001**, *396*, 187–198. [CrossRef] [PubMed]
34. Jain, M.; Ngoy, S.; Sheth, S.A.; Swanson, R.A.; Rhee, E.P.; Liao, R.; Clish, C.B.; Mootha, V.K.; Nilsson, R. A systematic survey of lipids across mouse tissues. *Am. J. Physiol. Metab.* **2014**, *306*, E854–E868. [CrossRef] [PubMed]
35. Yen, C.L.E.; Stone, S.J.; Koliwad, S.; Harris, C.; Farese, R.V. Thematic Review Series: Glycerolipids. DGAT enzymes and triacylglycerol biosynthesis. *J. Lipid Res.* **2008**, *49*, 2283–2301. [CrossRef] [PubMed]
36. Wu, C.; Jin, X.; Tsueng, G.; Afrasiabi, C.; Su, A.I. BioGPS: Building your own mash-up of gene annotations and expression profiles. *Nucleic Acids Res.* **2016**, *44*, D313–D316. [CrossRef] [PubMed]
37. Hashidate-Yoshida, T.; Harayama, T.; Hishikawa, D.; Morimoto, R.; Hamano, F.; Tokuoka, S.M.; Eto, M.; Tamura-Nakano, M.; Yanobu-Takanashi, R.; Mukumoto, Y.; et al. Fatty acyl-chain remodeling by LPCAT3 enriches arachidonate in phospholipid membranes and regulates triglyceride transport. *Elife* **2015**, *4*, e06328. [CrossRef]
38. Adam, O.; Wolfram, T.G. Impact of linoleic acid intake on arachidonic acid formation and eicosanoid biosynthesis in humans. *Prostaglandins Leukot. Essent. Fat. Acids* **2008**, *75*, 177–181. [CrossRef]
39. Akgoc, Z.; Sena-Esteves, M.; Martin, D.R.; Han, X.; d'Azzo, A.; Seyfried, T.N. Bis(monoacylglycero)phosphate: A secondary storage lipid in the gangliosidoses. *J. Lipid Res.* **2015**, *56*, 1006–1013. [CrossRef]
40. Cao, J.; Shen, W.; Chang, Z.; Shi, Y. ALCAT1 is a polyglycerophospholipid acyltransferase potently regulated by adenine nucleotide and thyroid status. *Am. J. Physiol. Metab.* **2008**, *296*, E647–E653. [CrossRef]
41. Yamashita, A.; Hayashi, Y.; Nemoto-Sasaki, Y.; Ito, M.; Oka, S.; Tanikawa, T.; Waku, K.; Sugiura, T. Acyltransferases and transacylases that determine the fatty acid composition of glycerolipids and the metabolism of bioactive lipid mediators in mammalian cells and model organisms. *Prog. Lipid Res.* **2014**, *53*, 18–81. [CrossRef]

42. Arnold, C.; Markovic, M.; Blossey, K.; Wallukat, G.; Fischer, R.; Dechend, R.; Konkel, A.; von Schacky, C.; Luft, F.C.; Muller, D.N.; et al. Arachidonic Acid-metabolizing Cytochrome P450 Enzymes Are Targets of ω-3 Fatty Acids. *J. Biol. Chem.* **2010**, *285*, 32720–32733. [CrossRef] [PubMed]
43. Igarashi, M.; Chang, L.; Ma, K.; Rapoport, S.I. Kinetics of eicosapentaenoic acid in brain, heart and liver of conscious rats fed a high n-3 PUFA containing diet. *Prostaglandins Leukot. Essent. Fat. Acids* **2013**, *89*, 403–412. [CrossRef] [PubMed]
44. Chen, C.T.; Bazinet, R.P. β-oxidation and rapid metabolism, but not uptake regulate brain eicosapentaenoic acid levels. *Prostaglandins Leukot. Essent. Fat. Acids* **2015**, *92*, 33–40. [CrossRef] [PubMed]
45. Scherer, M.; Schmitz, G. Metabolism, function and mass spectrometric analysis of bis(monoacylglycero)phosphate and cardiolipin. *Chem. Phys. Lipids* **2011**, *164*, 556–562. [CrossRef] [PubMed]
46. Pribasnig, M.A.; Mrak, I.; Grabner, G.F.; Taschler, U.; Knittelfelder, O.; Scherz, B.; Eichmann, T.O.; Heier, C.; Grumet, L.; Kowaliuk, J.; et al. α/β hydrolase domain-containing 6 (ABHD6) degrades the late Endosomal/Lysosomal lipid Bis(Monoacylglycero)phosphate. *J. Biol. Chem.* **2015**, *290*, 29869–29881. [CrossRef]
47. Matsuo, H.; Chevallier, J.; Mayran, N.; Le Blanc, I.; Ferguson, C.; Fauré, J.; Blanc, N.S.; Matile, S.; Dubochet, J.; Sadoul, R.; et al. Role of LBPA and Alix in Multivesicular Liposome Formation and Endosome Organization. *Science* **2004**, *303*, 531–534. [CrossRef]
48. Kobayashi, T.; Stang, E.; Fang, K.S.; De Moerloose, P.; Parton, R.G.; Gruenberg, J. A lipid associated with the antiphospholipid syndrome regulates endosome structure and function. *Nature* **1998**, *392*, 193–197. [CrossRef]
49. Kolter, T.; Sandhoff, K. Principles of lysosomal membrane digestion: Stimulation of sphingolipid degradation by sphingolipid activator proteins and anionic lysosomal lipids. *Annu. Rev. Cell Dev. Biol.* **2005**, *21*, 81–103. [CrossRef]
50. Bissig, C.; Gruenberg, J. Lipid sorting and multivesicular endosome biogenesis. *Cold Spring Harb. Perspect. Biol.* **2013**, *5*, a016816. [CrossRef]
51. Bouvier, J.; Zemski Berry, K.A.; Hullin-Matsuda, F.; Makino, A.; Michaud, S.; Geloën, A.; Murphy, R.C.; Kobayashi, T.; Lagarde, M.; Delton-Vandenbroucke, I. Selective decrease of bis(monoacylglycero)phosphate content in macrophages by high supplementation with docosahexaenoic acid. *J. Lipid Res.* **2009**, *50*, 243–255. [CrossRef]
52. Liebisch, G.; Ahrends, R.; Arita, M.; Arita, M.; Bowden, J.A.; Ejsing, C.S.; Griffiths, W.J.; Holcapek, M.; Kofeler, H.; Mitchell, T.W.; et al. Lipidomics needs more standardization. *Nat. Metab.* **2019**, *1*, 745–747.

11

Effects of Cyclic High Ambient Temperature and Dietary Supplementation of Orotic Acid, a Pyrimidine Precursor, on Plasma and Muscle Metabolites in Broiler Chickens

Saki Shimamoto [1,2], Kiriko Nakamura [1], Shozo Tomonaga [3], Satoru Furukawa [4], Akira Ohtsuka [1] and Daichi Ijiri [1,*]

1 Department of Agricultural Sciences and Natural Resources, Kagoshima University, Korimoto, Kagoshima 890-0065, Japan; shimamoto@agr.niigata-u.ac.jp (S.S.); k5023091@kadai.jp (K.N.); ohtsuka@chem.agri.kagoshima-u.ac.jp (A.O.)
2 Graduate School of Science and Technology, Niigata University, Nishi-ku, Niigata 950-2181, Japan
3 Division of Applied Biosciences, Graduate School of Agriculture, Kyoto University, Sakyo-ku, Kyoto 606-8502, Japan; tomonaga.shozo.4n@kyoto-u.ac.jp
4 Furukawa Research Office Co. Ltd., Setagaya-ku, Tokyo 157-0066, Japan; furukawa@furukawa-res.co.jp
* Correspondence: ijiri@chem.agri.kagoshima-u.ac.jp.

Abstract: The aim of this study was to evaluate the effects of high ambient temperature (HT) and orotic acid supplementation on the plasma and muscle metabolomic profiles in broiler chickens. Thirty-two 14-day-old broiler chickens were divided into four treatment groups that were fed diets with or without 0.7% orotic acid under thermoneutral (25 ± 1 °C) or cyclic HT (35 ± 1 °C for 8 h/day) conditions for 2 weeks. The chickens exposed to HT had higher plasma malondialdehyde concentrations, suggesting an increase in lipid peroxidation, which is alleviated by orotic acid supplementation. The HT environment also affected the serine, glutamine, and tyrosine plasma concentrations, while orotic acid supplementation affected the aspartic acid, glutamic acid, and tyrosine plasma concentrations. Untargeted gas chromatography–triple quadrupole mass spectrometry (GC-MS/MS)-based metabolomics analysis identified that the HT affected the plasma levels of metabolites involved in purine metabolism, ammonia recycling, pyrimidine metabolism, homocysteine degradation, glutamate metabolism, urea cycle, β-alanine metabolism, glycine and serine metabolism, and aspartate metabolism, while orotic acid supplementation affected metabolites involved in pyrimidine metabolism, β-alanine metabolism, the malate–aspartate shuttle, and aspartate metabolism. Our results suggest that cyclic HT affects various metabolic processes in broiler chickens, and that orotic acid supplementation ameliorates HT-induced increases in lipid peroxidation.

Keywords: chickens; heat stress; lipid peroxidation; metabolomics; orotic acid

1. Introduction

Ambient temperatures above the thermoneutral zone cause environmental heat stress. Chickens are more vulnerable to heat stress than other domestic animals, because they lack sweat glands and have higher body temperatures [1,2]. Under such high ambient temperature (HT) conditions, the generation of reactive oxygen species increases in various body tissues as the heat load increases [3]. This results in a range of physiological changes that can severely depress growth performance and meat yield [4,5] and reduce meat quality, accompanied by increasing lipid peroxidation levels [6,7] and drip loss [8,9], decreasing the share force value [10,11] and changing meat color [12–14].

Gas chromatography–mass spectrometry (GC-MS)-based, untargeted metabolomics analysis has been used to comprehensively characterize the effects of HT on the physiology of chickens [15], and has found that the plasma levels of 38 metabolites changed significantly when chickens were exposed to chronic heat (38 °C) for 4 days. These altered metabolites indicated that such heat exposure affected 35 metabolic processes, including the sulfur amino acid metabolic pathway, the kynurenine pathway in tryptophan metabolism, and nucleic acid metabolism [15]. In agreement with this observation, dietary supplementation with either a sulfur amino acid (methionine) or tryptophan, which is mainly metabolized to kynurenine, has been reported to alleviate the negative effects of HT [16,17]. However, the involvement of nucleic acids and their metabolites in the metabolic changes that occur in chickens kept in HT environments remains unclear.

Orotic acid, which is found in high concentrations in cow's milk, is a key intermediate in the pyrimidine biosynthesis pathway [18]. Pyrimidine nucleotides are important constituents of RNA and of the phospholipids present in cell membranes. Orotic acid can enter the de novo synthesis pathway for pyrimidines beyond the rate-limiting step, and thereby improve throughput. In chickens, the plasma level of orotic acid was found to significantly decrease, by approximately 60%, in response to short-term chronic heat exposure [15], suggesting that the plasma orotic acid concentration may be important under HT conditions.

In this study, we evaluated the effects of prolonged heat exposure and feeding orotic acid at 0.7% of the diet on the growth performance, plasma and muscle lipid peroxidation levels, and plasma and muscle metabolite concentrations of broiler chickens. To mimic realistic HT conditions, a cyclic HT environment (35 ± 1 °C for 8 h/day) was used. In addition, a food-grade orotic acid (Lactoserum; Matsumoto Trading Co., Ltd., Tokyo, Japan), which contains more than 98% of orotic acid monohydrate, was used for dietary supplementation.

2. Results

The final body weight, body weight gain, and feed conversion ratio were not affected by either the HT or orotic acid supplementation, whereas the feed intake was significantly depressed in the chickens kept under the HT conditions (Table 1). In addition, the body temperatures of the chickens kept in the HT environment were significantly increased compared with those of the chickens kept under thermoneutral conditions. However, orotic acid supplementation alleviated the increase in body temperature under the HT conditions.

Although the weights of the leg muscles were not affected by the rearing temperature, the weights of the breast muscles, breast tender muscles, livers, and hearts were lower in the chickens kept under the HT conditions (Table 2). The yield of the breast muscles (the ratio of breast muscles weight to body weight) was the highest in the chickens supplemented with orotic acid and kept in the thermoneutral environment compared to the other three groups of chickens (Table S1). However, the yield of neither the breast tender muscles nor the leg muscles were different among these four groups. Furthermore, the weight of the abdominal fat tissue was increased by the HT environment.

The malondialdehyde (MDA) concentration, which serves as an index for the lipid peroxidation level, was significantly affected by either temperature or orotic acid supplementation. The plasma MDA concentration of chickens fed a control diet and kept under the HT condition was significantly increased compared with the chickens kept in the thermoneutral environment (Figure 1). However, under the HT condition, the dietary supplementation of orotic acid decreased the plasma MDA concentration compared to the control diet. On the other hand, the HT condition did not significantly increase the muscle MDA concentration of chickens fed a control diet. Furthermore, dietary supplementation of orotic acid did not alleviate that of chickens under the HT condition.

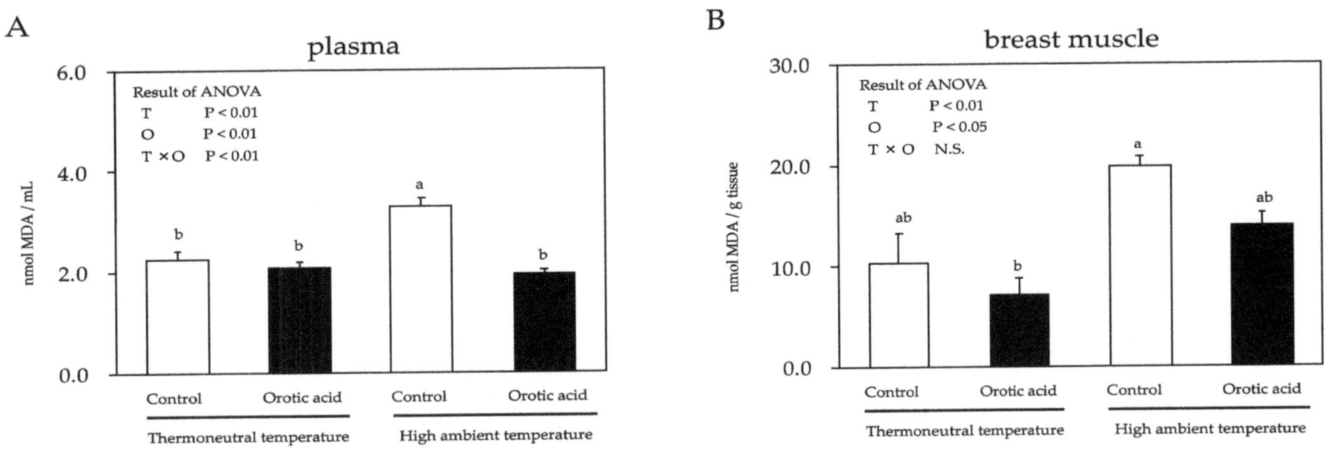

Figure 1. Effects of a cyclic high ambient temperature and feeding orotic acid on the plasma (**A**) and muscle (**B**) malondialdehyde (MDA) concentrations of broiler chickens. Results are expressed as mean ± standard error of the mean (SEM) ($n = 8$). Means with the same superscript letter within columns are not significantly different at $p < 0.05$. T: the effect of high ambient temperature; O: the effect of feeding orotic acid; T × O: the statistical interaction between high ambient temperature and feeding orotic acid.

Table 3 shows the plasma-free amino acid concentrations of the chickens kept under thermoneutral or HT conditions. There were significant effects of ambient temperature on the plasma serine, glutamine, and tyrosine concentrations, while orotic acid supplementation significantly affected the plasma aspartic acid, glutamic acid, and tyrosine concentrations. The two-way ANOVA revealed no significant interaction between the HT and orotic acid supplementation on these 18 plasma-free amino acids.

Untargeted gas chromatography–triple quadrupole mass spectrometry (GC-MS/MS)-based metabolomics analysis identified a total of 172 metabolites in the plasma of the chickens (Table 4 and Table S2). Of these metabolites, 23 were significantly affected by HT, of which 11 were significantly increased and 12 were decreased (Table 4). On the other hand, 12 were affected by orotic acid supplementation, of which 11 were significantly increased, and 1 was decreased. However, the two-way ANOVA revealed no significant interaction between the temperature and dietary treatments for the plasma metabolites.

The enrichment analysis indicated that nine and four metabolic pathways were affected by the HT environment and the orotic acid supplementation, respectively (Table 5).

The free amino acid and carnosine concentrations in the broiler chickens' breast muscles are shown in Table 6. The HT treatment significantly affected the serine, glutamine, arginine, methionine, and phenylalanine concentrations, while the orotic acid supplementation significantly affected the histidine, methionine, and carnosine concentrations. The two-way ANOVA revealed no significant interaction between the temperature and dietary treatments for the concentrations of the 18 free amino acids and carnosine in the muscle.

Table 1. Effects of a cyclic high ambient temperature and feeding orotic acid on the growth performance parameters of broiler chickens.

	Thermoneutral Temperature (25 ± 1 °C)		High Ambient Temperature (35 ± 1 °C for 8 h/day)		T	O	T × O
	Control	Orotic Acid	Control	Orotic Acid			
Final body weight (g)	1177.96 ± 64.68	1175.04 ± 70.06	1085.43 ± 33.39	1140.79 ± 40.13	N.S.	N.S.	N.S.
Body weight gain (g)	786.36 ± 64.27	777.49 ± 68.44	691.74 ± 36.24	746.31 ± 35.78	N.S.	N.S.	N.S.
Feed intake (g)	1216.59 ± 74.66	1268.61 ± 101.63	1022.19 ± 65.40	1118.18 ± 64.43	<0.05	N.S.	N.S.
Feed conversion ratio	1.56 ± 0.04	1.64 ± 0.05	1.48 ± 0.06	1.50 ± 0.05	N.S.	N.S.	N.S.
Body temperature (°C)	40.18 ± 0.07 ^{c}	40.68 ± 0.16 ^{bc}	41.37 ± 0.20 ^{a}	41.13 ± 0.15 ^{ab}	<0.001	N.S.	<0.05

Results are expressed as mean ± standard error of the mean (SEM) ($n = 8$). Means with the same superscript letter within rows are not significantly different at $p < 0.05$. T: the effect of temperature; O: the effect of feeding orotic acid; T × O: the statistical interaction between temperature and feeding orotic acid; N.S.: not significant.

Table 2. Effects of a cyclic high ambient temperature and feeding orotic acid on the weights of tissues of broiler chickens (g).

	Thermoneutral Temperature (25 ± 1 °C)		High Ambient Temperature (35 ± 1 °C for 8 h/day)		T	O	T × O
	Control	Otrotic Acid	Control	Orotic Acid			
Breast muscle	207.91 ± 11.01 ^{ab}	222.19 ± 13.22 ^{a}	182.65 ± 6.23 ^{b}	188.28 ± 4.70 ^{ab}	<0.01	N.S.	N.S.
Breast tender muscle	45.83 ± 2.54 ^{a}	45.98 ± 2.49 ^{a}	40.25 ± 1.21 ^{a}	41.52 ± 1.30 ^{a}	<0.05	N.S.	N.S.
Leg muscles	226.10 ± 16.26 ^{a}	215.90 ± 13.47 ^{a}	216.70 ± 6.56 ^{a}	220.00 ± 8.56 ^{a}	N.S.	N.S.	N.S.
Liver	22.20 ± 1.97 ^{ab}	23.41 ± 1.72 ^{a}	17.78 ± 0.65 ^{b}	19.35 ± 1.31 ^{ab}	<0.05	N.S.	N.S.
Heart	6.31 ± 0.54 ^{a}	6.44 ± 0.66 ^{a}	3.83 ± 0.16 ^{b}	4.38 ± 0.34 ^{b}	<0.001	N.S.	N.S.
Abdominal fat tissue	3.88 ± 1.25 ^{b}	4.35 ± 1.00 ^{b}	9.71 ± 1.55 ^{a}	9.83 ± 1.24 ^{a}	<0.001	N.S.	N.S.

Results are expressed as mean ± standard error of the mean (SEM) ($n = 8$). Means with the same superscript letter within rows are not significantly different at $p < 0.05$. T: the effect of temperature; O: the effect of feeding orotic acid; T × O: the statistical interaction between temperature and feeding orotic acid; N.S.: not significant.

Table 3. Effects of a cyclic high ambient temperature and feeding orotic acid on plasma-free amino acids of broiler chickens (μM).

	Thermoneutral Temperature (25 ± 1 °C)		High Ambient Temperature (35 ± 1 °C for 8 h/day)		T	O	T × O
	Control	Orotic Acid	Control	Orotic Acid			
Aspartic acid	15.82 ± 1.63	19.18 ± 3.13	12.00 ± 0.70	15.72 ± 2.10	N.S.	0.04	N.S.
Glutamic acid	28.52 ± 6.09	36.27 ± 3.15	32.13 ± 1.92	38.20 ± 3.82	N.S.	0.02	N.S.
Asparagine	17.56 ± 2.33	12.83 ± 3.05	14.04 ± 1.60	13.33 ± 2.06	N.S.	N.S.	N.S.
Serine	131.48 ± 8.88 ^{ab}	151.03 ± 11.88 ^{a}	98.78 ± 10.14 ^{b}	95.74 ± 6.92 ^{b}	<0.001	N.S.	N.S.
Glutamine	134.46 ± 8.69 ^{a}	116.84 ± 13.86 ^{a}	85.65 ± 10.75 ^{b}	81.62 ± 8.81 ^{b}	<0.05	N.S.	N.S.
Histidine	124.47 ± 11.15	151.57 ± 11.35	155.82 ± 12.55	161.53 ± 17.13	N.S.	N.S.	N.S.

Table 3. *Cont.*

	Thermoneutral Temperature (25 ± 1 °C)		High Ambient Temperature (35 ± 1 °C for 8 h/day)		T	O	T × O
	Control	Orotic Acid	Control	Orotic Acid			
Glycine	65.28 ± 6.59	85.55 ± 9.03	73.16 ± 7.27	69.77 ± 6.17	N.S.	N.S.	N.S.
Threonine	11.04 ± 1.05	17.10 ± 3.01	12.36 ± 1.93	14.60 ± 3.79	N.S.	N.S.	N.S.
Arginine	57.92 ± 5.42	58.30 ± 10.47	73.64 ± 8.20	60.60 ± 7.94	N.S.	N.S.	N.S.
Tyrosine	30.25 ± 2.00 ^{a}	22.66 ± 2.78 ^{ab}	24.09 ± 4.25 ^{ab}	15.67 ± 1.73 ^{b}	<0.05	<0.01	N.S.
Valine	21.28 ± 2.46	24.30 ± 2.38	20.02 ± 2.05	18.24 ± 1.49	N.S.	N.S.	N.S.
Methionine	13.89 ± 1.11	13.49 ± 2.35	12.27 ± 1.49	12.00 ± 1.12	N.S.	N.S.	N.S.
Tryptophan	19.69 ± 0.99	19.19 ± 1.17	19.77 ± 1.39	18.31 ± 1.01	N.S.	N.S.	N.S.
Phenylalanine	32.98 ± 3.11	33.37 ± 4.31	31.49 ± 3.99	27.36 ± 1.36	N.S.	N.S.	N.S.
Isoleucine	44.95 ± 4.43	44.59 ± 5.31	40.63 ± 5.29	35.77 ± 1.26	N.S.	N.S.	N.S.
Leucine	12.57 ± 1.20	12.83 ± 1.47	11.59 ± 1.87	9.68 ± 0.88	N.S.	N.S.	N.S.
Lysine	59.65 ± 10.69	75.17 ± 13.33	83.31 ± 14.47	72.88 ± 12.45	N.S.	N.S.	N.S.
Proline	51.70 ± 5.46	55.55 ± 4.10	64.98 ± 7.26	64.56 ± 6.29	N.S.	N.S.	N.S.

Results are expressed as mean ± standard error of the mean (SEM) ($n = 8$). Means with the same superscript letter within rows are not significantly different at $p < 0.05$. T: the effect of temperature; O: the effect of feeding orotic acid; T × O: the statistical interaction between temperature and feeding orotic acid; N.S.: not significant.

Table 4. Effects of a cyclic high ambient temperature and feeding orotic acid on plasma metabolites of broiler chickens.

	Thermoneutral Temperature (25 ± 1 °C)		High Ambient Temperature (35 ± 1 °C for 8 h/day)		T	O	T × O
	Control	Orotic Acid	Control	Orotic Acid			
Metabolites affected by temperature							
Nicotinic acid	100 ± 40 ^{ab}	76 ± 26 ^{b}	248 ± 95 ^{ab}	336 ± 94 ^{a}	< 0.01	N.S.	N.S.
Methionine	100 ± 25 ^{ab}	136 ± 27 ^{a}	39 ± 7 ^{b}	68 ± 17 ^{ab}	< 0.01	N.S.	N.S.
Galactosamine	100 ± 25 ^{a}	92 ± 28 ^{a}	30 ± 10 ^{ab}	20 ± 7 ^{b}	< 0.01	N.S.	N.S.
Uric acid	100 ± 28 ^{b}	148 ± 55 ^{ab}	521 ± 161 ^{ab}	621 ± 201 ^{a}	< 0.01	N.S.	N.S.
Xanthine	100 ± 23 ^{b}	269 ± 84 ^{ab}	585 ± 176 ^{a}	635 ± 112 ^{a}	< 0.01	N.S.	N.S.
Xanthosine monophosphate	100 ± 18	103 ± 26	61 ± 15	48 ± 11	< 0.05	N.S.	N.S.
Oleic acid	100 ± 29	70 ± 22	36 ± 13	29 ± 9	< 0.05	N.S.	N.S.
Thymine	100 ± 21	94 ± 18	155 ± 24	199 ± 55	< 0.05	N.S.	N.S.
Aspartic acid	100 ± 30	149 ± 40	49 ± 11	62 ± 16	< 0.05	N.S.	N.S.
Dihydrouracil	100 ± 21	119 ± 27	193 ± 48	189 ± 42	< 0.05	N.S.	N.S.
Ascorbic acid	100 ± 16	103 ± 18	64 ± 11	72 ± 13	< 0.05	N.S.	N.S.
Inosine	100 ± 22	100 ± 18	191 ± 62	192 ± 49	< 0.05	N.S.	N.S.
Ornithine	100 ± 26	112 ± 22	62 ± 11	59 ± 16	< 0.05	N.S.	N.S.
3-Phenyllactic acid	100 ± 17	114 ± 26	192 ± 65	257 ± 82	< 0.05	N.S.	N.S.

Table 4. *Cont.*

	Thermoneutral Temperature (25 ± 1 °C)		High Ambient Temperature (35 ± 1 °C for 8 h/day)		T	O	T × O
	Control	Orotic Acid	Control	Orotic Acid			
Cysteine	100 ± 21	124 ± 25	60 ± 12	79 ± 12	< 0.05	N.S.	N.S.
Glutaric acid	100 ± 25	135 ± 32	164 ± 44	258 ± 70	< 0.05	N.S.	N.S.
2-Hydroxyglutaric acid	100 ± 23	105 ± 17	169 ± 47	165 ± 33	< 0.05	N.S.	N.S.
Sucrose	100 ± 45	152 ± 70	35 ± 7	48 ± 11	< 0.05	N.S.	N.S.
Asparagine	100 ± 23	108 ± 27	56 ± 10	72 ± 12	< 0.05	N.S.	N.S.
Serine	100 ± 16	110 ± 17	77 ± 15	68 ± 14	< 0.05	N.S.	N.S.
Hypoxanthine	100 ± 18	123 ± 22	145 ± 34	182 ± 30	< 0.05	N.S.	N.S.
Metabolites affected by orotic acid							
Niacinamide	100 ± 17 [a]	66 ± 12 [ab]	95 ± 19 [ab]	42 ± 10 [b]	N.S.	< 0.01	N.S.
β-Alanine	100 ± 25 [ab]	191 ± 33 [a]	78 ± 18 [b]	164 ± 38 [ab]	N.S.	< 0.01	N.S.
Uridine	100 ± 16	167 ± 28	90 ± 19	171 ± 36	N.S.	< 0.05	N.S.
Guanine	100 ± 18	147 ± 21	110 ± 19	156 ± 22	N.S.	< 0.05	N.S.
3-Hydroxyanthranilic acid	100 ± 27 [b]	166 ± 46 [ab]	150 ± 22 [ab]	271 ± 65 [a]	N.S.	< 0.05	N.S.
Glycerol 3-phosphate	100 ± 28	151 ± 44	66 ± 14	140 ± 17	N.S.	< 0.05	N.S.
2′-Deoxyuridine	100 ± 15	152 ± 27	107 ± 30	171 ± 36	N.S.	< 0.05	N.S.
Cytosine	100 ± 23 [ab]	115 ± 38 [ab]	74 ± 13 [b]	195 ± 42 [a]	N.S.	< 0.05	N.S.
3-Aminoisobutyric acid	100 ± 34	189 ± 61	100 ± 16	269 ± 94	N.S.	< 0.05	N.S.
Phenylacetic acid	100 ± 29	180 ± 55	59 ± 9	127 ± 35	N.S.	< 0.05	N.S.
Metabolites affected by temperature and orotic acid							
Uracil	100 ± 18 [c]	273 ± 55 [ab]	173 ± 31 [bc]	379 ± 67 [a]	< 0.05	< 0.01	N.S.
Orotic acid	100 ± 45 [ab]	221 ± 30 [a]	28 ± 6 [b]	160 ± 34 [a]	< 0.05	< 0.01	N.S.

The relative quantities of the metabolites were means ± SEM ($n = 8$) and expressed as percentage of an arbitrary control set to 100%. Means with the same superscript letter within rows are not significantly different at $p < 0.05$. T: the effect of temperature; O: the effect of feeding orotic acid; T × O: the statistical interaction between temperature and feeding orotic acid; N.S.: not significant.

Table 5. Effects of a cyclic high ambient temperature and feeding orotic acid on metabolisms of broiler chickens.

Metabolism Name	*p*-Value
Metabolic pathways affected by high ambient temperature	
Purine metabolism	<0.01
Ammonia recycling	<0.01
Pyrimidine metabolism	<0.01
Homocysteine degradation	<0.05
Glutamate metabolism	<0.05

Table 5. *Cont.*

Metabolism Name	*p*-Value
Urea cycle	<0.05
β-Alanine metabolism	<0.05
Glycine and serine metabolism	<0.05
Aspartate metabolism	<0.05
Metabolic pathway affected by orotic acid	
Pyrimidine metabolism	<0.01
β-Alanine metabolism	<0.01
Malate–aspartate shuttle	<0.01
Aspartate metabolism	<0.05

Table 6. Effects of a cyclic high ambient temperature and feeding orotic acid on free amino acids and carnosine in the breast muscle of broiler chickens (mg/100 g).

	Thermoneutral Temperature (25 ± 1 °C)		High Ambient Temperature (35 ± 1 °C for 8 h/day)		T	O	T × O
	Control	Orotic Acid	Control	Orotic Acid			
Aspartic acid	5.91 ± 0.78	6.99 ± 0.48	4.99 ± 0.32	5.93 ± 0.73	N.S.	N.S.	N.S.
Glutamic acid	18.74 ± 3.10	19.30 ± 2.69	21.29 ± 1.39	21.91 ± 1.48	N.S.	N.S.	N.S.
Asparagine	5.98 ± 0.56	4.91 ± 0.21	5.25 ± 0.25	5.41 ± 0.26	N.S.	N.S.	N.S.
Serine	21.37 ± 1.60 ab	22.27 ± 1.59 a	16.49 ± 1.41 b	16.76 ± 1.45 b	< 0.05	N.S.	N.S.
Glutamine	25.59 ± 1.47 a	21.98 ± 2.18 ab	17.00 ± 1.50 b	18.84 ± 2.23 b	<0.05	N.S.	N.S.
Histidine	1.11 ± 0.32	0.57 ± 0.08	1.31 ± 0.27	0.79 ± 0.05	N.S.	<0.05	N.S.
Glycine	33.19 ± 11.45	31.47 ± 4.48	37.68 ± 7.35	35.29 ± 4.71	N.S.	N.S.	N.S.
Threonine	12.37 ± 1.82	15.21 ± 1.66	14.92 ± 0.53	13.97 ± 0.91	N.S.	N.S.	N.S.
Arginine	9.77 ± 1.25 b	11.83 ± 1.34 ab	16.63 ± 2.13 ab	13.82 ± 1.34 ab	<0.05	N.S.	N.S.
Tyrosine	5.65 ± 0.94	5.82 ± 0.47	6.74 ± 0.32	4.78 ± 0.24	N.S.	N.S.	N.S.
Valine	3.23 ± 0.35	4.19 ± 0.62	4.21 ± 0.28	3.82 ± 0.31	N.S.	N.S.	N.S.
Methionine	1.25 ± 0.27 b	2.19 ± 0.42 a	1.57 ± 0.13 b	1.28 ± 0.25 b	<0.05	<0.05	<0.05
Tryptophan	5.03 ± 0.27	5.62 ± 0.93	4.98 ± 0.41	4.74 ± 0.64	N.S.	N.S.	N.S.
Phenylalanine	3.60 ± 0.20	3.89 ± 0.38	4.62 ± 0.29	4.24 ± 0.28	<0.05	N.S.	N.S.
Isoleucine	2.66 ± 0.22	3.11 ± 0.50	2.76 ± 0.24	2.59 ± 0.26	N.S.	N.S.	N.S.
Leucine	4.03 ± 0.44	4.35 ± 0.59	4.02 ± 0.31	3.85 ± 0.34	N.S.	N.S.	N.S.
Lysine	6.30 ± 2.20	11.36 ± 1.56	11.44 ± 1.58	10.38 ± 1.01	N.S.	N.S.	N.S.
Proline	4.37 ± 0.53	4.40 ± 0.35	4.23 ± 0.27	4.71 ± 0.66	N.S.	N.S.	N.S.
Carnosine	364.01 ± 69.97 ab	446.29 ± 41.85 a	236.35 ± 26.80 b	437.63 ± 34.67 ab	N.S.	<0.05	N.S.

Results are expressed as mean ± standard error of the mean (SEM) ($n = 8$). Means with the same superscript letter within rows are not significantly different at $p < 0.05$. T: the effect of temperature; O: the effect of feeding orotic acid; T × O: the statistical interaction between temperature and feeding orotic acid; N.S.: not significant.

3. Discussion

Reactive oxygen species (ROS), such as singlet oxygen, hydrogen peroxide, and hydroxyl radicals, are highly reactive molecules produced by mitochondria [19]. Under high ambient temperature conditions, ROS generation increases in various body tissues of chickens as the heat load elevates [3], and consequently oxidizes and impairs lipids, proteins, and DNA [20]. Because chicken muscle has a high polyunsaturated fatty acid content, making it more sensitive to oxidative deterioration [21], the oxidation of such lipids negatively influences their industrial values (i.e., growth performance, meat yield, and meat quality) [4–14].

Under HT conditions, chickens have been known to show lower growth rate and feed efficiency, accompanied by decreasing meat yield [4,5]. In this study, although the HT environment did not affect the final body weight, body weight gain, or feed conversion ratio of the broiler chickens, it significantly depressed their feed intake and increased their average body temperature. These results concur with those of a previous study, which examined the effects of a similar HT environment [13]. In addition, the HT environment affected the weights of the breast muscles, breast tender muscles, livers, hearts, and abdominal fat tissue. In particular, the yield of breast muscle was significantly decreased by the HT environment. Furthermore, the plasma lipid peroxidation level was higher in the chickens kept under the HT conditions than those kept in the thermoneutral environment. Therefore, these results suggest that the HT environment used in this study could realistically induce the negative effects commonly observed in broiler chickens kept in heat stress-inducing environments.

In this study, most of the plasma amino acid concentrations were unaffected by HT, with the exception of the serine, tyrosine, and glutamine concentrations. This was particularly notable in the chickens fed the basal diet, in which the HT environment significantly decreased the plasma glutamine concentration. This concurred with the results of a previous study, which reported that chickens kept under HT conditions had lower plasma glutamine concentrations, in addition to poorer performance and carcass characteristics [22]. Although glutamine is a non-essential amino acid, it has been reported to promote enterocyte proliferation and survival, and regulate intestinal barrier function under a variety of stress conditions [23]. Furthermore, intestinal morphology and permeability were found to be disrupted under HT conditions, with this being accompanied by an increase in the plasma endotoxin concentration in broiler chickens [24]. These results suggest that glutamine is utilized to maintain intestinal integrity in broiler chickens under heat-stress conditions.

In addition, glutamine is known to be a precursor for the synthesis of purine and pyrimidine nucleotides, which are essential for DNA synthesis and the proliferation of cells [25]. In rats, GC-MS and liquid chromatography–mass spectrometry-based metabolomics analysis of the plasma indicated that short-term chronic heat exposure (37 °C for 48 h) altered pyrimidine and purine degradation [26]. To investigate the effects of an HT environment on metabolite profiles in the plasma of chickens, we performed untargeted GC-MS/MS-based metabolomics analysis. This analysis identified 172 metabolites in the plasma, of which 23 were significantly affected by the HT environment. The HT environment particularly affected the plasma levels of purine- (inosine, hypoxanthine, xanthine, xanthosine monophosphate, and uric acid) and pyrimidine-related (uracil, dihydrouracil, and thymine) metabolites. The changes in these metabolites are illustrated in Figure 2, with a general increase in concentration in response to the HT environment being apparent.

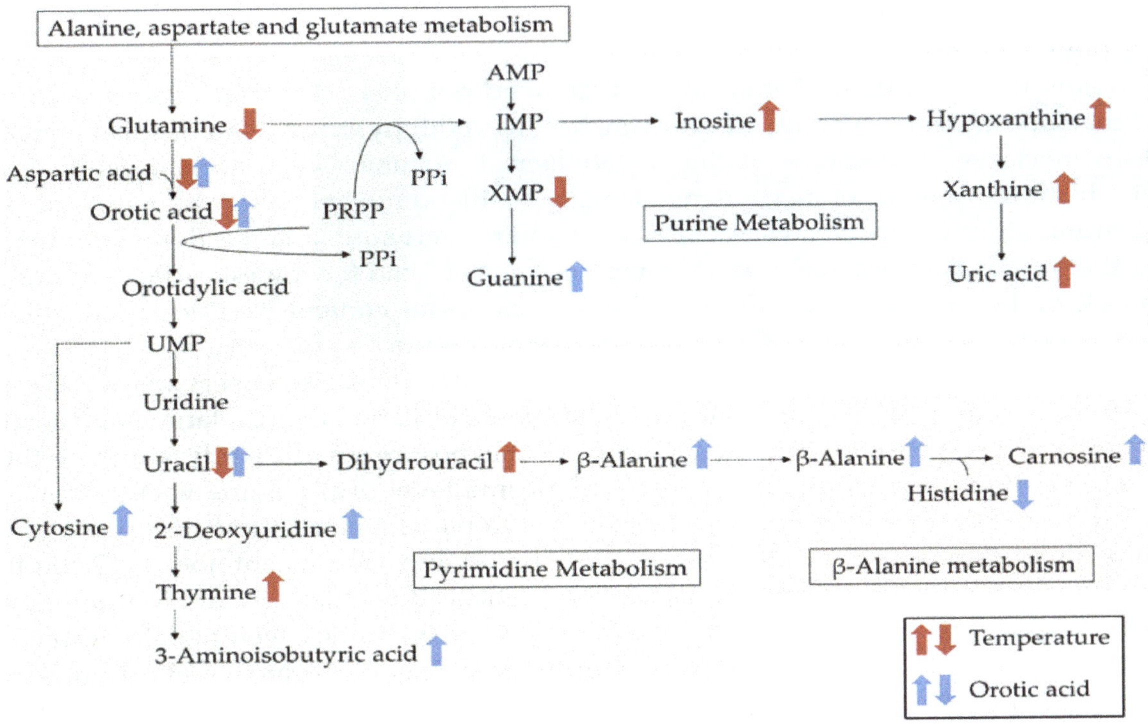

Figure 2. Integrated overview of the metabolic changes induced by either a cyclic high ambient temperature or by feeding orotic acid. Red arrows indicate high ambient temperature-induced changes in the metabolites, and blue arrows indicate orotic acid-induced changes. The metabolic scheme was based on information gathered from the KEGG PATHWAY Database (http://www.genome.jp/kegg/pathway.html).

One possible explanation for the changes in the metabolite concentrations observed may be the higher oxidative stress levels in the broiler chickens kept under the HT conditions. Uric acid is known to act as an antioxidant in the plasma [27], and is thought to stabilize ascorbic acid [28]. However, the concentration of ascorbic acid was lower in the plasma of the chickens kept in the HT environment than in the chickens kept under thermoneutral conditions. Another possibility is that these changes were linked the higher energy expenditure of the broiler chickens kept at HT. When broiler chickens are exposed to environmental temperatures above 31 °C, they increase heat production [29], suggesting an increase in adenosine triphosphate (ATP) production. Adenylate kinase, which acts as a regulator for phosphate nucleotide levels inside cells, converts two adenosine diphosphate (ADP) molecules to one ATP molecule and one adenosine monophosphate (AMP) molecule [30]. Since the concentration of AMP is maintained at a much lower level than that of either ADP or ATP [31], the generated AMP may be catabolized to form hypoxanthine, xanthine, or uric acid.

The enrichment analysis identified that the HT environment induced changes in purine metabolism, ammonia recycling, pyrimidine metabolism, homocysteine degradation, glutamate metabolism, the urea cycle, β-alanine metabolism, glycine and serine metabolism, and aspartate metabolism. Seven of these pathways (excluding homocysteine degradation and aspartate metabolism) concur with the findings of a previous study on the effects of short-term chronic heat exposure on the plasma metabolomic profiles of chicks [15]. Furthermore, the plasma homocysteine concentration is considered an oxidative stress marker [32]. In Japanese quail, HT increased both the homocysteine concentration and the lipid peroxidation level in the plasma [33,34], and it has been suggested that chickens kept under HT conditions have higher plasma homocysteine concentrations. Furthermore, the untargeted GC-MS/MS-based metabolomics analysis indicated that the HT environment decreased the plasma level of aspartic acid, which is a precursor for the synthesis of pyrimidines. These results suggest that although purine and pyrimidine synthesis pathways in chickens were enhanced in response to the HT conditions, the levels of related metabolites (e.g., glutamine, aspartic acid, and ascorbic acid) may have been disrupted.

It is noteworthy that the HT environment changed not only pyrimidine metabolism, but also aspartate metabolism and β-alanine metabolism, because both pyrimidine metabolism and aspartate metabolism are closely related to β-alanine metabolism. β-alanine is one of constituents of carnosine (β-alanyl-L-histidine) and anserine (β-alanyl-1-methyl-L-histidine), which are known to be present at high concentrations in the breast muscle of chickens and to exert antioxidant activity [35]. In this study, we found that the chickens fed the basal diet and kept at a HT had the lowest carnosine content in the breast muscle of the three treatment groups. This low carnosine content may explain the higher lipid peroxidation level in the breast muscle samples from this group.

An HT-induced decrease in the muscle carnosine content was also observed by Yang et al. [36], while Tomonaga et al. [15] reported that the constituents of carnosine (β-alanine and histidine) in chicks decreased in response to short-term chronic heat exposure. In this study, although the plasma dihydrouracil concentration was the highest, the plasma level of β-alanine was the lowest in the chickens fed the basal diet and kept under the HT conditions. Therefore, these results suggest that the conversion of dihydrouracil to β-alanine may be disrupted in chickens kept under HT conditions.

This study also evaluated the effects of feeding orotic acid to chickens, and found that it did not affect growth performance, as indicated by the final body weight, body weight gain, feed conversion ratio, feed intake, and tissue weights. However, orotic acid did ameliorate the HT-induced increases in lipid peroxidation level in the plasma of the chickens. This amelioration may have partially been due to the antioxidative activity of orotic acid [37], with the plasma level of orotic acid decreasing under the HT conditions and increasing in response to the orotic acid supplementation. The GC-MS/MS-based metabolomics analysis of the plasma indicated that 12 metabolites were affected by orotic acid supplementation, some of which have been reported to show antioxidative activity (e.g., uridine and 3-hydroxyanthranilic acid) [38,39] and appeared to be increased by orotic acid supplementation. Furthermore, the high-performance liquid chromatography analysis revealed that orotic acid significantly affected the plasma concentrations of aspartic acid, glutamic acid, and tyrosine. Of these three amino acids, aspartic acid and glutamic acid appeared to be increased by the orotic acid supplementation. Aspartic acid is converted into glutamic acid via the citric acid cycle [40], and glutamic acid is a component amino acid of glutathione, a well-known natural antioxidant [41,42]. These results suggest that alterations in the concentrations of these metabolites and amino acids may contribute to the amelioration of the HT environment-induced increase in the plasma lipid peroxidation level. Furthermore, orotic acid supplementation significantly affected the carnosine content of the breast muscle of the chickens. As previously mentioned, carnosine plays a role as an antioxidant [35], and the reduction of the lipid peroxidation level in the breast muscle tissue may therefore be linked to the orotic acid-induced increase in the carnosine content.

The GC-MS/MS-based metabolomics analysis indicated that orotic acid affected the plasma levels of pyrimidine-related metabolites (orotic acid, uracil, uridine, 2′-deoxyuridine, and β-alanine). This effect was likely linked to the entry of orotic acid into the de novo synthesis pathway for pyrimidines beyond the rate-limiting step. Furthermore, the de novo synthesis pathways for both pyrimidines and purines require 5-phosphoribosyl-1-pyrophosphate (PRPP), and the synthesis of PRPP by PRPP synthetase is the rate-limiting step for both pathways. In rats, it has been reported that orotic acid stimulates hepatic purine biosynthesis [43]. These results suggest that orotic acid may up-regulate the de novo synthesis of purines by providing additional PRPP. However, there was no significant interaction between the temperature and dietary treatments for the plasma metabolites. Further studies are required to understand the effects of orotic acid supplementation on the regulation of the de novo synthesis pathways for pyrimidines and purines in chickens.

The enrichment analysis identified changes in not only pyrimidine metabolism, but also in β-alanine metabolism, the malate–aspartate shuttle, and aspartate metabolism, in response to supplementing orotic acid. Of these, β-alanine metabolism and aspartate metabolism may be particularly important for the alleviation of HT-induced negative effects, as they were also found to be altered by HT conditions. Orotic acid supplementation was found to significantly increase the plasma level of β-alanine and the carnosine content of the breast muscles of the chickens. In chicks, orally administered β-alanine increased the carnosine content of the brain and muscles in a dose-dependent manner [44]. Therefore, these results suggest that orotic acid increases the muscle carnosine content by altering β-alanine metabolism, and thereby maintains the antioxidative capacity of broiler chickens under HT conditions. However, the mechanisms by which orotic acid supplementation increased the plasma level of β-alanine, even under HT conditions, remain unclear. Further research evaluating the effects of HT and orotic acid supplementation on the gene expression and activities of enzymes related to the conversion of dihydrouracil to β-alanine (i.e., dihydropyrimidinase and β-ureidopropionase) in chickens is necessary if we are to improve our insight into the reason for the HT-induced decrease in the muscle carnosine content and the alleviating effects of orotic acid supplementation.

As mentioned above, HT conditions severely increase lipid peroxidation levels [6,7], and consequently reduce meat quality (i.e., increase in drip loss, decrease in share force value, and change in meat color) [8–14]. The breast muscles of the chickens kept under cyclic HT conditions showed the highest lipid peroxidation levels, in conjunction with the lowest carnosine content, suggesting that their meat was of lower quality than that of the control chickens. In contrast, orotic acid increased the muscle carnosine content, possibly via pyrimidine metabolism and β-alanine metabolism, and consequently alleviated the HT environment-induced increase in the muscle lipid peroxidation levels. Cong et al. [45] reported that dietary supplementation of carnosine improved meat quality, antioxidant capacity, and lipid peroxidation status in broiler chickens. In addition, carnosine exerts antioxidant activity, even in cocked chicken meat [35]. These results suggest that orotic acid supplementation may have a positive effect on either carnosine content in meat or the meat quality of chickens.

4. Materials and Methods

4.1. Animals and Experimental Design

All experimental protocols and procedures were reviewed and approved by the Animal Care and Use Committee of Kagoshima University (approval number A18010). One hundred 1-day-old male broiler chicks (Chunky strain ROSS 308) were obtained from a commercial hatchery (Kumiai Hina Center, Kagoshima, Japan). Chicks were housed in an electrically-heated battery brooder and provided with water and a commercial diet (23% crude protein, 12.8 MJ/kg; Nichiwa Sangyou Company, Hyogo, Japan) until they were 14 days old. On day 14, 32 chicks were randomly selected from the group of 100. These chicks were housed individually in wire-bottomed aluminum cages (50 × 40 × 60 cm) and fed the basal diet (Table 7) for 3 days until the start of the main experimental period. The chicks were then randomly allocated to one of four groups, with the main experimental factors being diet and ambient temperature, in a 2 × 2 factorial design. The dietary treatments consisted of the basal diet or the basal diet supplemented with 0.7% of Lactoserum (Matsumoto Trading Co., Ltd., Tokyo, Japan), a food grade orotic acid that contains more than 98% of orotic acid as monohydrate form (i.e., more than 87.9% of orotic acid), and the temperature treatments consisted of either a thermoneutral environment at 25 ± 1 °C or an HT environment at 35 ± 1 °C. The experiment was conducted in a temperature-controlled room with 24 h of light and 50–70% relative humidity. The chicks assigned to the HT treatment groups were kept at 35 ± 1 °C for 8 h every day to mimic a realistic summer environment. At 32 days old, all the chickens were weighed, anesthetized by carbon dioxide, and killed by cervical dislocation. The chickens were then dissected, and the weights of the breast muscles (pectoralis major muscle), breast tender muscles (pectoralis minor muscle), leg muscles (thigh and drumstick), livers, hearts, and abdominal fat tissue depots were recorded. Blood samples were collected in heparinized test tubes, centrifuged at 5900× *g* for 10 min at 4 °C to separate the plasma, and stored at −30 °C until analysis.

Table 7. Composition and analysis of the basal diet.

	Ingredients (g/100 g)
Corn meal	57.90
Soybean meal	34.00
Corn oil	4.30
$CaCO_3$	0.66
$CaHPO_4$	2.00
NaCl	0.50
DL-Methionine	0.14
Mineral and vitamin premix [1]	0.50
Calculated analysis	
Crude protein (%)	20.00
Metabolizable energy (Mcal/kg)	3.10

[1] Content per kg of the vitamin and mineral premix: vitamin A = 90 mg, vitamin D3 = 1 mg, DL-alpha-tocopherol acetate = 2000 mg, vitamin K3 = 229 mg, thiamin nitrate = 444 mg, riboflavin = 720 mg, calcium d-pantothenate = 2174 mg, nicotinamide = 7000 mg, pyridoxine hydrochloride = 700 mg, biotin = 30 mg, folic acid = 110 mg, cyanocobalamine = 2 mg, calcium iodinate = 108 mg, MgO = 198,991 mg, $MnSO_4$ = 32,985 mg, $ZnSO_4$ = 19,753 mg, $FeSO_4$ = 43,523 mg, $CuSO_4$ = 4019 mg, and choline chloride = 299,608 mg.

4.2. Determination of MDA Concentration

To evaluate the oxidative stress levels, the MDA concentrations in the breast muscles and plasma were determined colorimetrically, using the thiobarbituric acid reactive substances assay, as described by Yagi [46] and Ohkawa et al. [47].

4.3. Determination of Free Amino Acid Concentrations

The analysis of the free amino acid and carnosine concentrations in the plasma and breast muscle tissue samples from the chickens was performed using a pre-column technique with liquid chromatography, according to previously reported methods [48]. A liquid chromatography system with automated pre-column derivatization functionality was used in the analysis (Nexera X2; Shimadzu Corporation, Kyoto, Japan). A total of 21 compounds were measured in the analysis, including the following basic amino acids and associated molecules: alanine, anserine, arginine, asparagine, aspartic acid, carnosine, glutamic acid, glutamine, glycine, histidine, isoleucine, leucine, lysine, methionine, phenylalanine, proline, serine, threonine, tryptophan, tyrosine, and valine. The concentrations of the amino acids in the plasma are expressed in μmol/L, and those in the breast muscle are expressed in mg/100 g. In this study, alanine and anserine were not separated, and carnosine was not detected in the plasma.

4.4. Sample Preparation for GC-MS/MS Analysis

Fifty μL aliquots of plasma were suspended in 250 μL of methanol/chloroform/water (5:2:2), with 5 μL of 1 mg/mL 2-isopropylmalic acid as the internal standard. The samples were then mixed in a shaker at 1200 rpm at 37 °C for 30 min, and then centrifuged at 16,000× *g* at 4 °C for 5 min. Next, 225 μL of the supernatant was mixed with 200 μL of distilled water and vortex-mixed, followed by centrifugation at 16,000× *g* at 4 °C for 5 min. Subsequently, 250 μL of the supernatant was dried under a vacuum using a centrifugal evaporator (RD-400; Yamato Scientific, Tokyo, Japan), after cooling at −80 °C for 10 min. Methoxyamine hydrochloride in pyridine (20 mg/mL, 40 μL) was then added to the tubes, and they were vortex-mixed, then shaken at 1200× *g* at 30 °C for 90 min in the dark to allow oximation. N-methyl-N-trimethylsilyltrifluoroacetamine (20 μL) was then added to each tube, and the contents were vortex-mixed. To prepare trimethylsilyl derivatives, the tubes were shaken at 1200× *g* at 37 °C for 45 min in the dark.

4.5. GC-MS/MS Analysis and Data Processing

GC-MS/MS analysis was performed as previously described [49], using a GCMS-TQ8050 (Shimadzu Corporation, Kyoto, Japan). A 30 m × 0.25 mm (internal diameter) BPX-5 column (SGE, Melbourne, Australia) with a 0.25 µm film thickness was used, according to the method described in the Smart Metabolites Database (Shimadzu, Kyoto, Japan).

Data processing was performed using the Smart Metabolites Database (Shimadzu, Kyoto, Japan), MS-DIAL version 3.08 [50], and the MRMPROBS program version 2.42 [51]. Peaks were recorded for the 45–600 m/z mass range, and were automatically detected via MS-DIAL using the peak detection option of a minimum peak height of 2000. A data quality check was conducted using the thresholds of $-10 < RI < 10$, dot production > 0.8, and presence > 0.6, and the remaining data was then manually checked. Ultimately, 172 metabolites were identified in the plasma samples. The relative quantities of the metabolites were calculated using the peak areas of each metabolite relative to that of the internal standard (2-isopropylmalic acid), and expressed as a percentage of an arbitrary control set to 100%.

4.6. Statistical Analysis

The data were analyzed using two-way ANOVA, with individual comparisons being made using Tukey's multiple comparison test. All analyses were performed using R [52]. Statistical significance was set at $p < 0.05$, and the data are expressed as the means ± standard error of the mean (SEM).

Quantitative enrichment analysis using the pathway-associated metabolite sets included in MetaboAnalyst 4.0 [53], which is an established tool for metabolite set enrichment analysis, was performed using the plasma metabolite compounds that were significantly affected by either the HT treatment or orotic acid supplementation. Differences were considered significant at $p < 0.05$.

5. Conclusions

A cyclic HT environment altered nucleic acid metabolism and increased lipid peroxidation levels in the plasma and breast muscle of broiler chickens. Feeding the pyrimidine precursor orotic acid to the chickens altered the plasma concentrations of metabolites mainly related to pyrimidine metabolism and β-alanine metabolism, and ameliorated the increases in lipid peroxidation levels caused by rearing the broiler chickens under HT conditions.

Supplementary Materials: Table S1. Effects of a cyclic high ambient temperature and feeding orotic acid on the muscle yields of broiler chickens. Table S2. Metabolites identified in the plasma of broiler chickens reared under either thermoneutral or high ambient temperature (HT) conditions, and fed either a control or orotic acid-supplemented diet.

Author Contributions: Conceptualization, S.S., K.N., S.F., A.O., and D.I.; methodology, D.I. and S.T.; software, S.T. and D.I.; validation, S.S. and S.T.; formal analysis, S.S. and K.N; investigation, S.S. and K.N.; resources, S.F; original draft preparation, S.S. and K.N; manuscript review and editing, S.S., S.T., S.F., A.O., and D.I.; supervision, A.O.; project administration, D.I. All authors have read and agreed to the published version of the manuscript.

References

1. Ensminger, M.E.; Oldfield, J.E.; Heineman, W.W. *Feeds and Nutrition*, 2nd ed.; Ensminger Pub Co: Prentice Hall: Upper Saddle River, NJ, USA, 1990; pp. 8–110.
2. Sahin, K.; Sahin, N.; Kucuk, O.; Hayirili, A.; Prasad, A.S. Role of dietary zinc in heat stressed poultry: A review. *Poult. Sci.* **2009**, *88*, 2176–2183. [CrossRef] [PubMed]
3. Khan, R.U.; Naz, S.; Nikousefat, Z.; Selvaggi, M.; Laudadio, V.; Tufarelli, V. Effect of ascorbic acid in heat-stressed poultry. *Worlds Poult. Sci. J.* **2012**, *68*, 477–489. [CrossRef]
4. Yalçin, S.; Settar, P.; Ozkan, S.; Cahaner, A. Comparative evaluation of three commercial broiler stocks in hot versus temperate climates. *Poult. Sci.* **1997**, *76*, 921–929. [CrossRef] [PubMed]
5. Yunis, R.; Cahaner, A. The effects of naked neck (Na) and frizzle (F) genes on growth and meat yields of broilers and their interactions with ambient temperatures and potential growth rate. *Poult. Sci.* **1999**, *78*, 1347–1352. [CrossRef]

6. Whitehead, C.C.; Keller, T. An update on ascorbic acid in poultry. *Worlds Poult. Sci. J.* **2003**, *59*, 161–184. [CrossRef]
7. Zeferino, C.P.; Komiyama, C.M.; Pelícia, V.C.; Fascina, V.B.; Aoyagi, M.M.; Coutinho, L.L.; Sartori, J.R.; Moura, A.S.A.M.T. Carcass and meat quality traits of chickens fed diets concurrently supplemented with vitamins C and E under constant heat stress. *Animal* **2016**, *10*, 163–171. [CrossRef]
8. Zhu, X.; Ruusunen, M.; Gusella, M.; Zhou, G.; Puolanne, E. High post-mortem temperature combined with rapid glycolysis induces phosphorylase denaturation and produces pale and exudative characteristics in broiler pectoralis major muscles. *Meat Sci.* **2011**, *89*, 181–188. [CrossRef]
9. Wang, R.H.; Liang, R.R.; Lin, H.; Zhu, L.X.; Zhang, Y.M.; Mao, Y.W.; Dong, P.C.; Niu, L.B.; Zhang, M.H.; Luo, X.X. Effect of acute heat stress and slaughter processing on poultry meat quality and postmortem carbohydrate metabolism. *Poult. Sci.* **2017**, *96*, 738–746. [CrossRef]
10. Sosnicki, A.A.; Greaser, M.L.; Pieterzak, M.; Pospiech, E.; Sante, V. PSE-like syndrome in breast muscle of domestic turkeys: A review. *J. Muscle Foods* **1998**, *9*, 13–23. [CrossRef]
11. Hashizawa, Y.; Kubota, M.; Kadowaki, N.; Fujimura, S. Effect of dietary vitamin E on broiler meat qualities, color, water-holding capacity and shear force value, under heat stress conditions. *Anim. Sci. J.* **2013**, *84*, 732–736. [CrossRef]
12. Petracci, M.; Betti, M.; Cavani, C. Color variation and characterization of broiler breast meat during processing in Italy. *Poult. Sci.* **2004**, *83*, 2085–2092. [CrossRef]
13. Inoue, H.; Simamoto, S.; Takahashi, H.; Kawashima, Y.; Sato, W.; Ijiri, D.; Ohtsuka, A. Effects of astaxanthin-rich dried cell powder from paracoccus carotinifaciens on carotenoid composition and lipid peroxidation in skeletal muscle of broiler chickens under thermo-neutral or realistic high temperature conditions. *Anim. Sci. J.* **2019**, *90*, 229–236. [CrossRef] [PubMed]
14. Al-Sagan, A.A.; Khalil, S.; Hussein, E.O.S.; Attia, Y.A. Effects of fennel seed powder supplementation on growth performance, carcass characteristics, meat quality, and economic efficiency of broilers under thermoneutral and chronic heat stress conditions. *Animals (Basel)* **2020**, *10*, 206. [CrossRef]
15. Tomonaga, S.; Okuyama, H.; Tachibana, T.; Makino, R. Effects of high ambient temperature on plasma metabolomic profiles in chicks. *Anim. Sci. J.* **2018**, *89*, 448–455. [CrossRef] [PubMed]
16. Tabiri, H.Y.; Sato, K.; Takahashi, K.; Toyomizu, M.; Akiba, Y. Effects of heat stress and dietary tryptophan on performance and plasma amino acid concentrations of broiler chickens. *Asian Austral. J. Anim.* **2002**, *15*, 247–253. [CrossRef]
17. Del Vesco, A.P.; Gasparino, E.; Grieser, D.O.; Zancanela, V.; Voltolini, D.M.; Khatlab, A.S.; Guimarães, S.E.; Soares, M.A.; Oliveira Neto, A.R. Effects of methionine supplementation on the expression of protein deposition-related genes in acute heat stress-exposed broilers. *PLoS ONE* **2015**, *25*, e0115821. [CrossRef] [PubMed]
18. Löffler, M.; Carrey, E.A.; Zameitat, E. Orotic acid, more than just an intermediate of pyrimidine *de novo* synthesis. *J Genet. Genom.* **2015**, *42*, 207–219. [CrossRef]
19. Freeman, B.A.; Crapo, J.D. Biology of disease: Free radicals and tissue injury. *Lab. Investig.* **1982**, *47*, 412–426.
20. Liu, D.; Wen, J.; Liu, J.; Li, L. The roles of free radicals in amyotrophic lateral sclerosis: Reactive oxygen species and elevated oxidation of protein, DNA, and membrane phospholipids. *FASEB J.* **1999**, *13*, 2318–2328. [CrossRef]
21. Grashorn, M.A. Functionality of poultry meat. *J. Appl. Poult. Res.* **2007**, *16*, 99–106. [CrossRef]
22. Dai, S.F.; Gao, F.; Zhang, W.H.; Song, S.X.; Xu, X.L.; Zhou, G.H. Effects of dietary glutamine and gamma-aminobutyric acid on performance, carcass characteristics and serum parameters in broilers under circular heat stress. *Anim. Feed Sci. Technol.* **2011**, *168*, 51–60. [CrossRef]
23. Wang, B.; Wu, G.; Zhou, Z.; Dai, Z.; Sun, Y.; Ji, Y.; Li, W.; Wang, W.; Liu, C.; Han, F.; et al. Glutamine and intestinal barrier function. *Amino Acids* **2015**, *47*, 2143–2154. [CrossRef] [PubMed]
24. Nanto-Hara, F.; Kikusato, M.; Ohwada, S.; Toyomizu, M. Heat stress directly affects intestinal integrity in broiler chickens. *J. Poult. Sci.* **2019**. [CrossRef]
25. Wu, G. Intestinal mucosal amino acid catabolism. *Nutr. J.* **1998**, *128*, 1249–1252. [CrossRef] [PubMed]
26. Ippolito, D.L.; Lewis, J.A.; Yu, C.; Leon, L.R.; Stallings, J.D. Alteration in circulating metabolites during and after heat stress in the conscious rat: Potential biomarkers of exposure and organ-specific injury. *BMC Physiol.* **2014**, *14*, 14. [CrossRef]
27. Ames, B.N.; Cathcart, R.; Schwiers, E.; Hochstein, P. Uric acid provides an antioxidant defense in humans

against oxidant- and radical-caused aging and cancer: A hypothesis. *Proc. Natl. Acad. Sci. USA* **1981**, *78*, 6858–6862. [CrossRef]
28. Sevanian, A.; Davies, K.J.; Hochstein, P. Serum urate as an antioxidant for ascorbic acid. *Am. J. Clin. Nutr.* **1991**, *54*, 1129S–1134S. [CrossRef]
29. Yunianto, V.D.; Hayashi, K.; Kaneda, S.; Ohtsuka, A.; Tomita, Y. Effect of environmental temperature on muscle protein turnover and heat production in tube-fed broiler chickens. *Br. J. Nutr.* **1997**, *77*, 897–909. [CrossRef]
30. Dzeja, P.P.; Zeleznikar, R.J.; Goldberg, N.D. Adenylate kinase: Kinetic behavior in intact cells indicates it is integral to multiple cellular processes. *Mol. Cell. Biochem.* **1998**, *184*, 169–182. [CrossRef]
31. Beis, I.; Newsholme, E.A. The contents of adenine nucleotides, phosphagens and some glycolytic intermediates in resting muscles from vertebrates and invertebrates. *Biochem. J.* **1975**, *152*, 23–32. [CrossRef]
32. Miller, J.W.; Nadeau, M.R.; Smith, J.; Smith, D.; Selhub, J. Folate-deficiency-induced homocysteinaemia in rats: Disruption of S-adenosylmethionine's co-ordinate regulation of homocysteine metabolism. *Biochem. J.* **1994**, *298*, 415–419. [CrossRef] [PubMed]
33. Sahin, K.; Onderci, M.; Sahin, N.; Gursu, M.F.; Kucuk, O. Dietary vitamin C and folic acid Supplementation ameliorates the detrimental effects of heat stress in Japanese quail. *J. Nutr.* **2003**, *133*, 1882–1886. [CrossRef] [PubMed]
34. Sahin, K.; Onderci, M.; Sahin, N.; Gursu, M.F.; Khachik, F.; Kucuk, O. Effects of lycopene supplementation on antioxidant status, oxidative stress, performance and carcass characteristics in heat-stressed Japanese quail. *J. Therm. Biol.* **2006**, *31*, 307–312. [CrossRef]
35. O'Neill, L.M.; Galvina, K.; Morrisseya, P.A.; Buckleyb, D.J. Inhibition of lipid oxidation in chicken by carnosine and dietary α-tocopherol supplementation and its determination by derivative spectrophotometry. *Meat Sci.* **1998**, *50*, 479–488. [CrossRef]
36. Yang, P.; Hao, Y.; Feng, J.; Lin, H.; Feng, Y.; Wu, X.; Yang, X.; Gu, X. The expression of carnosine and its effect on the antioxidant capacity of *longissimus dorsi muscle* in finishing pigs exposed to constant heat stress. *Asian-Australas. J. Anim. Sci.* **2014**, *27*, 1763–1772. [CrossRef]
37. Prachayasittikul, S.; Wongsawatkul, O.; Worachartcheewan, A.; Ruchirawat, S.; Prachayasittikul, V. Vasorelaxation and superoxide scavenging activities of orotic acid. *Int. J. Pharmacol.* **2010**, *6*, 413–418.
38. Chadha, R.; Mahal, H.S.; Mukherjee, T.; Kapoor, S. Evidence for a possible role of 3-hydroxyanthranilic acid as an antioxidant. *J. Phys. Org. Chem.* **2008**, *22*, 349–354. [CrossRef]
39. Lu, X.; Li, N.; Qiao, X.; Qiu, Z.; Liu, P. Composition analysis and antioxidant properties of black garlic extract. *J. Food Drug Anal.* **2017**, *25*, 340–349. [CrossRef]
40. Simon, G.; Drori, J.B.; Cohen, M.M. Mechanism of conversion of aspartate into glutamate in cerebral-cortex slices. *Biochem. J.* **1967**, *102*, 153–162. [CrossRef]
41. Hopkins, G.F.; Dixon, M. On glutathione. II. A thermostable oxidation-reduction system. *J. Biol. Chem.* **1992**, *54*, 527–563.
42. Hunter, G.; Eagles, B.A. Glutathione: A critical study. *J. Biol. Chem.* **1927**, *72*, 147–166.
43. Windmueller, H.G.; Spaeth, A.E. Stimulation of hepatic purine biosynthesis by orotic acid. *J. Biol. Chem.* **1965**, *240*, 4398–4405. [PubMed]
44. Tomonaga, T.; Matsumoto, M.; Furuse, M. β-Alanine enhances brain and muscle carnosine levels in broiler chicks. *J. Poult. Sci.* **2012**, *4*, 308–312. [CrossRef]
45. Cong, J.; Zhang, L.; Li, J.; Wang, S.; Gao, F.; Zhou, G. Effects of dietary supplementation with carnosine on meat quality and antioxidant capacity in broiler chickens. *Br. Poult. Sci.* **2017**, *58*, 69–75. [CrossRef]
46. Yagi, K. A simple fluorometric assay for lipoperoxide in blood plasma. *Biochem. Med.* **1976**, *15*, 212–216. [CrossRef]
47. Ohkawa, H.; Ohishi, N.; Yagi, K. Assay for lipid peroxides in animal tissues by thiobarbituric acid reaction. *Anal. Biochem.* **1979**, *95*, 351–358. [CrossRef]
48. Azuma, K.; Hirao, Y.; Hayakawa, Y.; Murahata, Y.; Osaki, T.; Tsuka, T.; Imagawa, T.; Okamoto, Y.; Ito, N. Application of pre-column labeling liquid chromatography for canine plasma-free amino acid analysis. *Metabolites* **2016**, *6*, 3. [CrossRef]
49. Goto, T.; Mori, H.; Shiota, S.; Tomonaga, S. Metabolomics approach reveals the effects of breed and feed on the composition of chicken eggs. *Metabolites* **2019**, *9*, 224. [CrossRef]
50. Lai, Z.; Tsugawa, H.; Wohlgemuth, G.; Mehta, S.; Mueller, M.; Zheng, Y.; Ogiwara, A.; Meissen, J.;

Showalter, M.; Takeuchi, K.; et al. Identifying metabolites by integrating metabolome databases with mass spectrometry cheminformatics. *Nat. Methods* **2018**, *15*, 53–56. [CrossRef]

51. Tsugawa, H.; Kanazawa, M.; Ogiwara, A.; Arita, M. MRMPROBS suite for metabolomics using large-scale MRM assays. *Bioinformatics* **2014**, *30*, 2379–2380. [CrossRef]
52. R Core Team. *R: A Language and Environment for Statistical Computing*; R Foundation for Statistical Computing: Vienna, Austria, 2019; Available online: https://www.R-project.org/ (accessed on 17 December 2019).
53. Chong, J.; Soufan, O.; Li, C.; Caraus, I.; Li, S.; Bourque, G.; Wishart, D.S.; Xia, J. MetaboAnalyst 4.0: Towards more transparent and integrative metabolomics analysis. *Nucleic Acids Res.* **2018**, *46*, 486–494. [CrossRef] [PubMed]

12

Integrative Analysis of Metabolomic and Transcriptomic Profiles Uncovers Biological Pathways of Feed Efficiency in Pigs

Priyanka Banerjee, Victor Adriano Okstoft Carmelo and Haja N. Kadarmideen *

Quantitative Genomics, Bioinformatics and Computational Biology Group, Department of Applied Mathematics and Computer Science, Technical University of Denmark, 2800 Kongens Lyngby, Denmark; priyankabnrj@gmail.com (P.B.); vaocar@dtu.dk (V.A.O.C.)
* Correspondence: hajak@dtu.dk.

Abstract: Feed efficiency (FE) is an economically important trait. Thus, reliable predictors would help to reduce the production cost and provide sustainability to the pig industry. We carried out metabolome-transcriptome integration analysis on 40 purebred Duroc and Landrace uncastrated male pigs to identify potential gene-metabolite interactions and explore the molecular mechanisms underlying FE. To this end, we applied untargeted metabolomics and RNA-seq approaches to the same animals. After data quality control, we used a linear model approach to integrate the data and find significant differently correlated gene-metabolite pairs separately for the breeds (Duroc and Landrace) and FE groups (low and high FE) followed by a pathway over-representation analysis. We identified 21 and 12 significant gene-metabolite pairs for each group. The valine-leucine-isoleucine biosynthesis/degradation and arginine-proline metabolism pathways were associated with unique metabolites. The unique genes obtained from significant metabolite-gene pairs were associated with sphingolipid catabolism, multicellular organismal process, cGMP, and purine metabolic processes. While some of the genes and metabolites identified were known for their association with FE, others are novel and provide new avenues for further research. Further validation of genes, metabolites, and gene-metabolite interactions in larger cohorts will elucidate the regulatory mechanisms and pathways underlying FE.

Keywords: feed efficiency; linear model; metabolomics; pigs; transcriptomics

1. Introduction

Feed represents about 60–70% of total pork production costs in modern pig production. Thus, to decrease costs and increase profitability, it is pivotal to identify feed efficient (FE) animals [1]. However, due to the polygenic architecture of FE, individual pigs in a herd exhibit considerable variation in FE despite belonging to similar genetic background and environment [2]. Considering this variation, different methods have been proposed and widely used to measure the FE, including feed conversion ratio (FCR) and residual feed intake (RFI) [1,3]. FCR is the ratio of feed intake (FI) per unit body weight gain and is affected by many factors such as breed, sex, diet, and environmental conditions [4,5]. Pigs with low FCR are considered high FE and vice-versa. RFI estimates the difference between actual and expected FI predicted on production traits as average daily gain (ADG) [6]. Since FCR considers both FI and weight gain, and FCR is also one of the critical predictors of FE, it suggests that feed efficient pigs may possess different physiological-biochemical profiles compared to the inefficient ones [2].

Based on the advances in omics technologies, several approaches have been adopted to shed light on the genetic mechanisms underlying FE in pigs. Among these omics technologies, transcriptomics and metabolomics have provided tools to elucidate the molecular basis of FE. While transcriptomics allows us to have a transcriptional snapshot of genes underpinning the phenotype under investigation, metabolomics bridges the gap between genotype and phenotype. Recently, an increasing number of metabolomic studies have reported the role of metabolites in economically important traits [7], such as meat quality [8,9], pre-slaughter stress [10], and FE [3]. Likewise, several transcriptomic studies have pointed out candidate genes underpinning FE and other related-traits such as immune response, growth, and metabolism in pigs [11–13].

Recently, we have investigated RNA-seq data on the 41 Danish production pigs that underwent feed efficiency and performance testing trials to identify differentially expressed genes and gene networks and reported 13 genes as potential biomarkers for feed efficiency [14]. Despite the new insights into key genes and molecular mechanisms reported in these studies, these approaches rely solely on data from a single biological layer. It has been shown that the integration of transcriptomics datasets with genomic and other omics datasets (systems genomics) increases the power to detect causal and regulatory factors and molecular pathways underlying complex phenotypes or diseases in animals [15,16].

To gain further insights into biochemical aspects of complex traits, data integration analysis has emerged as a fruitful tool, unveiling potential biomarkers via integration of metabolomics and transcriptomics [17]. By the development of analytical technologies for data integration, the assessment of system-wide changes of transcript levels as surrogate measurements of metabolic rearrangements can be widely assessed. Metabolite-transcript co-responses using combined profiling can yield vital information on the complex biological regulation of the trait. Transcriptome-metabolome integration is a powerful combination as the metabolome is affected by the phenotypic measurements to which the global measures of transcriptome can be anchored [18]. Therefore, herein, we integrated data from metabolome-transcriptome approaches to unveil the unique gene-metabolite pairs. To this end, we adopted a two-step framework, as follows: (1) we first employed the numerical integration of gene-metabolite levels to identify gene-metabolite interaction pairs separately for the breeds (Duroc and Landrace) and FE groups (low and high FE) using IntLIM R-package; (2) next, a knowledge-based integration approach based on pathway over-representation analysis was used to reveal the underlying pathways in each group (breed-specific and FE-specific). To the best of our knowledge, this is the first study of its kind to ever combine high throughput metabolomics data with RNA sequence based gene expression data in pigs to unravel the missing links between genes and metabolites and to shed light on the molecular basis that characterizes the specific differences based on breed and feed efficiency.

2. Results

2.1. Descriptive Statistics and Linear Model Analysis for genes and Metabolites

The data on 749 metabolites and 25,880 genes from 40 samples were generated using untargeted metabolomics and transcriptomic approaches, respectively. We utilized data of 405 annotated metabolites (see methodology) for further analysis. For the transcriptomic data generated on 25,880 genes, we analyzed the data for each of the two groups (breed-specific and FE-specific), as described in the methodology. The genes with a gene count <1 were removed, resulting in 20,233 genes for both the groups. The gene count data for each group (breed-specific and FE-specific) was normalized, and the linear model was fitted into the data as given in the methodology. The genes were also screened for their chromosomal information from the Ensembl *Sus scrofa* database. After normalization, removal of values < 0 and obtaining the gene chromosomal location information, 17,726 (breed-specific), and 17,697 (FE-specific) genes were retained in each group. As a quality control for IntLIM, we filtered out genes with the lowest 5% of the variation, which gave 16,839 genes (breed-specific) and 16,812 genes (FE-specific) that were subjected to IntLIM analysis. A schematic representation of the study design and analysis steps are given in Figure 1. We performed the PCA analysis (Figure S1) on the filtered metabolome-transcriptome data, which included 405 metabolites, and 16,839 genes (breed-specific), and 16,812 genes (FE-specific). The results of PCA analysis for the metabolites-genes based on breed and FE groups are shown in Figure S1.

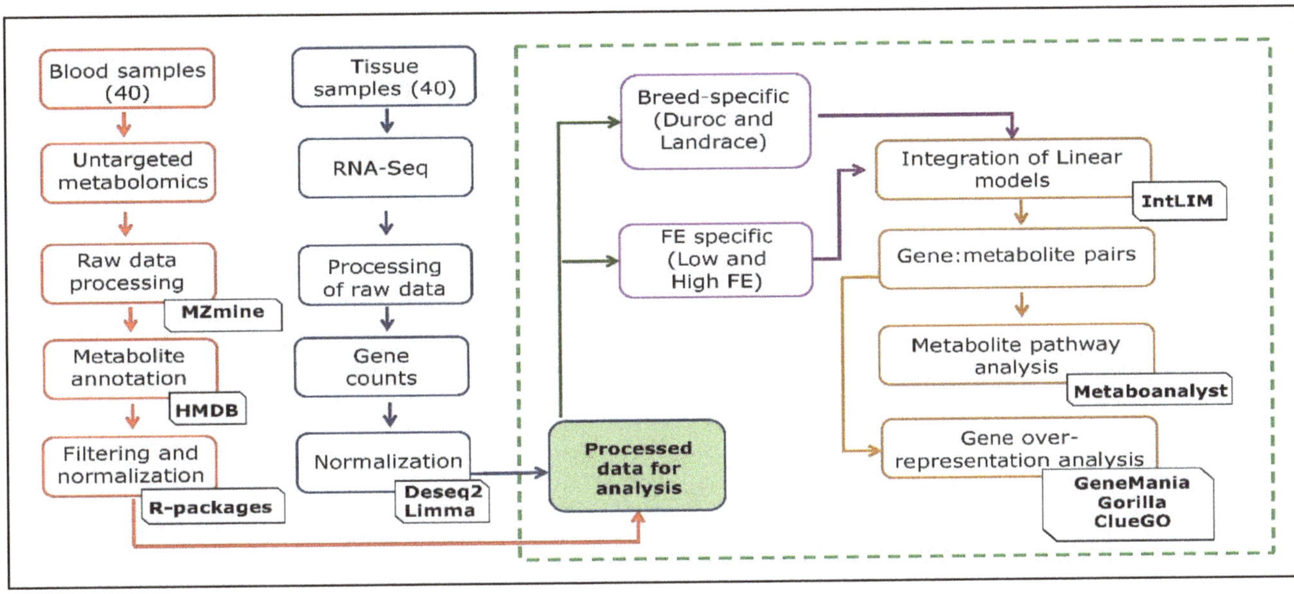

Figure 1. Schematic representation of the study design and analysis steps.

2.2. *Gene Metabolite Interaction of Breed-Specific and FE-Specific Groups*

From the IntLIM analysis, we identified gene-metabolite pairs that have a strong association with respect to the breed (Duroc and Landrace) and FE (low and high FE) groups. For the breed-specific groups, all possible combinations of gene-metabolite pairs (6,819,795 model runs) were evaluated, using Duroc and Landrace as the breed-group. Based on this approach, we identified 21 gene-metabolite associations (false discovery rate—FDR adjusted p-value ≤ 0.1 and correlation difference effect size > 0.1) (Table 1). Clustering these pairs by the direction of association (positive and negative correlations) within each breed group revealed two major clusters (Figure 2a) in each breed. First, the Duroc correlated/Landrace anti-correlated cluster consists of seven gene-metabolite pairs (three unique metabolites and five unique genes) with a high positive correlation in Duroc and low or negative correlation in Landrace (Figure 2a). Second, the Duroc anti-correlated/Landrace correlated cluster consists of 14 gene-metabolite pairs (10 unique metabolites and nine unique genes) with relatively high negative correlations in Duroc and positive correlations in Landrace. The two top-ranked gene-metabolite pairs (ranked in descending order of absolute value of Spearman correlation difference between Duroc and Landrace) in the Duroc correlated and anti-correlated clusters were ENSSSCG00000028124 (*SNRPN*)—Rhodamine B (Figure 2b) and ENSSSCG00000000401 (*GLS2*)—cystathionine ketimine (Figure 2c) respectively. *SNRPN* and Rhodamine B are positively correlated in Duroc ($r = 0.7$) but negatively correlated in Landrace ($r = -0.5$) (Figure 2b). *GLS2* and cystathionine ketimine are negatively correlated in Duroc ($r = -0.9$), but positively correlated in Landrace ($r = 0.2$) (Figure 2c).

Similarly, we used IntLIM for the FE-specific group and evaluated all possible combinations of gene-metabolite pairs (6,808,860 models), with low and high FE as a binary phenotype. With this approach, we identified 12 FE-specific gene-metabolite correlations (FDR adjusted interaction p-value ≤ 0.1, and a Spearman correlation difference > 0.1) involving eight unique gene and metabolites each (Table 2). The heat map with gene-metabolite Spearman correlation for low and high FE group showed a clear separation between the two groups (Figure 3a). The high FE-correlated cluster of 12 gene-metabolite pairs (eight unique genes and metabolites with high correlations in high-FE groups) were negatively correlated with the low-FE group. The two gene-metabolite pairs (ranked in descending order of absolute value Spearman correlation difference between high and low FE group) in high-FE correlated clusters were ENSSSCG00000025106 (*THNSL2*)—pyrocatechol (Figure 3b) and ENSSSCG00000036609 (*TBXT*)—ketoleucine (Figure 3c), respectively. Both pairs showed a positive correlation in high-FE group (r = 0.6, r = 0.5) while showed a negative correlation in the low-FE group (r = −0.7, r = −0.3) (Figure 3b,c).

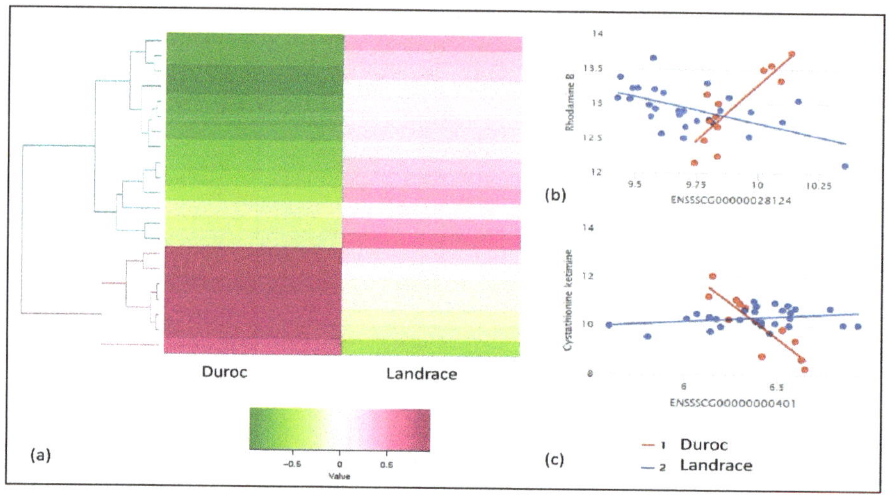

Figure 2. Results of IntLIM applied to breed-specific groups. (**a**) Clustering of 21 identified gene-metabolite pairs (FDR adjusted *p*-value of interaction coefficient < 0.1, Spearman correlation difference > 0.1 in Duroc and Landrace breeds, (**b**) Gene-metabolite difference in ENSSSCG00000028124 (*SNRPN*)—rhodamine B (FDR adjusted *p*-value = 0.1, Duroc Spearman correlation = 0.7, Landrace Spearman correlation = −0.5), (**c**) Gene-metabolite difference in ENSSSCG00000000401 (*GLS2*)—cystathionine ketimine (FDR adjusted *p*-value = 0.01, Duroc Spearman correlation = −0.9, Landrace Spearman correlation = 0.2).

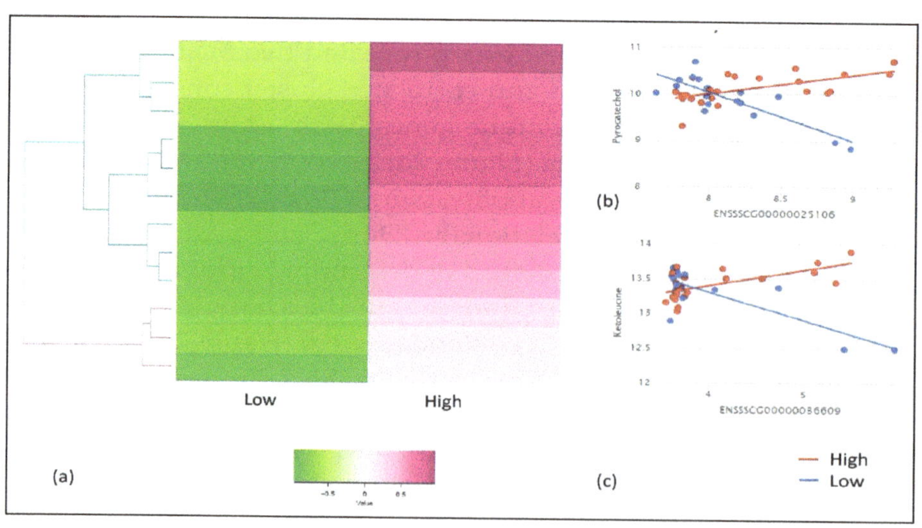

Figure 3. Results of IntLIM applied to FE-specific groups. (**a**) Clustering of 12 identified gene-metabolite pairs (FDR adjusted *p*-value of interaction coefficient < 0.1, Spearman correlation difference > 0.1 in high and low FE groups, (**b**) Gene-metabolite difference in ENSSSCG00000025106 (*THNSL2*)—Pyrocatechol (FDR adjusted *p*-value = 0.06, High-FE Spearman correlation = 0.6, Low-FE Spearman correlation = −0.7), (**c**) Gene-metabolite difference in ENSSSCG00000036609 (*TBXT*)—ketoleucine (FDR adjusted *p*-value = 0.08, High-FE Spearman correlation = 0.5, Low-FE Spearman correlation = −0.3).

Table 1. Gene-metabolite interaction pairs from IntLIM for the breed-specific groups.

Metabolite_Name	Ensembl ID	Gene Name	Duroc_cor	Landrace_cor	Abs diff.corr	Pval	FDRadjPval
Rhodamine B	ENSSSCG00000028124	*SNRPN*	0.776224	−0.54242	−1.31864	3.11×10^{-7}	0.1
L-glutamic acid 5-phosphate	ENSSSCG00000010274	*SGPL1*	0.888112	−0.21456	−1.10267	1.57×10^{-7}	0.09
Cystathionine ketimine	ENSSSCG00000010274	*SGPL1*	0.874126	−0.18719	−1.06132	2.67×10^{-7}	0.1
Cystathionine ketimine	ENSSSCG00000040110	*Novel_gene*	0.923077	−0.11385	−1.03692	1.08×10^{-8}	0.02
L-glutamic acid 5-phosphate	ENSSSCG00000038948	*ETS2*	0.895105	−0.11002	−1.00512	2.66×10^{-8}	0.02
L-glutamic acid 5-phosphate	ENSSSCG00000040110	*Novel_gene*	0.937063	0.010947	−0.92612	1.48×10^{-9}	0.01
L-glutamic acid 5-phosphate	ENSSSCG00000014632	*FAM160A2*	0.951049	0.229885	−0.72116	1.09×10^{-7}	0.08
Aloesol	ENSSSCG00000004128	*ZC2HC1B*	−0.22767	0.048714	0.276385	4.33×10^{-7}	0.1
Theogallin	ENSSSCG00000026442	*FAM163B*	−0.29772	0.45156	0.749284	1.69×10^{-7}	0.09
Fenamiphos	ENSSSCG00000018649	*Novel_gene*	−0.68652	0.093596	0.780112	3.44×10^{-7}	0.1
Taraxacolide 1-o-b-d-glucopyranoside	ENSSSCG00000026442	*FAM163B*	−0.27321	0.648057	0.921262	1.55×10^{-7}	0.09
Proanthocyanidin a2	ENSSSCG00000033688	*ZDHHC22*	−0.63047	0.32567	0.956144	4.20×10^{-7}	0.1
Fenamiphos	ENSSSCG00000040467	*Novel_gene*	−0.67251	0.288998	0.961504	2.25×10^{-8}	0.02
L-glutamic acid 5-phosphate	ENSSSCG00000000401	*GLS2*	−0.85315	0.109469	0.962616	1.80×10^{-7}	0.09
Paracetamol sulfate	ENSSSCG00000000401	*GLS2*	−0.94406	0.038314	0.98237	2.87×10^{-7}	0.1
L-glutamic acid 5-phosphate	ENSSSCG00000034989	*LRRTM2*	−0.83916	0.15052	0.989681	2.17×10^{-8}	0.02
Cystathionine ketimine	ENSSSCG00000034989	*LRRTM2*	−0.79021	0.204707	0.994917	2.53×10^{-8}	0.02
Ketoleucine	ENSSSCG00000026442	*FAM163B*	−0.5289	0.466886	0.995783	1.19×10^{-8}	0.02
Ganoderenic acid e	ENSSSCG00000034200	*SEC22C*	−0.85315	0.288998	1.142145	3.47×10^{-7}	0.1
Cystathionine ketimine	ENSSSCG00000000401	*GLS2*	−0.93007	0.249042	1.179112	2.92×10^{-9}	0.01
Ganoderenic acid e	ENSSSCG00000037595	*Novel_gene*	−0.85315	0.449371	1.302517	2.26×10^{-7}	0.1

Duroc_cor: Spearman correlation in Duroc; Landrace_cor: Spearman correlation in Landrace; Abs diff.corr: the absolute difference in correlation between Duroc and Landrace; Pval: *p*-value ($p < 10^{-7}$ was selected as the cut-off value); FDRadjPval: FDR adjusted *p*-value.

Table 2. Gene-metabolite interaction pairs from IntLIM for the FE-specific groups.

Metabolite_Name	Ensembl ID	Gene Name	High_cor	Low_cor	Abs diff.corr	Pval	FDRadjPval
Pyrocatechol	ENSSSCG00000025106	*THNSL2*	0.66996	−0.73284	−1.4028	4.72×10^{-8}	0.06
2-pyrocatechuic acid	ENSSSCG00000025106	*THNSL2*	0.645257	−0.68137	−1.32663	7.13×10^{-8}	0.08
Ketoleucine	ENSSSCG00000017043	*RNF145*	0.548419	−0.75735	−1.30577	2.17×10^{-7}	0.1
Ketoleucine	ENSSSCG00000025106	*THNSL2*	0.498024	−0.58088	−1.07891	1.19×10^{-7}	0.10
Theogallin	ENSSSCG00000036609	*TBXT*	0.727273	−0.35049	−1.07776	2.87×10^{-8}	0.05
Neodiospyrin	ENSSSCG00000029077	*TUBAL3*	0.608696	−0.46814	−1.07683	2.63×10^{-8}	0.05
Theogallin	ENSSSCG00000025106	*THNSL2*	0.4417	−0.6152	−1.0569	2.52×10^{-7}	0.1
Proanthocyanidin a2	ENSSSCG00000008938	*ENAM*	0.37954	−0.60539	−0.98493	1.99×10^{-7}	0.1
Ketoleucine	ENSSSCG00000036609	*TBXT*	0.557312	−0.3701	−0.92741	8.78×10^{-8}	0.08
Adrenochrome	ENSSSCG00000038441	*Novel_gene*	0.171485	−0.58088	−0.75237	1.71×10^{-8}	0.05
Proanthocyanidin a2	ENSSSCG00000019329	*U2*	0.052384	−0.66176	−0.71415	1.11×10^{-8}	0.05
Levulinic acid	ENSSSCG00000009250	*PRKG2*	0.075117	−0.54902	−0.62414	1.93×10^{-7}	0.1

High_cor: Spearman correlation in high FE group; Low_cor: Spearman correlation in low FE group; Abs diff.corr: the absolute difference in correlation between the high and low FE group; Pval: *p*-value ($p < 10^{-7}$ was selected as the cut-off value); FDRadjPval: FDR adjusted *p*-value.

2.3. Pathway and Gene Ontology Over-Representation Analysis

We identified the pathways associated with the unique metabolites in each cluster identified from breed-specific (21) and FE-specific (12) interactions. The three unique metabolites from Duroc correlated/Landrace anti-correlated clusters were associated with arginine and proline metabolism (p-value = 0.02). Furthermore, the ten unique metabolites from Duroc anti-correlated/Landrace correlated cluster were associated with valine-leucine-isoleucine biosynthesis (p-value = 0.01) and valine-leucine-isoleucine degradation (p-value = 0.07) along with arginine and proline metabolism (p-value = 0.07). The eight unique metabolites from high-FE correlated/low-FE anti-correlated cluster were associated with valine-leucine-isoleucine biosynthesis (p-value = 0.01) and degradation (p-value = 0.07). The pathways associated with the metabolites in breed-specific and FE-specific clusters for unique metabolites are given in Supplementary Table S1a.

Unique and mappable genes from each group (breed-specific—each cluster, and FE-specific) were screened by using GeneMania to generate a composite functional association network that includes all the evidence of co-functionality. From the breed-specific group, unique genes (4 genes) from Duroc correlated/Landrace anti-correlated cluster (Table 1) mapped to 20 genes based on co-functionality from GeneMania (Table S1b). The gene-ontology enrichment analysis of the identified 24 genes (unique genes from Table 1 and co-functional genes from GeneMania) revealed enrichment of the regulation of hemopoiesis, response to thyroid hormone, and the sphingolipid catabolic process (Table S1c). These genes were enriched for the sphingolipid metabolism KEGG pathway (adjusted p-value corrected with Bonferroni step down ≤ 0.05) (Supplementary Table S1d). Unique and mappable genes (6 genes) identified from Duroc anti-correlated/Landrace correlated cluster (Table 1) were co-functional with 20 genes based on GeneMania (Table S1b). The gene ontology enrichment analysis of these 26 genes (unique genes from Table 1 and co-functional genes from GeneMania) revealed the ER to Golgi vesicle-mediated transport and membrane fusion (Table S1c) as an enriched biological process. Butanoate metabolism, alanine-aspartate-glutamate metabolism, and valine-leucine-isoleucine degradation were significantly enriched KEGG pathways from the Duroc anti-correlated/Landrace correlated cluster (Supplementary Table S1d). Similarly, from the FE-specific group, unique mapped genes (7 genes) from high-FE correlated/low-FE anti-correlated clusters (Table 2) were co-functional with 20 genes identified from GeneMania (Table S1b). These genes were involved with the cGMP metabolic process, purine nucleotide biosynthesis, and phosphorus-oxygen lyase activity pathways (Table S1c). The top significant KEGG pathway enriched was the cGMP-PKG signaling pathway (Supplementary Table S1d).

3. Discussion

FE is an important quantitative trait, which quantifies the efficiency of nutrient conversion from the feed to a tissue that is of nutritional and economic significance [19]. Understanding the molecular mechanisms underlying FE will be advantageous in the efficient selection of pigs and benefit the pig producers. In the Danish pig industry, Duroc is the most popularly used terminal sires in combination with crossbred Landrace X Yorkshire breeds [20], so the selection pressures for FE in Duroc is higher as compared to Landrace. Thus, in the current study, we attempted to identify the gene-metabolite interactions specific to each breed. FE is a complex trait influenced by environmental and health factors and involves many organs. Skeletal muscle, being the largest organ in the body, is an essential location for the metabolism of carbohydrates and lipids [21,22]. It plays a significant role in the utilization and storage of energy acquired from the feed. Thus, understanding the difference in the regulatory processes from a divergent FE group will add a layer of knowledge to the understanding of biological mechanisms involved with this complex trait.

A plethora of metabolome and transcriptome studies for FE in pigs are reported [3,9,10,12]. However, to the best of our knowledge, markers from the integration of metabolome and transcriptome in Duroc and Landrace pigs has not been done before. Herein, we unveiled the gene-metabolite relationships that are phenotype dependent. This approach highlighted molecular functions and pathways that are strongly evidenced by the integration study. Evaluating phenotype-specific relationships between metabolites and genes assists us to elucidate novel co-regulation patterns that would not be identified by single approaches. In the current study, we integrated untargeted metabolomic and transcriptomic data. We used a numerical data integration approach that employed the integration of a linear model (IntLIM package) to predict metabolite levels from gene-expression in a phenotype dependent manner [23].

We attempted to identify the breed-specific and FE-specific gene-metabolite pairs in the current study. The PCA analysis showed a difference in the expression of genes in Duroc and Landrace. However, PCA for the FE group did not exhibit significant clusters between groups, which may be due to the small sample size evaluated here. With our current metabolome-transcriptome analysis, we identified 21 gene-metabolite breed-specific pairs and 12 gene-metabolite FE-specific pairs.

3.1. Breed-Specific Pathway Analysis

In the breed-specific analysis, two clusters were identified between Duroc and Landrace breeds. The Duroc correlated/Landrace anti-correlated cluster associated L-glutamic acid 5-phosphate metabolite with the *FAM160A2*, ENSSSCG00000040110, *ETS2*, and *SGPL1* genes; cystathionine ketimine metabolite with ENSSSCG00000040110 and *SGPL1* genes and Rhodamine B with the *SNRPN* genes. The arginine and proline metabolism pathways were associated with the unique metabolites from this cluster. The gene ontology enrichment analysis of the unique genes identified from this cluster with the co-functional genes found enrichment for the multicellular organismal process, sphingolipid catabolic process, regulation of hemostasis, and coagulation pathways. Sphingolipid metabolism associated with the *SGPL1* and *GBA* genes was identified as the top KEGG pathway. *SGPL1* (sphingosine-1-phosphate) catalyzes the final step of the sphingolipid pathway by irreversibly converting sphingosine-1-phosphate (*S1P*) to its by-products [24], thereby regulating *S1P*. *S1P* plays a role as a muscle trophic factor by activating quiescent muscle stem cells (satellite cells) for efficient skeletal muscle repair and regeneration [25]. The role of FE on skeletal muscle mass was well established from previous studies wherein the improvement of muscle properties and an increase in muscle mass is attributed by selection for low RFI in pigs [26]. *S1P* is also reported to trigger glutamate secretion and potentiates depolarization-evoked glutamate secretion [27]. Glutamic acid has been found to play a crucial role in FE as it improves the FE of weaned piglets [28]. This supports the results in the current study where *SGPL1* was associated with L glutamic acid 5-phosphate as a significant gene-metabolite pair. A brief overview of the role of *SGPL1* in sphingolipid-metabolism regulating FE is given in Figure 4. Therefore, further investigation of *SGPL1*-L-glutamic acid 5-phosphate gene-metabolite pair, which was positively correlated with Duroc while negatively correlated with Landrace concerning FE traits, could be a major avenue for breed-specific research and its effect on FE and meat quality traits.

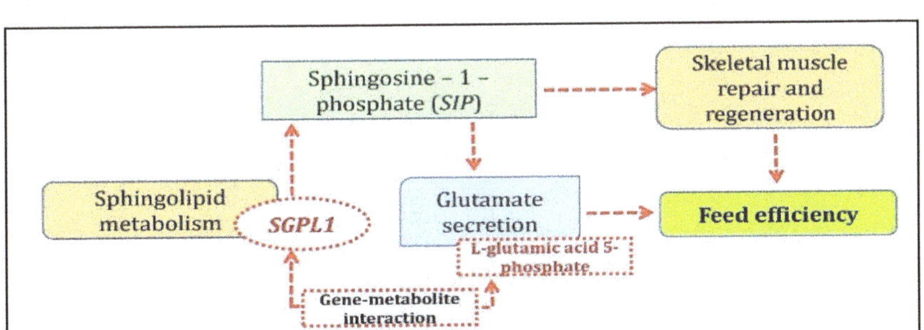

Figure 4. Biological mechanism showing the involvement of *SGPL1*-L-glutamic acid 5-phosphate gene-metabolite pair underlying sphingolipid metabolism identified from Duroc correlated/Landrace anti-correlated cluster.

We also identified the *SNRPN* gene and Rhodamine B relation in the Duroc correlated cluster. A previous study showed the *SNRPN* gene (small nuclear ribonucleoprotein polypeptide N) was ubiquitously expressed in pigs [29]. Small nuclear ribonucleoproteins and heterogeneous small nuclear riboproteins play roles in nucleolar ribosomal RNA (rRNA) and messenger RNA (mRNA) synthesis in conjunction with spliceosome activity responsible for cleaving on introns from the pre mRNA molecule [30]. Furthermore, in a study of FE in broiler chickens, a high FE phenotype exhibited enrichment of ribosome assembly including small nuclear ribonucleoprotein, as well as nuclear transport and protein translation processes than low FE phenotype [31].

The Duroc anti-correlated/Landrace correlated cluster identified aloesol—*ZC2HC1B*, Proanthocyanidin a2—*ZDHHC22*; fenamiphos metabolite with ENSSSCG00000040467, ENSSSCG00000018649 gene; ganoderenic acid e metabolite with ENSSSCG00000037595 and *SEC22C* genes; *FAM163B* gene with taraxacolide 1-o-b-d-glucopyranoside, theogallin, and ketoleucine metabolites; *LRRTM2* gene with cystathionine ketimine and L-glutamic acid 5-phosphate metabolites and *GLS2* gene with L-glutamic acid 5-phosphate, cystathionine ketimine, and paracetamol sulfate metabolites. One of the significant pathways in Landrace correlated cluster was valine, leucine, and isoleucine degradation which included the *ABAT* and *ACADS* genes (co-functional genes) identified in the current study. Valine, leucine, and isoleucine are branched-chain amino acids (BCAA) and have a crucial role in protein synthesis and energy production [32]. The degradation of BCAA can be glucogenic (valine), ketogenic (leucine and isoleucine), or both (isoleucine). The end products from their degradation, succinyl-CoA and/or acetyl-CoA can enter the tricarboxylic acid (TCA) cycle for energy generation and gluconeogenesis or may act as precursors for lipogenesis and ketone body production through acetyl-CoA and acetoacetate [33]. Glucose metabolism and the TCA pathway in the skeletal muscle is a key pathway regulating FE traits in pigs [34]. In an interesting proteomic study involving glucose metabolism and the TCA cycle reported earlier, the proteins catalyzing the conversion of glucose to pyruvate and oxaloacetate were up-regulated in high-FE pigs while those involved in the conversion of pyruvate to lactate or acetyl-CoA were down-regulated in high-FE pigs [34]. This resulted in inhibition of the TCA cycle in high-FE pigs due to the down-regulation of key catalytic proteins [34]. Thus, the pathway identified with BCAA in the current study may cause differences in FE concerning specific breed as evident from the indirect link with TCA and FE (Figure 5). The BCAA also affects protein synthesis, as reported earlier in a study with reduced degradation of rat skeletal muscle proteins [35]. Additionally, leucine activates mTOR signaling, one of the central regulators of cell growth and metabolism along with an increase in fatty acid oxidation. With all these supporting facts of BCAA regulating cellular metabolism, protein, and fatty acid degradation, which are also key factors influencing FE, the role of the valine-leucine-isoleucine pathway in FE cannot be overlooked. Furthermore, this pathway has been found over-represented in a GWAS study for RFI in beef cattle [36] and pigs [3].

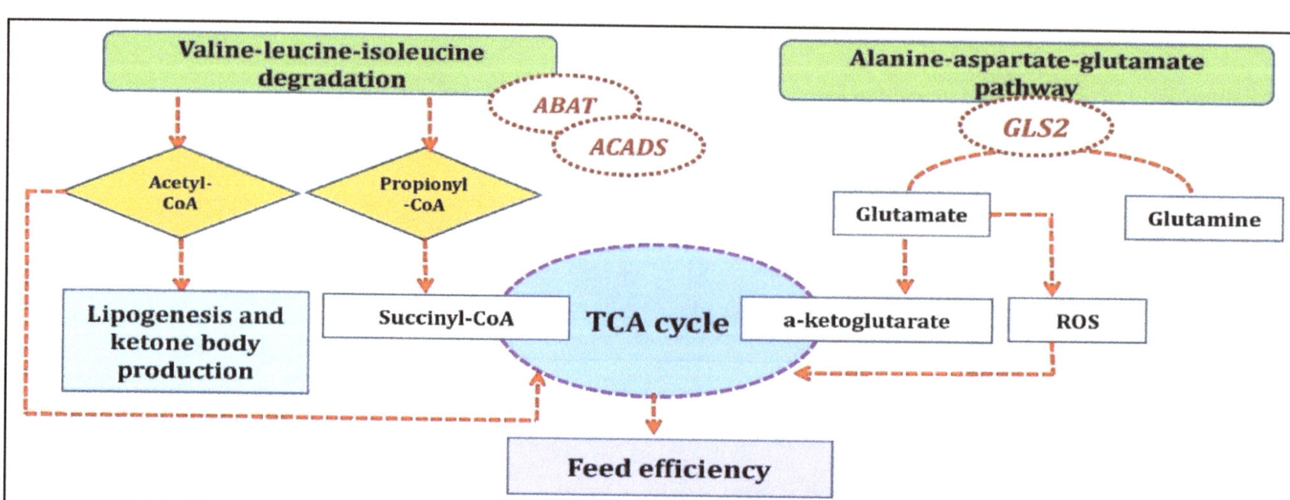

Figure 5. Biological mechanism showing the involvement of genes identified from Duroc anti-correlated/Landrace correlated clusters underlying feed efficiency.

We identified the alanine-aspartate-glutamate metabolism KEGG pathway involving the *GLS2* and *ABAT* genes within the Duroc anti-correlated/Landrace correlated cluster. *GLS2* is a mitochondrial glutaminase that catalyzes glutamine to glutamate which is further converted to a-ketoglutarate, an important substrate for the citric acid cycle to produce ATP in mitochondria. Glutamate is also a precursor of reduced glutathione (*GSH*), an important antioxidant molecule, and a scavenger for ROS (Reactive oxygen species) [37]. ROS are regulated by FE related traits as reported from the previous studies, higher levels of ROS production and oxidized mitochondrial proteins have been found in the muscle of low FE chickens [38] while in pigs, ROS production in mitochondria was higher in semitendinosus muscle of less efficient pigs selected for high RFI compared to high efficient pigs (low RFI) [39]. *GLS2* interacted with L-glutamic acid 5-phosphate and cystathionine ketimine metabolites as identified from Duroc anti-correlated/Landrace correlated clusters (Figure 5).

The unique metabolites identified in the breed-specific clusters were also previously reported in another study for the metabolomic analysis of FE related traits in Duroc and Landrace [3]. The breed-specific unique metabolites such as aloesol and ketoleucine affected FE in Duroc [3]. In contrast, rhodamine B, taraxacolide 1-o-b-d-glucopyranoside, and ganoderenic acid were underlying testing daily gain (TDG) in Duroc [3]. Theogallin and ketoleucine were involved with TDG and daily gain (DG) in Duroc and Landrace and RFI in Duroc [3]. L-glutamic acid 5-phosphate, cystathionine ketimine, and paracetamol sulfate were associated with FE and RFI in Landrace [3]. It is worth highlighting the interaction of metabolites L-glutamic acid 5-phosphate and cystathionine ketimine identified in this study. While these metabolites interact with the *SGPL1* gene as identified in the Duroc correlated cluster, on the contrary, they interact with the *GLS2* gene in the Landrace correlated cluster. Both *SGPL1* and *GLS2* were in the top significant pathways in their respective cluster. Therefore, these gene-metabolite interactions which are highly specific to breed differences open up the avenues for further research to extrapolate differences in FE related traits concerning breeds.

3.2. cGMP-PKG Pathway Involved with FE-Specific Analysis

In the FE-specific analysis, we found 12 significant gene-metabolite pairs. The gene-metabolite pairs in high-FE correlated/low-FE anti-correlated cluster were *TBXT* gene with theogallin and ketoleucine metabolites; *THNSL2* gene with pyrocatechol, 2-pyrocatechuic acid, ketoleucine, and theogallin metabolites; *TUBAL3* gene with neodispyrin metabolite, *RNF145* gene with ketoleucine metabolite, *ENAM* gene and U2 snRNA with proanthocyanidin a2 metabolite, ENSSSCG00000038441 gene with adrenochrome metabolite, and *PRKG2* gene with levulinic acid metabolite. The pathway analysis with the unique metabolites identified from high-FE correlated/low-FE anti-correlated clusters was over-represented for valine-leucine-isoleucine biosynthesis and degradation pathway. The unique mapped genes and co-functional genes were enriched for the following biological processes: lyase activity, cGMP metabolic process, phosphorus-oxygen lyase activity, and cyclic purine nucleotide metabolic process. cGMP-PKG, purine metabolism, and renin secretion were the KEGG pathways identified in this cluster. The cGMP pathways were also identified in the studies reported earlier with FE related traits with pigs [40] and beef cattle [41]. The *PRKG2* gene, one of the main predictors for cGMP pathways and also identified in this study, encodes the serine/threonine-protein, which binds to inhibits the activation of several receptor tyrosine kinases and is a regulator of the intestinal secretion, bone growth, and renin secretion (https://www.ncbi.nlm.nih.gov/gene/5593). *PRKG2* encodes for CGKII (guanosine 3,5-cyclic monophosphate (cGMP)-dependent protein kinase II) and is abundantly expressed in intestinal epithelium. *CGKII* relays signaling through a membrane-associated, cGMP-producing enzyme, guanylyl cyclase (GC). The catalytic activity of this receptor-enzyme is triggered by two locally produced ligands, the peptides guanylin and uroguanylin [42]. The GC is activated by nitric oxide (NO) and catalyzes the conversion of intracellular guanosine-5′-triphosphate (GTP) to cyclic guanosine-3′,5′-monophosphate (cGMP). This enzyme has two forms: a membrane protein and a soluble form with specific kinetic properties and tissue distributions. The soluble GC (sGC) form is a heterodimeric protein consisting of α (α_1 and α_2,) and β (β_1 and β_2) subunits encoded by distinct genes [43]. An alpha subunit of guanylyl cyclase, *GUCY1A2* and a beta subunit *GUCY1B3* was identified as co-functional genes in the current study and were involved with cGMP, phosphorus metabolic process, nitrogen metabolic process pathways as identified in the current study. Uroguanylin is a gastrointestinal hormone primarily involved in fluid and electrolyte handling. It has recently been reported that prouroguanylin, secreted postprandially, is converted to uroguanylin in the brain and activates the receptor guanylate cyclase-C (GC-C) to reduce food intake in mice [44]. Reduced FI is a characteristic feature for the selection of the pigs known for high FE [1]. The overview of the mechanism involving the cGMP-PKG pathway and its role in FE is given in Figure 6.

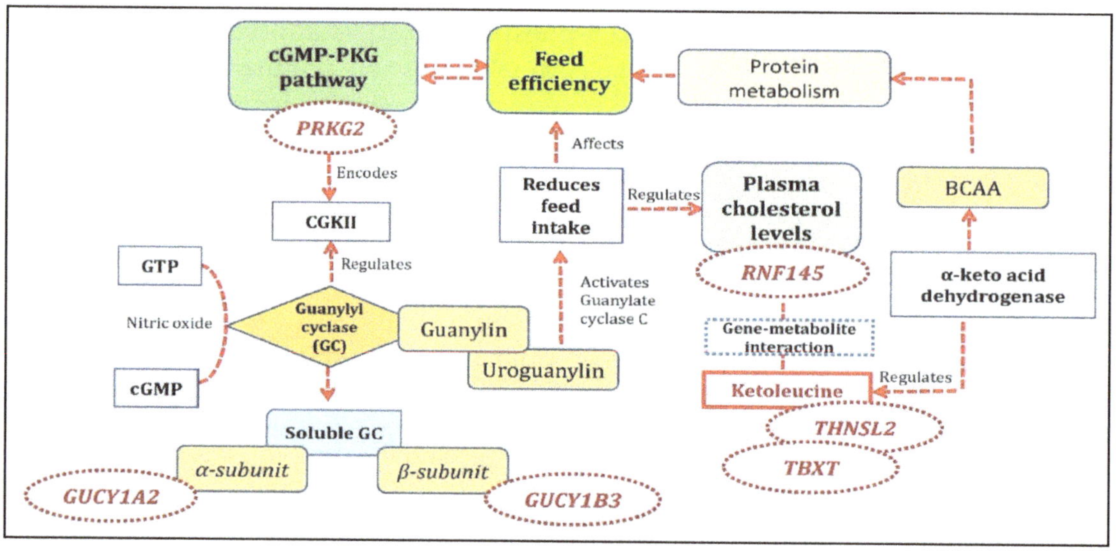

Figure 6. An overview of gene-metabolite pairs affecting the cGMP-PKG pathway involved with feed efficiency.

A positive correlation between FI and plasma cholesterol levels is previously established in pigs [45]. *RNF145*, which was positively correlated in the high-FE cluster participates in key signaling pathways of cholesterol homeostasis [46]. *RNF145* expression is induced in response to LXR activation and high-cholesterol diet feeding [46]. Transduction of *RNF145* into mouse liver inhibits the expression of genes involved in cholesterol biosynthesis and reduces plasma cholesterol levels. On the other hand, its inactivation increases cholesterol levels both in the liver and plasma [46]. In this study, *RNF145* was identified with ketoleucine as a significant gene-metabolite pair in this cluster (Figure 6). Ketoleucine is an abnormal metabolite that arises from the incomplete breakdown of branched-chain amino acids (https://hmdb.ca/metabolites/HMDB0000695). Ketoleucine is regulated by branched-chain α-keto acid dehydrogenase. Studies reported that the branched-chain α-keto acid dehydrogenase catalyzes the irreversible oxidative decarboxylation of all three branched-chain keto acids (BCKA) derived from branched-chain amino acids (BCAA), i.e., α-ketoisocaproate (ketoleucine) [47]. They demonstrated changes in BCKA activity that showed a significant alteration in BCAA and protein metabolism during starvation in rats [47]. BCAA also plays a crucial role in FE by regulating energy homeostasis in addition to lipid and protein metabolism as reported in pigs [48].

Apart from *RNF145*, gene-metabolite interaction of ketoleucine with *TBXT* and *THNSL2* was also identified in this study. *THNSL2* was reported among the top 40 significantly differentially expressed genes of characterized proteins between high- and low-ADG steers from a liver transcriptome profiling of beef cattle [49]. This gene has also been associated with abdominal and visceral fat in humans based on GWAS [50]. An interaction of proanthocyanidin a2 with ENSSSCG00000019329 (U2 snRNA) was also identified with low but positive correlation with high-FE cluster while a negative correlation with the low-FE cluster. U2 spliceosomal snRNAs are the molecules found in the major spliceosomal machinery of all eukaryotic organisms and affect their gene expression [51]. U2 snRNA plays a central role in the splicing of mRNA precursors by regulating a dynamic set of

RNA-RNA base-pairing interactions [52]. From the previously reported studies, the role of precursor mRNA in gene expression has been established as it removes the intronic sequence from immature RNA, leading to a production of mature mRNA that might differ in function [53]. Regulation of pre-mRNA splicing by nutrients modulates the carbohydrate and lipid metabolism [53]. U2 interacts with proanthocyanidin a2 in the current study. Proanthocyanidin a2 is an antioxidant and has a broad spectrum of biologic properties against oxidative stress [54]. Proanthocyanidin significantly increased the activity of antioxidant enzymes such as superoxide dismutase, glutathione peroxidase, and catalase [54]. The role of antioxidant activity with FE was reported earlier in beef cattle as low feed efficient steers had greater superoxide dismutase and glutathione peroxidase activity than the high feed efficient ones, potentially using a greater proportion of energy [55]. Thus, as evident from all these facts, the potential role of proanthocyanidin a2–U2 interaction in high-FE pigs identified in the current study might be interesting to explore.

The gene-metabolite pairs identified in the present study over-represented some pathways that have been reported to have a role in FE related traits. Some of the genes identified are novel and were not included in the pathway analysis. Since these gene-metabolite pairs selected have a highly significant correlation with respect to each study group, a detailed study of these genes and metabolites are needed to better understand their role in FE related-traits. Further studies on the identified gene-metabolite pairs may assist in the discovery of biomarkers as these significant pairs identified directly reflect the phenotype as revealed by the candidate gene-expression with the downstream metabolite differences in pigs with low and high FE groups.

4. Materials and Methods

4.1. Data Resource and Phenotype Generation

The pigs used in this experiment were raised at the pig testing station "Bøgildgård" operated by SEGES within Landbrug and Fødevarer (L&F: Danish Agriculture and Food Council). Pigs were *ad libitum* fed and free water supplied. The authors of this study were not responsible for animal husbandry, diet, and care as the testing station is a facility within the Danish breeding program run by SEGES. The initial bodyweight of the pigs before the testing period was approximately 7 kg, followed by a 5-week acclimatization phase. For the phenotypic traits, we calculated FCR and RFI, as reported in our previous study [40]. We considered the same classification of animals in this study as efficient and inefficient (low and high FCR, respectively), as previously reported [40]. The classification was done by selecting pigs that were one standard deviation above or below the mean FCR for each breed as previously reported.

4.2. Gene Expression Profile, Metabolite Profile, and Data Analyses

For transcriptome analysis, we collected *psoas major* muscle from 40 purebred uncastrated male pigs from two breeds comprising of 12 Danbred Duroc and 28 Danbred Landrace. The tissue samples were preserved in RNAlater (Ambion, Austin, TX, USA) immediately post-slaughter and stored at −25 °C until subsequent analysis. Total RNA isolation and sequencing were carried out by BGI Genomics. Paired-end sequencing (100 bp) was performed on the BGISEQ-500 platform after Oligo dT library preparation. Read quality control, mapping, and gene counts were reported elsewhere [14]. Lowly expressed genes were filtered out, and the gene counts normalization was carried out by applying the variance stabilizing transformation (*VST*) function from DeSeq2 [56].

To identify significant gene-metabolite pairs, we analyzed the data considering two approaches, i.e., breed-specific (Duroc-Landrace) and FE-specific (low-high FE groups). Thus, we fitted a linear model for adjusting the read counts with the covariates using the Limma R package [57]. For adjusting the data to identify breed-specific differences, we adopted the following model:

$$y_{ijkl} = \mu + F_i + R_j + A_k + P_l + \varepsilon_{ijkl}$$

where y_{ijkl}: is the normalized read counts; μ: is the intercept; F_i: is the fixed effect of the FE group (two levels, high and low); R_j: is the covariate for the RIN values; A_k: is the covariate for the animal's slaughter age, in days; P_l: is the fixed effect of the pen (8 levels); ε_{ijkl}: is the random residual effect associated with each observation.

To identify differences between high and low FE groups, the breed effect was added in the linear model, as follows:

$$y_{ijkl} = \mu + B_i + R_j + A_k + P_l + \varepsilon_{ijkl}$$

where y_{ijkl}: is the normalized read counts; μ: is the intercept; B_i: is the fixed effect of the breed (two levels, Duroc and Landrace); R_j: is the covariate for the RIN values; A_k: is the covariate for the animal's slaughter age, in days; P_l: is the fixed effect of the pen (eight levels); ε_{ijkl}: is the random residual effect associated with each observation.

Regarding the identification of the metabolites, we used an untargeted metabolomic approach, as reported elsewhere [3]. In summary, 5 mL of blood samples at two-time points were collected from jugular venipuncture of each non-fasted pig into the EDTA tubes and immediately placed on ice. The plasma samples extracted from 109 animals (59 Duroc and 50 Landrace) were subjected to metabolomics analysis, as described in a previous study [3]. The metabolite data from this study were accessed using MetaboLights accession ID MTBLS1384 with a link: https://www.ebi.ac.uk/metabolights/MTBLS1384. Due to the need for paired data to carry the integrative analysis, only those samples with both metabolite and RNA-Seq data were used herein. The metabolite data from time-point two in 40 pigs were log-normalized before fitting into a linear model. Only those with the relative standard deviation > 0.15 were used based on the raw counts. As proposed for the RNA-seq, we adjusted the log-normalized metabolite data considering both approaches. First, the following model was employed for the breed-specific analysis:

$$m_{ijkl} = \mu + F_i + D_j + A_k + P_l + \varepsilon_{ijkl}$$

where m_{ijkl}: is the is the log-normalized concentration of each metabolite (n = 749); μ: is the intercept; F_i: is the fixed effect of the FE group (two levels, high and low); D_j: is the fixed effect of the batch for metabolomic analysis (two levels); A_k: is the covariate for the sampling age, in days; P_l: is the fixed effect of the pen (8 levels); ε_{ijkl}: is the random residual effect associated with each observation.

For the FE-specific group approach, we fitted the data as follows:

$$m_{ijkl} = \mu + B_i + D_j + A_k + P_l + \varepsilon_{ijkl}$$

where m_{ijkl}: is the is the log-normalized concentration of each metabolite (n = 749); μ: is the intercept; B_i: is the fixed effect of the breed (two levels, Duroc and Landrace); D_j: is the fixed effect of the batch for metabolomic analysis (two levels); A_k: is the covariate for the sampling age, in days; P_l: is the fixed effect of the pen (8 levels); ε_{ijkl}: is the random residual effect associated with each observation.

The metabolites were annotated with HMDB (Human metabolome database) based on library search of the masses in HMDB with a mass uncertainty of 0.005 Da or 5 ppm. Those metabolites that did not correspond to HMDB entries were left unannotated and removed from the analysis.

4.3. Integration of Transcriptomic and Metabolomic Data Based on the Linear Model

To uncover the complex relationship between metabolites and genes, we adopted a linear model framework using the IntLIM (Integration of Linear model) R-package (version 0.1.0) (https://github.com/mathelab/IntLIM) [23]. The IntLIM approach allows us to integrate the metabolomic-transcriptomic data considering a case-control design. Thus, as initially proposed, we compared the breeds (Duroc vs. Landrace) and the FE groups (low and high FCR animals). As a quality control step from IntLIM, we filtered out genes with the lowest 5% of the variation. Gene and metabolite exploratory analyses were performed by applying Principal Component Analysis to identify breed- and FE-specific clusters.

The linear model for data integration is given as described in the following equation:

$$m = \beta_1 + \beta_2 g + \beta_3 p + \beta_4 (g : p) + \varepsilon$$

where m: is the log-normalized metabolite abundance; β_1 : is the intercept; $\beta_2 g$: is the normalized and adjusted gene expression level; $\beta_3 p$: is the phenotype (FE group—high and low FE; or breed—Duroc and Landrace); $\beta_4(g : p)$: is the interaction between gene expression and phenotype; ε: is the residual effect associated with each observation ($\varepsilon = N(0, \sigma)$).

A statistically significant two-tailed p-value of the gene-phenotype (g-p) interaction indicates the difference in the phenotype of FE groups (high and low) or breed (Duroc and Landrace) calculated by the slope relating gene-expression and metabolite abundance [23]. The two-tailed p-value indicates that the slope relating gene-expression and metabolite abundance is different from one phenotype compared to the other. Thus, it was used to identify gene-metabolite associations that are specific to a particular phenotype (breed—Duroc and Landrace, FE—low and high). We calculated the absolute difference in the Spearman correlation identified from IntLIM between the FE and breed groups to find the significant ($p < 10^{-7}$) gene-metabolite pairs. The absolute difference between the FE group was estimated as ($r_{Low\_cor} - r_{High\_cor}$) where ($r_{High\_cor}$) is the correlation given for a gene-metabolite pair in high-FE group (Table 2) while (r_{Low_cor}) is the correlation given for a gene-metabolite pair in the low-FE group (Table 2). The absolute difference in the correlation between the breeds was estimated as ($r_{Landrace\_cor} - r_{Duroc\_cor}$), where, ($r_{Duroc\_cor}$) is the correlation for a given gene-metabolite pair in Duroc (Table 1) while ($r_{Landrace\_cor}$) is the correlation for a gene-metabolite pair in Landrace (Table 1).

4.4. Pathway Over-Representation Analysis

In the breed-specific and FE-specific group, the same metabolite can be related to more than one gene and vice-versa. So, we screened for the common metabolites and genes in the group and referred them as unique metabolites and unique genes respectively in this study. We analyzed the unique metabolites in each group (breed-specific and FE-specific) using Metaboanalyst 4.0 (www.metaboanalyst.ca) [58]. We used three parameters for the pathway analysis: the pathway library, algorithm for pathway over-representation analysis, and algorithm for topological analysis. For the current study, we selected the *Homo sapiens* (KEGG) pathway library to estimate the importance of the compound in a given metabolic pathway. For pathway over-representation and topology analysis, we used the hypergeometric test and relative-betweenness centrality algorithm, respectively, to measure the connections with the other nodes, including the number of shortest paths going through the node of interest.

Regarding the unique mapped (with chromosomal location information) genes, we carried out a co-functionality analysis using GeneMANIA [59] (www.genemania.org). GeneMANIA considers our query list of unique genes identified in each cluster (breed-specific and FE-specific) and allows us to predict the co-functional genes underlying similar functions. Thus, we analyzed the unique genes in each cluster (breed-specific—Duroc correlated cluster, and Duroc anti-correlated cluster, FE-specific—High-FE correlated cluster) to identify the co-functional genes in each cluster. Next, we used the unique genes, as well as the co-functional genes in each cluster of each group, to identify the GO terms using GOrilla (Gene ontology enrichment analysis and visualization tool) (http://cbl-gorilla.cs.technion.ac.il/) [60]. To this end, the *Homo sapiens* were used as the reference, and the entire set of identified and annotated genes in this study (n = 15,187 genes) was used as a background. Over-representation KEGG pathway analysis with the unique and co-functional genes in each cluster was performed using ClueGO version 2.5.4 [61] to cluster redundant terms with a kappa score of 0.4 and *S. scrofa* annotation as the background. The pathways were selected after filtering for group p-value corrected with Bonferroni step down ≤ 0.05 and those with at-least two over-represented genes.

5. Conclusions

This study applied a novel approach for metabolome-transcriptome data integration using the linear model unveiling potential gene-metabolite pairs affecting the biological processes related to FE in pigs. To the best of our knowledge, this is the first study to report the gene-metabolite interaction mechanisms that may determine nutrient partitioning and energy utilization and hence affect FE in pigs. The approach followed here provided many interesting genes and metabolites with significant p-values. While some of the metabolites and genes identified were known with their association for FE, others are novel and provide new avenues for further research. The unique metabolites were associated with valine-leucine-isoleucine biosynthesis/degradation and arginine-proline metabolism. The unique genes enriched for sphingolipid metabolism, valine-leucine-isoleucine degradation, alanine-aspartate-glutamate pathway (breed-specific), and cGMP-PKG signaling pathway (FE-specific). Further validation of genes, metabolites, and gene-metabolite interactions in a cohort with more animals with additional features such as alteration in dietary components, farm variations, and other environmental effects would help to establish a framework for future FE prediction using metabolomics biomarker profiles that could be practical to use in large populations other than genomic profiling. More data would also make it possible to model the complex relations in gene-metabolite profiles over time more accurately and will help to elucidate the regulatory mechanisms affecting the pathways underlying FE.

Supplementary Materials:
Figure S1: The principal component analysis of metabolites and genes. A and B: PCA plot of genes and metabolites, respectively, in (1) Duroc and (2) Landrace; C and D: PCA plot of genes and metabolites, respectively, in high and low feed efficient groups, Table S1: Pathways enriched by unique gene-metabolite pairs. Spreadsheet tabs are divided into (a) metabolite enrichment analysis results by Metaboanalyst; (b) list of the co-functional genes by GeneMania; (c) gene ontology pathways enriched by Gorilla; (d) KEGG pathways enriched by ClueGO.

Author Contributions: H.N.K. conceived and designed this "FeedOMICS" project, obtained funding as the main applicant. V.A.O.C. and H.N.K designed blood sampling experiments, phenotype data collection, metabolite profiling, and RNA-Seq experiments. P.B. carried out biostatistical and bioinformatic data analysis and wrote the manuscript. All authors collaborated in the interpretation of results, discussion, and write up of the manuscript. All authors have read, reviewed, and approved the final manuscript. All authors have read and agreed to the published version of the manuscript.

Acknowledgments: This study was funded by the Independent Research Fund Denmark (DFF)—Technology and Production (FTP) grant (grant number: 4184-00268B). V.A.O.C. received partial Ph.D. stipends from the University of Copenhagen and Technical University of Denmark. Authors thank Wellison Jarles Da Silva Diniz for helping in the interpretation of results. Authors thank SEGES—Pig Research Centre (VSP) Denmark for access to blood samples and phenotype datasets used in this study.

References

1. Patience, J.F.; Rossoni-Serão, M.C.; Gutiérrez, N.A. A review of feed efficiency in swine: biology and application. *J. Anim. Sci. Biotechnol.* **2015**, *6*, 33. [CrossRef]
2. He, B.; Li, T.; Wang, W.; Gao, H.; Bai, Y.; Zhang, S.; Zang, J.; Li, D.; Wang, J. Metabolic characteristics and nutrient utilization in high-feed-efficiency pigs selected using different feed conversion ratio models. *Sci. China Life Sci.* **2019**, *62*, 959–970. [CrossRef]
3. Carmelo, V.A.O.; Banerjee, P.; da Silva Diniz, W.J.; Kadarmideen, H.N. Metabolomic networks and pathways associated with feed efficiency and related-traits in Duroc and Landrace pigs. *Sci. Rep.* **2020**, *10*, 1–14. [CrossRef] [PubMed]
4. Godinho, R.M.; Bergsma, R.; Silva, F.F.; Sevillano, C.A.; Knol, E.F.; Lopes, M.S.; Lopes, P.S.; Bastiaansen, J.W.M.; Guimarães, S.E.F. Genetic correlations between feed efficiency traits, and growth performance and carcass traits in purebred and crossbred pigs. *J. Anim. Sci.* **2018**, *96*, 817–829. [CrossRef]
5. Ren, P.; Yang, X.J.; Cui, S.Q.; Kim, J.S.; Menon, D.; Baidoo, S.K. Effects of different feeding levels during three short periods of gestation on gilt and litter performance, nutrient digestibility, and energy homeostasis in gilts. *J. Anim. Sci.* **2017**, *95*, 1232–1242. [CrossRef] [PubMed]
6. Do, D.N.; Strathe, A.B.; Ostersen, T.; Pant, S.D.; Kadarmideen, H.N. Genome-wide association and pathway analysis of feed efficiency in pigs reveal candidate genes and pathways for residual feed intake. *Front. Genet.* **2014**, *5*, 307. [CrossRef] [PubMed]

7. Novais, F.J.; Dromms, R.A.; Alexandre, P.A.; Pires, P.R.L.; Styczynski, M.P.-W.; Ferraz, J.B.S.; Fukumasu, H.; Iglesias, A.H. Identification of a metabolomic signature associated with feed efficiency in beef cattle. *BMC Genom.* **2019**, *20*, 1–10. [CrossRef] [PubMed]
8. Rohart, F.; Paris, A.; Laurent, B.; Canlet, C.; Molina, J.; Mercat, M.J.; Tribout, T.; Muller, N.; Iannuccelli, N.; Villa-Vialaneix, N.; et al. Phenotypic prediction based on metabolomic data for growing pigs from three main european breeds. *J. Anim. Sci.* **2012**, *90*, 4729–4740. [CrossRef]
9. D'Alessandro, A.; Marrocco, C.; Zolla, V.; D'Andrea, M.; Zolla, L. Meat quality of the longissimus lumborum muscle of Casertana and Large White pigs: Metabolomics and proteomics intertwined. *J. Proteom.* **2011**, *75*, 610–627. [CrossRef]
10. Bertram, H.C.; Oksbjerg, N.; Young, J.F. NMR-based metabonomics reveals relationship between pre-slaughter exercise stress, the plasma metabolite profile at time of slaughter, and water-holding capacity in pigs. *Meat Sci.* **2010**, *84*, 108–113. [CrossRef]
11. Jing, L.; Hou, Y.; Wu, H.; Miao, Y.; Li, X.; Cao, J.; Michael Brameld, J.; Parr, T.; Zhao, S. Transcriptome analysis of mRNA and miRNA in skeletal muscle indicates an important network for differential Residual Feed Intake in pigs. *Sci. Rep.* **2015**, *5*, 11953. [CrossRef] [PubMed]
12. Vincent, A.; Louveau, I.; Gondret, F.; Tréfeu, C.; Gilbert, H.; Lefaucheur, L. Divergent selection for residual feed intake affects the transcriptomic and proteomic profiles of pig skeletal muscle. *J. Anim. Sci.* **2015**, *93*, 2745–2758. [CrossRef]
13. Horodyska, J.; Hamill, R.M.; Reyer, H.; Trakooljul, N.; Lawlor, P.G.; Mccormack, U.M.; Wimmers, K. RNA-Seq of Liver From Pigs Divergent in Feed Efficiency Highlights Shifts in Macronutrient Metabolism, Hepatic Growth and Immune Response. *Front. Genet.* **2019**, *10*. [CrossRef]
14. Carmelo, V.A.O.; Kadarmideen, H.N. Genome Regulation and Gene Interaction Networks Inferred From Muscle Transcriptome Underlying Feed Efficiency in Pigs. *Front. Genet.* **2020**, *11*. [CrossRef]
15. Kadarmideen, H.N. Genomics to systems biology in animal and veterinary sciences: Progress, lessons and opportunities. *Livest. Sci.* **2014**, *166*, 232–248. [CrossRef]
16. Suravajhala, P.; Kogelman, L.J.A.; Kadarmideen, H.N. Multi-omic data integration and analysis using systems genomics approaches: methods and applications in animal production, health and welfare. *Genet. Sel. Evol.* **2016**, *48*, 38. [CrossRef]
17. Carrillo, J.A.; He, Y.; Li, Y.; Liu, J.; Erdman, R.A.; Sonstegard, T.S.; Song, J. Integrated metabolomic and transcriptome analyses reveal finishing forage affects metabolic pathways related to beef quality and animal welfare. *Sci. Rep.* **2016**, *6*, 25948. [CrossRef]
18. Cavill, R.; Jennen, D.; Kleinjans, J.; Briedé, J.J. Transcriptomic and metabolomic data integration. *Brief. Bioinform.* **2016**, *17*, 891–901. [CrossRef]
19. Wilkinson, J.M. Re-defining efficiency of feed use by livestock. *Animal* **2011**, *5*, 1014–1022. [CrossRef]
20. Do, D.N.; Strathe, A.B.; Jensen, J.; Mark, T.; Kadarmideen, H.N. Genetic parameters for different measures of feed efficiency and related traits in boars of three pig breeds. *J. Anim. Sci.* **2013**, *91*, 4069–4079. [CrossRef]
21. Morales, P.E.; Bucarey, J.L.; Espinosa, A. Muscle Lipid Metabolism: Role of Lipid Droplets and Perilipins. *J. Diabetes Res.* **2017**, *2017*, 1–10. [CrossRef] [PubMed]
22. Pedersen, B.K. Muscle as a Secretory Organ. In *Comprehensive Physiology*; John Wiley & Sons, Inc.: Hoboken, NJ, USA, 2013.
23. Siddiqui, J.K.; Baskin, E.; Liu, M.; Cantemir-Stone, C.Z.; Zhang, B.; Bonneville, R.; McElroy, J.P.; Coombes, K.R.; Mathé, E.A. IntLIM: integration using linear models of metabolomics and gene expression data. *BMC Bioinform.* **2018**, *19*, 81. [CrossRef] [PubMed]
24. Taylor, V.A.; Stone, H.K.; Schuh, M.P.; Zhao, X.; Setchell, K.D.; Erkan, E. Disarranged Sphingolipid Metabolism From Sphingosine-1-Phosphate Lyase Deficiency Leads to Congenital Nephrotic Syndrome. *Kidney Int. Rep.* **2019**, *4*, 1763–1769. [CrossRef] [PubMed]
25. Donati, C.; Cencetti, F.; Bruni, P. Sphingosine 1-phosphate axis: a new leader actor in skeletal muscle biology. *Front. Physiol.* **2013**, *4*. [CrossRef]
26. Lefaucheur, L.; Lebret, B.; Ecolan, P.; Louveau, I.; Damon, M.; Prunier, A.; Billon, Y.; Sellier, P.; Gilbert, H. Muscle characteristics and meat quality traits are affected by divergent selection on residual feed intake in pigs1. *J. Anim. Sci.* **2011**, *89*, 996–1010. [CrossRef]
27. Kajimoto, T.; Okada, T.; Yu, H.; Goparaju, S.K.; Jahangeer, S.; Nakamura, S. Involvement of Sphingosine-1-Phosphate in Glutamate Secretion in Hippocampal Neurons. *Mol. Cell. Biol.* **2007**, *27*, 3429–3440. [CrossRef]

28. Santos, L.S.; Miassi, G.M.; Tse, M.L.P.; Gomes, L.M.; Berto, P.N.; Denadai, J.C.; Caldara, F.R.; Dalto, D.B.; Berto, D.A. Growth performance and intestinal replacement time of 13C in newly weaned piglets supplemented with nucleotides or glutamic acid. *Livest. Sci.* **2019**, *227*, 160–165. [CrossRef]
29. Wang, M.; Zhang, X.; Kang, L.; Jiang, C.; Jiang, Y. Molecular characterization of porcine NECD, SNRPN and UBE3A genes and imprinting status in the skeletal muscle of neonate pigs. *Mol. Biol. Rep.* **2012**, *39*, 9415–9422. [CrossRef]
30. Wahl, M.C.; Will, C.L.; Lührmann, R. The Spliceosome: Design Principles of a Dynamic RNP Machine. *Cell* **2009**, *136*, 701–718. [CrossRef]
31. Bottje, W.G.; Lassiter, K.; Piekarski-Welsher, A.; Dridi, S.; Reverter, A.; Hudson, N.J.; Kong, B.-W. Proteogenomics Reveals Enriched Ribosome Assembly and Protein Translation in Pectoralis major of High Feed Efficiency Pedigree Broiler Males. *Front. Physiol.* **2017**, *8*. [CrossRef]
32. Zhang, S.; Zeng, X.; Ren, M.; Mao, X.; Qiao, S. Novel metabolic and physiological functions of branched chain amino acids: a review. *J. Anim. Sci. Biotechnol.* **2017**, *8*, 10. [CrossRef] [PubMed]
33. Harper, A.E.; Miller, R.H.; Block, K.P. Branched-chain amino acid metabolism. *Annu. Rev. Nutr.* **1984**, *4*, 409–454. [CrossRef] [PubMed]
34. Fu, L.; Xu, Y.; Hou, Y.; Qi, X.; Zhou, L.; Liu, H.; Luan, Y.; Jing, L.; Miao, Y.; Zhao, S.; et al. Proteomic analysis indicates that mitochondrial energy metabolism in skeletal muscle tissue is negatively correlated with feed efficiency in pigs. *Sci. Rep.* **2017**, *7*, 45291. [CrossRef]
35. Freund, H.R.; Hanani, M. The metabolic role of branched-chain amino acids. *Nutrition* **2002**, *18*, 287–288. [CrossRef]
36. Duarte, D.A.S.; Newbold, C.J.; Detmann, E.; Silva, F.F.; Freitas, P.H.F.; Veroneze, R.; Duarte, M.S. Genome-wide association studies pathway-based meta-analysis for residual feed intake in beef cattle. *Anim. Genet.* **2019**, *50*, 150–153. [CrossRef]
37. Hu, W.; Zhang, C.; Wu, R.; Sun, Y.; Levine, A.; Feng, Z. Glutaminase 2, a novel p53 target gene regulating energy metabolism and antioxidant function. *Proc. Natl. Acad. Sci. USA* **2010**, *107*, 7455–7460. [CrossRef] [PubMed]
38. Zampiga, M.; Flees, J.; Meluzzi, A.; Dridi, S.; Sirri, F. Application of omics technologies for a deeper insight into quali-quantitative production traits in broiler chickens: A review. *J. Anim. Sci. Biotechnol.* **2018**, *9*, 61. [CrossRef] [PubMed]
39. Grubbs, J.K.; Fritchen, A.N.; Huff-Lonergan, E.; Dekkers, J.C.M.; Gabler, N.K.; Lonergan, S.M. Divergent genetic selection for residual feed intake impacts mitochondria reactive oxygen species production in pigs1. *J. Anim. Sci.* **2013**, *91*, 2133–2140. [CrossRef] [PubMed]
40. Banerjee, P.; Carmelo, V.A.O.; Kadarmideen, H.N. Genome-Wide Epistatic Interaction Networks Affecting Feed Efficiency in Duroc and Landrace Pigs. *Front. Genet.* **2020**, *11*. [CrossRef]
41. Higgins, M.G.; Fitzsimons, C.; McClure, M.C.; McKenna, C.; Conroy, S.; Kenny, D.A.; McGee, M.; Waters, S.M.; Morris, D.W. GWAS and eQTL analysis identifies a SNP associated with both residual feed intake and GFRA2 expression in beef cattle. *Sci. Rep.* **2018**, *8*, 14301. [CrossRef]
42. Bijvelds, M.J.C.; Tresadern, G.; Hellemans, A.; Smans, K.; Nieuwenhuijze, N.D.A.; Meijsen, K.F.; Bongartz, J.-P.; Ver Donck, L.; de Jonge, H.R.; Schuurkes, J.A.J.; et al. Selective inhibition of intestinal guanosine 3′,5′-cyclic monophosphate signaling by small-molecule protein kinase inhibitors. *J. Biol. Chem.* **2018**, *293*, 8173–8181. [CrossRef]
43. Valente, T.S.; Baldi, F.; Sant'Anna, A.C.; Albuquerque, L.G.; Paranhos da Costa, M.J.R. Genome-Wide Association Study between Single Nucleotide Polymorphisms and Flight Speed in Nellore Cattle. *PLoS ONE* **2016**, *11*, e0156956. [CrossRef] [PubMed]
44. Valentino, M.A.; Lin, J.E.; Snook, A.E.; Li, P.; Kim, G.W.; Marszalowicz, G.; Magee, M.S.; Hyslop, T.; Schulz, S.; Waldman, S.A. A uroguanylin-GUCY2C endocrine axis regulates feeding in mice. *J. Clin. Investig.* **2011**, *121*, 3578–3588. [CrossRef]
45. Rauw, W.M.; Portolés, O.; Corella, D.; Soler, J.; Reixach, J.; Tibau, J.; Prat, J.M.; Diaz, I.; Gómez-Raya, L. Behaviour influences cholesterol plasma levels in a pig model. *Animal* **2007**, *1*, 865–871. [CrossRef]
46. Zhang, L.; Rajbhandari, P.; Priest, C.; Sandhu, J.; Wu, X.; Temel, R.; Castrillo, A.; de Aguiar Vallim, T.Q.; Sallam, T.; Tontonoz, P. Inhibition of cholesterol biosynthesis through RNF145-dependent ubiquitination of SCAP. *Elife* **2017**, *6*. [CrossRef] [PubMed]
47. Holeček, M. Effect of starvation on branched-chain α-keto acid dehydrogenase activity in rat heart and skeletal muscle. *Physiol. Res.* **2001**, *50*, 19–24. [PubMed]

48. Duan, Y.; Duan, Y.; Li, F.; Li, Y.; Guo, Q.; Ji, Y.; Tan, B.; Li, T.; Yin, Y. Effects of supplementation with branched-chain amino acids to low-protein diets on expression of genes related to lipid metabolism in skeletal muscle of growing pigs. *Amino Acids* **2016**, *48*, 2131–2144. [CrossRef] [PubMed]
49. Mukiibi, R.; Vinsky, M.; Keogh, K.; Fitzsimmons, C.; Stothard, P.; Waters, S.M.; Li, C. Liver transcriptome profiling of beef steers with divergent growth rate, feed intake, or metabolic body weight phenotypes1. *J. Anim. Sci.* **2019**, *97*, 4386–4404. [CrossRef] [PubMed]
50. Sung, Y.J.; Pérusse, L.; Sarzynski, M.A.; Fornage, M.; Sidney, S.; Sternfeld, B.; Rice, T.; Terry, J.G.; Jacobs, D.R.; Katzmarzyk, P.; et al. Genome-wide association studies suggest sex-specific loci associated with abdominal and visceral fat. *Int. J. Obes.* **2016**, *40*, 662–674. [CrossRef]
51. Valadkhan, S.; Gunawardane, L.S. Role of small nuclear RNAs in eukaryotic gene expression. *Essays Biochem.* **2013**, *54*, 79–90. [CrossRef]
52. Sun, J.S.; Manley, J.L. A novel U2-U6 snRNA structure is necessary for mammalian mRNA splicing. *Genes Dev.* **1995**, *9*, 843–854. [CrossRef] [PubMed]
53. Ravi, S.; Schilder, R.J.; Kimball, S.R. Role of Precursor mRNA Splicing in Nutrient-Induced Alterations in Gene Expression and Metabolism. *J. Nutr.* **2015**, *145*, 841–846. [CrossRef] [PubMed]
54. Mansouri, E.; Khorsandi, L.; Moaiedi, M.Z. Grape seed proanthocyanidin extract improved some of biochemical parameters and antioxidant disturbances of red blood cells in diabetic rats. *Iran. J. Pharm. Res.* **2015**, *14*, 329–334. [CrossRef] [PubMed]
55. Russell, J.R.; Sexten, W.J.; Kerley, M.S.; Hansen, S.L. Relationship between antioxidant capacity, oxidative stress, and feed efficiency in beef steers. *J. Anim. Sci.* **2016**, *94*, 2942–2953. [CrossRef] [PubMed]
56. Anders, S.; Huber, W. Differential expression analysis for sequence count data. *Genome Biol.* **2010**, *11*, R106. [CrossRef] [PubMed]
57. Ritchie, M.E.; Phipson, B.; Wu, D.; Hu, Y.; Law, C.W.; Shi, W.; Smyth, G.K. limma powers differential expression analyses for RNA-sequencing and microarray studies. *Nucleic Acids Res.* **2015**, *43*, e47. [CrossRef]
58. Chong, J.; Soufan, O.; Li, C.; Caraus, I.; Li, S.; Bourque, G.; Wishart, D.S.; Xia, J. MetaboAnalyst 4.0: Towards more transparent and integrative metabolomics analysis. *Nucleic Acids Res.* **2018**, *46*, W486–W494. [CrossRef]
59. Mostafavi, S.; Ray, D.; Warde-Farley, D.; Grouios, C.; Morris, Q. GeneMANIA: a real-time multiple association network integration algorithm for predicting gene function. *Genome Biol.* **2008**, *9*, S4. [CrossRef]
60. Eden, E.; Navon, R.; Steinfeld, I.; Lipson, D.; Yakhini, Z. GOrilla: a tool for discovery and visualization of enriched GO terms in ranked gene lists. *BMC Bioinform.* **2009**, *10*, 48. [CrossRef]
61. Bindea, G.; Mlecnik, B.; Hackl, H.; Charoentong, P.; Tosolini, M.; Kirilovsky, A.; Fridman, W.H.; Pagès, F.; Trajanoski, Z.; Galon, J. ClueGO: A Cytoscape plug-in to decipher functionally grouped gene ontology and pathway annotation networks. *Bioinformatics* **2009**, *25*, 1091–1093. [CrossRef]

13

First Insights into the Urinary Metabolome of Captive Giraffes by Proton Nuclear Magnetic Resonance Spectroscopy

Chenglin Zhu [1,†], Sabrina Fasoli [2,†], Gloria Isani [2] and Luca Laghi [1,*]

1 Department of Agro-Food Science and Technology, University of Bologna, 47521 Cesena, Italy; chenglin.zhu2@unibo.it
2 Department of Veterinary Medical Sciences, University of Bologna, Ozzano Emilia, 40064 Bologna, Italy; sabrina.fasoli2@unibo.it (S.F.); gloria.isani@unibo.it (G.I.)
* Correspondence: l.laghi@unibo.it.
† These authors contributed equally to this work.

Abstract: The urine from 35 giraffes was studied by untargeted ^{1}H-NMR, with the purpose of obtaining, for the first time, a fingerprint of its metabolome. The metabolome, as downstream of the transcriptome and proteome, has been considered as the most representative approach to monitor the relationships between animal physiological features and environment. Thirty-nine molecules were unambiguously quantified, able to give information about diet, proteins digestion, energy generation, and gut-microbial co-metabolism. The samples collected allowed study of the effects of age and sex on the giraffe urinary metabolome. In addition, preliminary information about how sampling procedure and pregnancy could affect a giraffe's urinary metabolome was obtained. Such work could trigger the setting up of methods to non-invasively study the health status of giraffes, which is utterly needed, considering that anesthetic-related complications make their immobilization a very risky practice.

Keywords: captive giraffes; urine; metabolomics; ^{1}H-NMR

---

1. Introduction

According to the International Union for Conservation of Nature (IUCN), giraffe (*Giraffa camelopardalis*) is declared a vulnerable species [1]. Moreover, different measures have been taken to monitor and protect giraffe population. For example, the International Union for Conservation of Nature (IUCN) Species Survival Commission (SSC) Giraffe and Okapi Specialist Group (GOSG) was established with the aim of studying and guaranteeing the conservation needs of this species (https://www.giraffidsg.org/). In addition, from November 26, 2019, giraffes are included in Appendix II of the CITES (Convention on International Trade in Endangered Species of Wild Fauna and Flora) to improve its protection, subjecting it to strict regulation (https://www.cites.org/).

Zoos represent a significant part of the protection strategy for giraffes, with projects explicitly aimed at protecting endangered species and pursuing high standards of animal welfare [2]. In these structures, however, giraffes may be subjected to sources of stress that reverberate negatively on individual and social behaviors [3]. Causes of stress could be represented by the presence of visitors and attendants [4]. Among the efforts that have been made to reduce the stressors, some are devoted to developing protocols to evaluate their general health status that do not involve immobilization, but are based on indirect methods [3]. In fact, giraffes are particularly prone to anesthetic-related complications and death, due to their unique cardiovascular system, making immobilization a risky practice [5,6].

The possibility of obtaining information from urine collected from the ground seems particularly attractive from this point of view, but the literature on this type of sampling is absent for giraffes and it has been only reported in okapi [7]. Indeed giraffes have been studied more for their iconic height and the mechanisms existing at the cardiovascular level to counterbalance the consequent state of primary hypertension [8–10].

Among the completely unexplored characteristics of giraffe urine is its metabolome, the ensemble of low weight molecules produced by the cellular metabolism. Studies carried out by liquid chromatography–mass spectrometry (LC/MS) or by proton nuclear magnetic resonance spectroscopy (^{1}H-NMR) on humans and other animals suggest that the giraffe's urinary metabolome may be particularly informative about the general health of the animal. In horse urine, molecules revealing the action of the intestinal microbiota were found in micromolar concentrations [11,12]. Molecular patterns of the urinary metabolome linked to inflammatory processes have been identified in humans [13]. Urinary profile responses to the calorie content of the diet were identified in rat [14]. The effects of heat stress were studied in cattle by metabolomic profiling of urine [15]. Indeed, the use of urine as a source of biological data in giraffes could be a suitable alternative, due to its non-invasive approach that could avoid the immobilization of animals.

Among the analytical platforms capable of fulfilling the requirements, proton nuclear magnetic resonance spectroscopy (^{1}H-NMR) has been widely used for the investigation of urine metabolomes, taking advantage of its high reproducibility and minimal sample preparation.

In the present study, we wanted to verify the feasibility of ^{1}H-NMR based metabolomic studies focusing on the urine of giraffes. For this purpose, we characterized the molecular profile of healthy giraffes held in captivity to obtain preliminary quantitative values that could be applied for the diagnosis of diseases affecting this animal. Moreover, the samples collected gave the opportunity to have a first insight about the influence of important physiological factors, such as the sex and age of the subjects, on the urinary metabolomic profile.

2. Results

2.1. Urinary Metabolites Identification by Untargeted ^{1}H-NMR

A representative spectrum of the metabolites identified in the giraffe's urine is reported in Figure 1. In this study, we identified 39 molecules (Table S1). These molecules mainly pertain to the classes of amino acids and derivatives and organic acids and derivatives. Hippurate (30.63%), creatinine (25.17%), and phenylacetylglycine (12.64%) were the most represented metabolites.

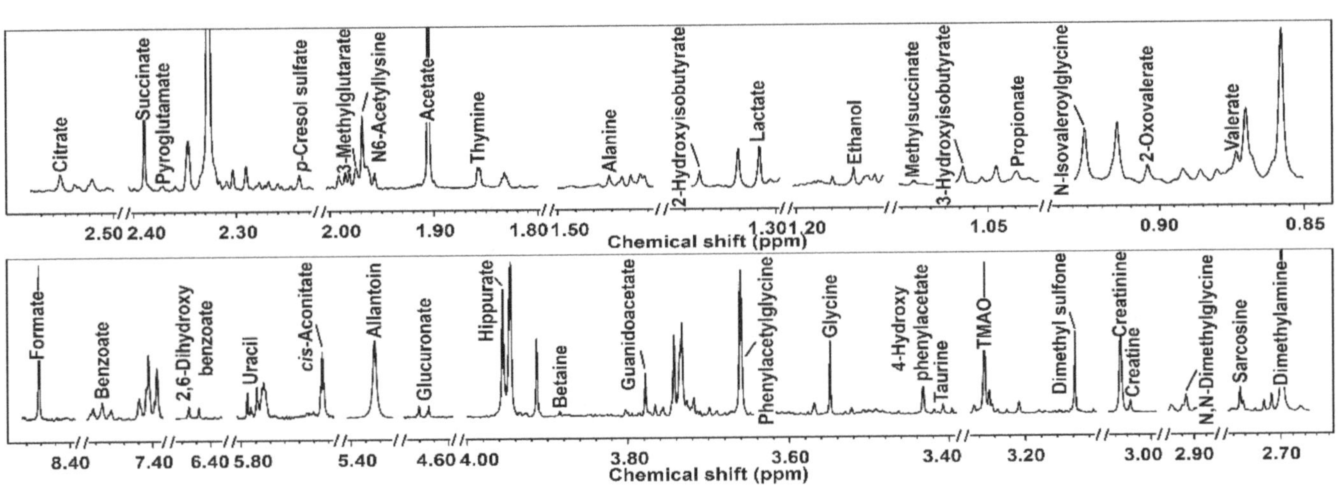

Figure 1. Portions of ^{1}H-NMR spectra, representative of all the spectra obtained in this study. Each molecule's name appears over the NMR peak used for its quantification. To ease the visual inspection of each portion, a different spectrum with a convenient signal-to-noise ratio has been selected.

2.2. Effects of Sampling Procedure and Location

To check the potential influence of the different sampling methods, we wanted to collect pairs of samples during the same voiding, one directly and one from the ground. Unfortunately, we only succeeded in this task for one individual (Ronny). Among 39 quantified compounds, four molecules showed a variation of concentration higher than 50%, namely *p*-cresol sulfate, citrate, glycine, and benzoate. ^{1}H-NMR signals for these compounds are reported in Figure 2. In detail, benzoate and glycine were more concentrated in the urine collected from the ground, while citrate and *p*-cresol sulfate showed the opposite trend. Overall, the 39 molecules showed a median difference between the two samples of 4.8%. As these observations were based only on one pair of samples from a single individual, we decided not to exclude these molecules from the subsequent analyses.

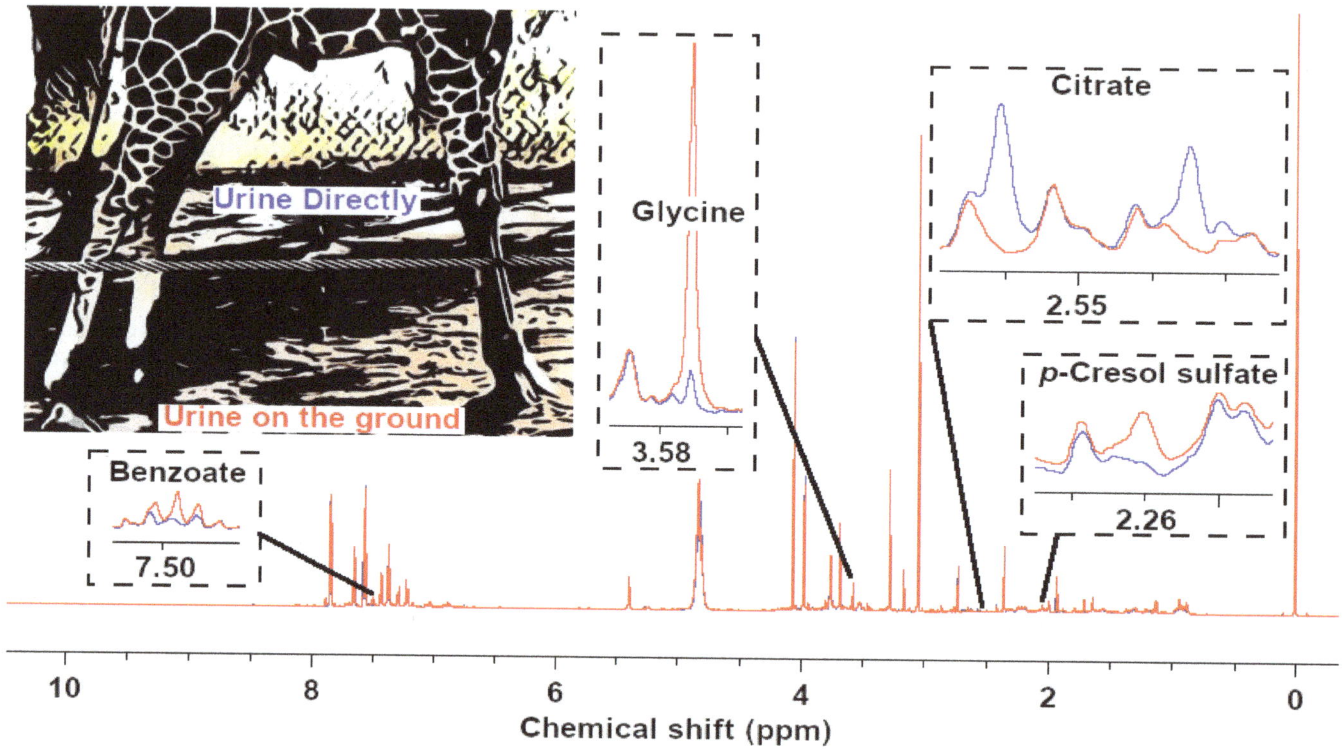

Figure 2. Representative sections of two spectra obtained from analyzing urine from the same giraffe (Ronny), collected directly (**blue line**) and from the ground (**red line**) during one urination, respectively.

To obtain hints about the potential effects of location on the metabolome of giraffe urine, we selected the samples from the locations BG (Parco Faunistico Le Cornelle) and FA (Zoosafari Fasanolandia), where most of the samples had been collected, and we set up a three-way ANOVA analysis aiming at excluding any effect related to gender or age. None of the molecules quantified appeared as significantly different in relation to zoo, so this variable was not considered in the subsequent analyses.

2.3. Sex Affects the Giraffe Urine Molecular Profile

To obtain preliminary data on the effect of sex on the urinary metabolome, we focused on samples collected from adult, non-pregnant individuals. Six molecules were found to be significantly ($p < 0.05$) affected by sex, as shown in Table 1.

Table 1. Metabolite concentrations (mmol/L, median (IQR)) in the adult group were significantly ($p < 0.05$) affected by sex, as assessed by *t*-test.

	Females (6)	Males (7)	Trend	*p* Value
Acetate	$2.04\ (5.23 \times 10^{-1})$	$1.33\ (9.04 \times 10^{-1})$	↓	0.034
Hippurate	13.50 (10.70)	19.30 (19.50)	↑	0.047
Lactate	$2.77 \times 10^{-1}\ (8.90 \times 10^{-2})$	$1.28 \times 10^{-1}\ (7.35 \times 10^{-2})$	↓	0.003
Phenylacetylglycine	7.82 (2.41)	15.20 (5.53)	↑	0.014
Succinate	$2.48 \times 10^{-1}\ (3.00 \times 10^{-2})$	$1.66 \times 10^{-1}\ (8.80 \times 10^{-2})$	↓	0.006
Thymine	$1.77 \times 10^{-1}\ (4.94 \times 10^{-2})$	$2.86 \times 10^{-1}\ (1.79 \times 10^{-1})$	↑	0.043

To have an overall view of the data, a robust principal component analysis (rPCA) model was calculated on their concentration, as shown in Figure 3.

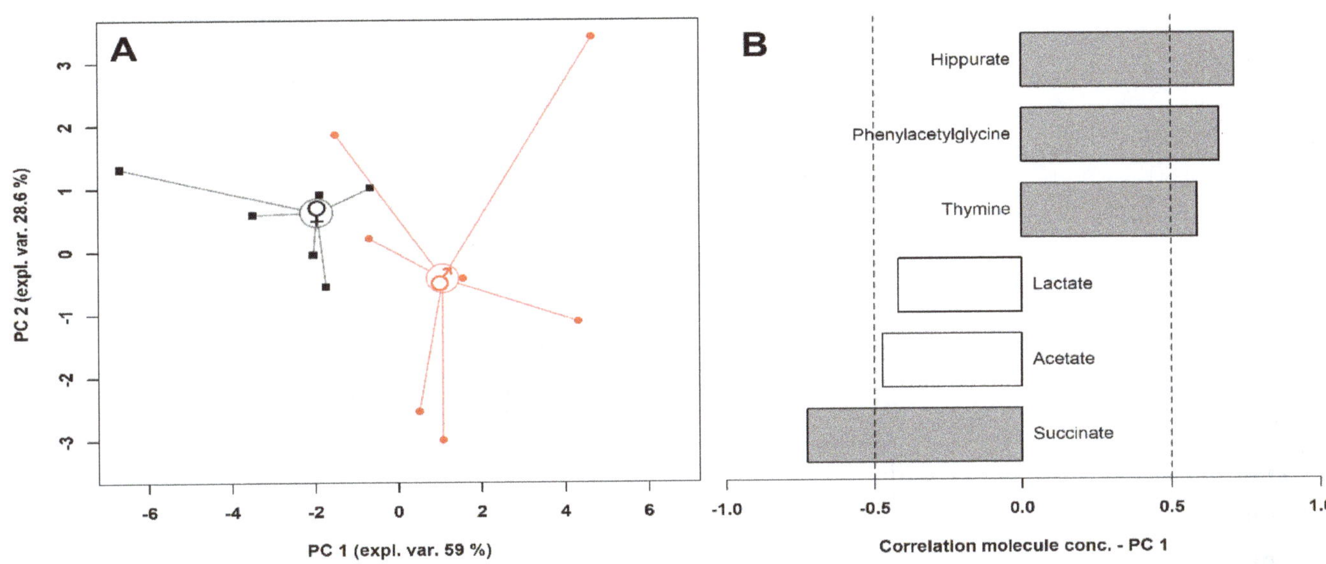

Figure 3. rPCA model calculated on the concentration of the significantly different molecules between male and female giraffes. The scoreplot (**A**) represents with squares and circles females and males, respectively. The median of each sample group is represented by wide circles. The loading plot (**B**) reports the correlation between the importance of each substance over principal component 1 and its concentration. Gray bars highlight significant correlations ($p < 0.05$).

Three principal components (PCs) were accepted by the algorithm to depict the overall data features. PC 1, accounting for 59% of the variance thus represented, indeed significantly summarized the peculiarities connected to sex ($p < 0.05$), with female and male individuals appearing respectively at low and high PC scores. Among these molecules, hippurate, phenylacetylglycine, and thymine were more abundant in the urine of male individuals, while lactate, acetate, and succinate were more concentrated in the females' urine.

2.4. Effect of Age on the Urinary Metabolome

Age was found to significantly affect ($p < 0.05$) the concentration of three urinary metabolites, namely formate, alanine, and valerate, (Figure 4). To understand if their evolution was part of a trend spanning over the entire life of the giraffe, these molecules were used as a base for an rPCA model (Figure 5).

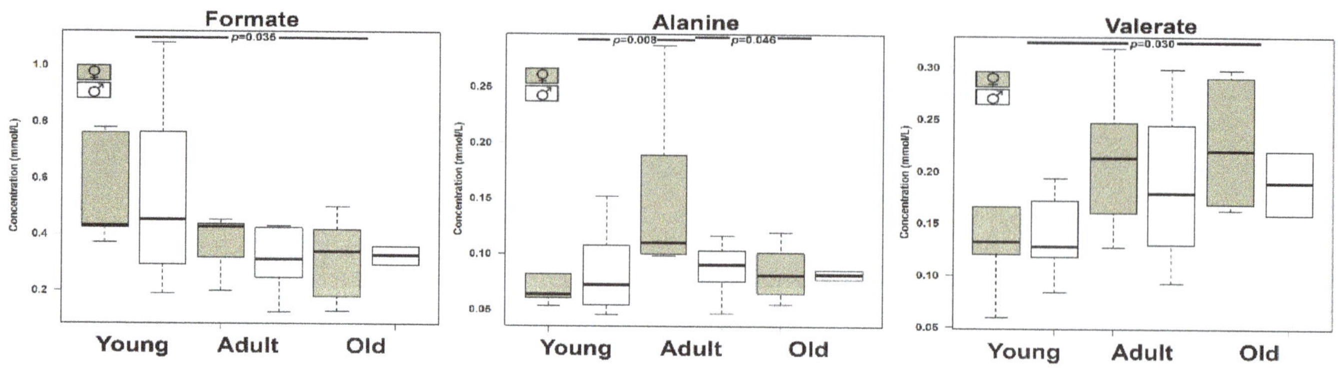

Figure 4. Boxplots showing the concentration of molecules significantly ($p < 0.05$) affected by age, as assessed by two-way ANOVA followed by Tukey post-hoc test.

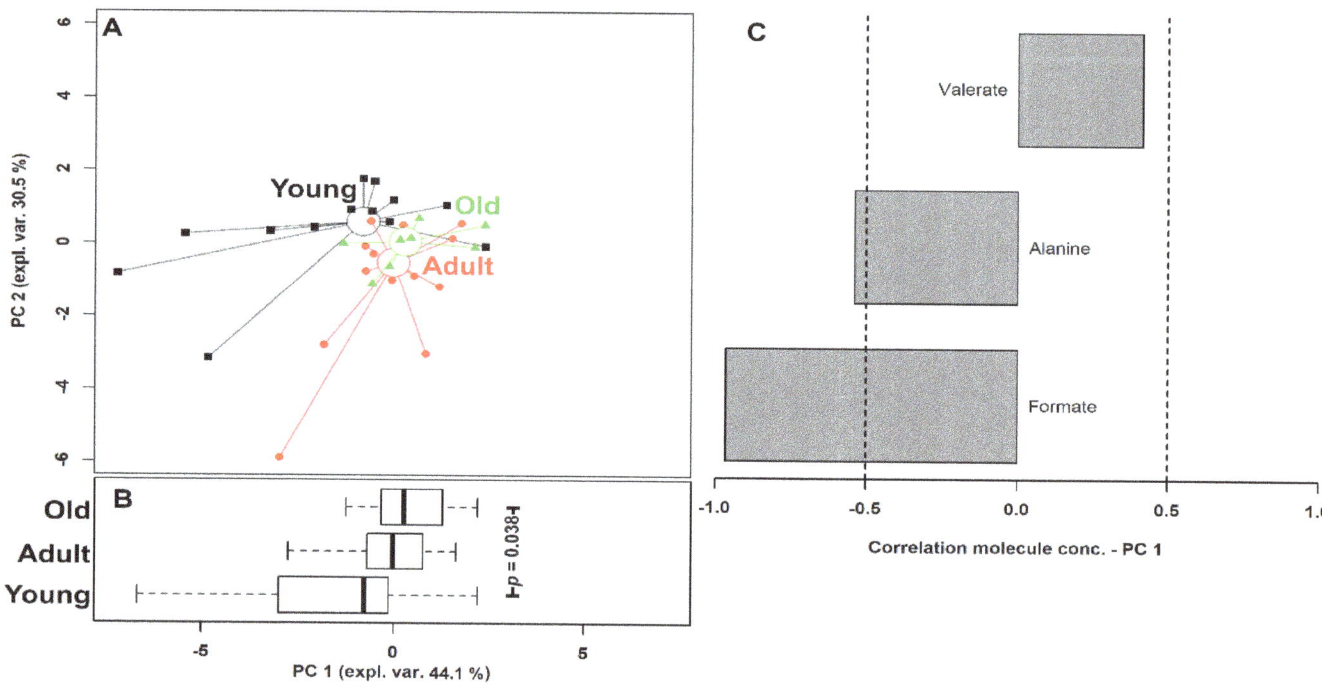

Figure 5. rPCA model of the concentration of the molecules showing a significant difference among the giraffes grouped by age. The scoreplot (**A**) shows the samples from the three groups with squares (Young), circles (Adult), and triangles (Old). The median of each sample group is represented by wide circles. The boxplot (**B**) summarizes the positions of the samples along PC1 and compares them by two-way ANOVA, followed by Tukey post-hoc test. The loading plot (**C**) reports the correlation between the importance of each substance over PC 1 and its concentration. Gray bars highlight significant correlations ($p < 0.05$).

Three PCs were accepted by the algorithm to depict the overall data features. PC 1, accounting for 44.1% of the variance thus represented, summarized effectively the peculiarities connected to age ($p < 0.05$), with Young, Adult, and Old individuals appearing respectively at low, intermediate, and high PC scores. Among these molecules, formate and alanine were more abundant in young individuals, while valerate showed an opposite trend.

2.5. Pregnancy Related Urinary Metabolome

Urine samples were obtained from two female giraffes during and after pregnancy (Table S2). Despite the limited number of samples, it was possible to observe a variation of five metabolites during the pregnancy. These molecules showed consistent trends in the samples from both giraffes.

All these molecules showed a relevant increase in concentration during the pregnancy, except for phenylacetylglycine, as shown in Table 2.

Table 2. Urinary metabolites (mmol/L) affected by pregnancy consistently across the two giraffes observed.

	Giulietta		Nicole	
	Not Pregnant	Pregnant [1]	Not Pregnant	Pregnant
Phenylacetylglycine	10.20	3.52 ↓	10.40	5.02 ↓
Benzoate	2.14	3.88 ↑	2.46	12.22 ↑
Glycine	1.06	3.04 ↑	1.79	11.65 ↑
Taurine	1.75×10^{-1}	2.93×10^{-1} ↑	7.98×10^{-2}	1.33×10^{-1} ↑
p-Cresol sulfate	1.46×10^{-2}	2.15×10^{-2} ↑	6.37×10^{-2}	3.50×10^{-1} ↑

[1] For readability, only molecules changing by more than 40% for both giraffes are shown.

3. Discussion

The present paper describes one of the first studies ever devoted to the urinary metabolome of nonfarmed animals, and the very first focusing on the giraffe metabolome. Due to such paucity of studies on the topic, a key point that needs to be addressed before giraffe urine can be used for metabolomics studies is the possibility of relying on samples collected from the ground. Several aspects, in fact, make the collection of urine directly from the individual during urination highly impractical. To obtain a first insight on this point, we managed to collect the same urine sample either at the start of a spontaneous voiding or from the ground with a syringe at the end the voiding. The corresponding ^{1}H-NMR spectra were highly superimposable, except for four molecules, namely benzoate, citrate, *p*-cresol sulfate, and glycine. The fact that the non-volatile glycine showed the greatest differences gave hints that the discrepancies could be mainly connected to dynamic variations in composition during urination, in agreement with Sink and Weinstein [16]. Modifications induced by the collection method could therefore be considered a confounding factor of lower entities than inhomogeneity in the composition of urine during voiding.

The 39 molecules identified give information about protein digestion, diet, gut-microbial co-metabolism, and energy production. Their quantitative observation therefore offers a handy perspective of the health status of giraffes, through a quintessentially non-invasive sampling method.

Comparisons with the urinary metabolome of other animals are also possible, giving indirect information about the differences in metabolism. An example of this possibility is offered by allantoin. This molecule is the fourth most concentrated in giraffe urine (Table S1), identically to yak (*Bos grunniens*) [17] and horse [18]. Differently from these strictly herbivorous animals, this molecule is the most concentrated in the urine of the giant panda [19], even if the giant panda consumes an amount of vegetables in relation to body weight (as much as 30%) much higher than ruminants or horses, which should lead to the lowest concentration of urinary allantoin [20]. This apparent contradiction leads to speculate that the main mechanism determining the concentration of allantoin in the urine of the above-mentioned animals is likely to be its renal reabsorption, which is very effective in strictly herbivorous animals [20].

3.1. *Sex Affects the Giraffe Urine Molecular Profile*

In the current study acetate, succinate, and lactate concentrations appeared to be significantly higher in female giraffe urine, while hippurate, phenylacetylglycine, and thymine were more concentrated in male urine. For acetate, two of the authors of the present paper identified a similar situation in horse urine [18]. For the other molecules, indirect connections with published findings can be devised. There is an abundance of references, focusing on humans, showing that exercise leads to higher concentrations of acetate, succinate, and lactate in urine, and lower concentrations of thymine and hippurate [21–23]. Ginnett et al. showed that female giraffes spend more time walking, foraging, feeding, and traveling than males [24]. The two observations seem to suggest that the sex-related differences observed in the urine of males and females may be partly due to the different daily activities. Contrary to the previously reported molecules, phenylacetylglycine is mainly a co-metabolite of gut microorganisms, derived from valine, leucine, phenylalanine, lysine, or ornithine [25]. Its different concentration in relation to sex may therefore reflect peculiarities in gut microbiota profiles or different foraging behaviors, similarly to what was recently observed in the giant panda [19]. Ginnett et al., in fact, demonstrated that males prefer larger bites than females, with potential consequences on the food, and in turn urine, metabolome profile [24]. It is tantalizing to speculate that the length of the neck, which is higher in males [5], may play a role too. In fact, Schüßler and Greven [26] found an allometric direct relationship between rumen-to-mouth distance and the duration of rumination intercycles, influencing in turn the digestive action of ruminal microorganisms.

3.2. *Effect of Age*

By removing the gender effect by two-way ANOVA, it was possible to focus on the effect of age. In parallel with previous studies in rats and humans [27,28], formate and alanine were negatively related to age. The trend observed for formate is very likely related to the gut microbiome. In fact, in the gut microbiota of the juvenile giraffes there is a prevalence of *Bacteroides* and *Acinetobacter* genera, responsible for the degradation of starch and cellulose to formate [29], while in the gut of adult giraffes other genera tend to prevail, such as *Treponema* [30].

The concentration of amino acids in urine has been consistently linked to the turnover of muscle amino acids [18,31], with urinary concentration of alanine specifically related to exercise [32]. Therefore, the difference in the concentration of alanine could be ascribed to a variation of daily activity intensity along age.

3.3. *Effect of Pregnancy*

Early identification of pregnant giraffes with maximum accuracy is an important issue for optimizing their management. Although some diagnostic methods (e.g., ultrasonography) have been described in domestic animals [33], their application to wild or captive animals is hindered by practical reasons. Metabolomics approaches seem in principle promising for setting up diagnostic methods that might be more convenient in specific contexts, due to the possibility to quantify a high number of molecules at the same time. However, previous studies performed in domestic animals were focused on serum [34,35], a sub-optimal sample from the point of view of non-invasivity. Therefore, despite the restricted number of samples analyzed in the present study, the obtained data can provide a preliminary urinary fingerprint of pregnancy in giraffes.

Taurine is an important amino acid during pregnancy and lactation, because it satisfies the needs of both the fetus and suckling infant. In our research, taurine excretion through urine increased during early pregnancy, consistent with human studies [36]. Taurine is rarely found in plants [37], so that herbivores cannot obtain a sufficient amount taurine from the diet. Remarkably, in ruminants the urinary taurine concentration is strongly diet-dependent, as can be inferred from the works of Bristow et al. on cows fed with maize silage compared to free grazing cows [38]. Diet is therefore likely to trigger biosynthetic pathways, such as the one leading to taurine from methionine [39]. Moreover, a specific pathway, converting homocysteine to taurine and glycine through cysteine, is known to become effective in early pregnancy [39]. This latter mechanism is a likely reason for the increasing trend of taurine excretion we found in the present work.

A further contribution to urine metabolome profile modifications may be due to changes in the gut microbiota. In fact, among the molecules showing the greatest changes we found *p*-cresol sulfate and phenylacetylglycine, mainly described as gut microorganism co-metabolites [11,25], absorbed at the intestinal level and then expelled through urine. Interestingly, the change in the concentration of both has been related, in humans, with alterations in the microbiota profile linked to inflammatory states [13,40], in which pregnancy is known to play a role [41]. Despite the very limited number of cases here, these observations support the compelling possibility to use the urine metabolome to gain specific information about giraffe inflammatory status during pregnancy, as modulated by the gut microbiota.

4. Materials and Methods

4.1. Compliance with Ethical Requirements

All the procedures related to animals respected the Directive 2010/63/EU of the European Parliament and of the Council of September 22, 2010 on the protection of animals used for scientific purposes (Article 1, Paragraph 1, Letter b) and the Italian legislation (D. Lgs. n. 26/2014, Article 2, Paragraph 1, Letter b).

4.2. Sample Collection

A total of 35 captive giraffes (*Giraffa camelopardalis*) were involved in the current study. Based on physical examinations, giraffes did not show symptoms of diseases both before and during the urine sampling period. The giraffes were housed in five Italian zoos: Zoosafari Fasanolandia (FA) (N = 11), Safari Ravenna (RA) (N = 4), Giardino Zoologico di Pistoia (PT) (N = 1), Parco Natura Viva (VR) (N = 4), and Parco Faunistico Le Cornelle (BG) (N = 15).

The details for each giraffe are reported in Table 3. Their age ranged from a minimum of 6 months to a maximum of 20 years. The giraffes were categorized in 3 age classes: Young (from 6 months to 6 years old, N = 14), Adult (from 6 to 15 years old, N = 16), and Old (older than 15, N = 9), according to the following information. In female giraffes the first birth is at about 6.4 years old, even if sexual maturity is reached at 3–4 years [42,43]. Giraffe males are considered as adults when older than 6 years old, according to Lee et al. [44].

Table 3. Animal information.

Sample ID	Name	Sex	Age (years)	Zoo
N.01	Ronny	Male	14	FA
N.02	Nicole	Female	14	FA
N.03	Giulietta	Female	17	FA
N.04	Marcello	Male	9	FA
N.05	Italia	Female	8	FA
N.06	Carlos	Male	2	RA
N.07	Daniele	Male	11	RA
N.08	Cleopatra	Female	20	PT
N.09	Alto	Male	2	FA
N.10	Congo	Male	0.3	FA
N.11	Roberto	Male	0.6	RA
N.12	Martina	Female	0.6	RA
N.13	Linda	Female	16	BG
N.14	Sandy	Female	16	BG
N.15	Raffa	Female	7	BG
N.16	Telete	Female	2	BG
N.17	Rusman	Male	16	BG
N.18	Akuna	Female	10	BG
N.19	Ciokwe	Male	5	BG
N.20	Miro	Male	9	BG
N.21	Lucia	Female	16	BG
N.22	Nuvola	Female	7	BG
N.23	Sahel	Female	2	BG
N.24	Russel	Male	16	BG
N.25	Ramiro	Male	3	BG
N.26	Madiba	Male	6	BG
N.27	Nasanta	Female	2	BG

Table 3. *Cont.*

N.28	Macchia	Male	5	VR
N.29	Secondo	Male	11	VR
N.30	Akasha	Male	7	VR
N.31	Quarto	Male	9	VR
N.32	Luna	Female	15	FA
N.33	Kenya	Female	20	FA
N.34	Alessia	Female	4	FA
N.35	Mina	Female	14	FA

The samples were collected between 10:00 a.m. and 2:00 p.m., in connection to the daily activities of the keepers. Urine samples were collected with a syringe from the ground. To limit the soil contaminants, only the upper part of the urine was collected, immediately after the spontaneous voiding, before it was absorbed by the soil. A sample from one male was also collected directly into a sterile beaker, preventing the sample from touching the ground. Four urine samples were collected from two females during and after pregnancy. After collection, the urine samples were centrifuged at 1500× *g* for 10 min, to further remove potential ground contaminants, and the supernatants were frozen at −80 °C.

4.3. Metabolomic Analysis

We prepared urine samples for NMR by thawing and centrifuging them for 15 min at 18,630× *g* at 4 °C. We added the supernatant (350 μL) to bi-distilled water (350 μL) and to a D_2O solution (200 μL) of TSP (3-(trimethylsilyl)-propionic-2,2,3,3-d4 acid) 10 mM and of NaN_3 2 mM. A 1M phosphate buffer had been used to set the D_2O solution to pH of 7.00 ± 0.02. After a further centrifugation, we recorded ^{1}H-NMR spectra at 298 K with an AVANCE III spectrometer (Bruker, Milan, Italy), at a frequency of 600.13 MHz, equipped with Topspin software (Ver. 3.5).

According to Zhu et al. [17], we suppressed the signals from broad resonances using a CPMG-(Carr-Purcell-Meiboom-Gill) filter composed of 400 echoes with a of 400 s and a 180° pulse of 24 s, for a total filter of 330 ms. We also applied pre-saturation, to reduce the signal from water. We employed Topspin software to apply a line broadening of 0.3 Hz and to adjust the phase of each spectrum. We set the recycle delay to 5 s, by considering the relaxation time of the protons under investigation. We employed R computational language [45] for any further processing of spectra, quantification of molecules, and data mining, with custom scripts.

We aligned the spectra by using the TSP signal as a reference (−0.017 ppm). We adjusted the baseline of each spectrum by distinguishing irregularities of the baseline from genuine signals, according to the "rolling ball" idea [46], implemented in the R package "baseline" [47]. We performed the assignment of the signals by comparing chemical shift and multiplicity with the libraries (Ver. 10) of Chenomx software (Chenomx Inc., Canada, v. 8.3).

According to Dieterle et al. [48], water intake behavior can change the dilution of urine as much as five times, obscuring any trend in metabolite concentrations. We removed this confounding factor by calculating, for each sample, the ratio between the area of TSP peak and the intensity of the spectrum. This allowed us to estimate the dilution of each sample and to select the one with the mostly representative dilution. We used this sample as a reference by quantifying the molecules from the added TSP. We then normalized the other samples towards the reference by probabilistic quotient normalization (PQN) [48].

4.4. Statistical Analysis

We conducted the statistical analysis in R computational language [45] and we refined the artwork by GIMP (version 2.10, www.gimp.org). Prior to univariate analysis, we transformed the data to normality by BoxCox transformation [49]. To investigate the effects of sex on urinary metabolites, we considered only adult, non-pregnant giraffes. This allowed us to reduce potential interferences due to different age classes. We then highlighted any difference by t-test. To investigate age related effects, by removing sex effect, we applied a two-way ANOVA test followed by Tukey-HSD, by taking advantage of the "aov" function of the R package "stats" [50]. For the above statistical tests, we accepted a cut-off p-value of 0.05.

In agreement with Bazzano et al. [51], we highlighted any trend characterizing the samples with robust principal component analysis (rPCA) models [52], using the molecules accepted by univariate analysis as a base. We took advantage of the PcaHubert algorithm implemented in the "rrcov" package. The algorithm grants robustness with a two-steps approach. In the first step outlying samples are detected according to their distance from the others along and orthogonally to the PCA plane. A second step determines the optimal number of principal components (PCs). The main features of each rPCA model are summarized by a scoreplot and by a Pearson correlation plot. The former is the projection of the samples in the PC space and highlights the underlying structure of the data. The latter relates the concentration of each variable to the components of the model.

5. Conclusions

This work represents a primer in giving quantitative information about the urinary metabolome of captive giraffes, as detected by untargeted ^{1}H-NMR. Foraging behaviors and daily activity could be considered as one of the main reasons for the differences we highlighted that are linked to sex and age. A preliminary observation conducted on two female giraffes suggests that ^{1}H-NMR based metabolomics could be conveniently applied to monitor modifications occurring during pregnancy, some of which are potentially related to inflammatory status triggered by modification of the microbiota profile.

Author Contributions: C.Z., L.L., S.F., and G.I. conceived and designed the research. C.Z. and L.L. performed metabolomics analysis. S.F. collected the samples. C.Z., L.L., S.F., and G.I. wrote the manuscript. All authors have read and agreed to the published version of the manuscript.

Acknowledgments: Chenglin Zhu gratefully acknowledges financial support from Chinese Scholarship Council (grant n° 201606910076). Part of the samples used in this study was collected with the financial Grant won by Sabrina Fasoli and donated by the Zebra Foundation for Veterinary Zoological Education in 2019. The authors would like to thank Bandoli Francesca, Cordon Rossana, Cotignoli Chiara, Laguardia Daniele, Laricchiuta Pietro, Sandri Camillo, Schneider Rainer, and Spiezio Caterina for their support. The authors also would like to thank all staff of the zoos involved in this study for their precious help during the sampling and Vito Barnaba, Pietro Ciaccia, Giacomo Melani, and Fulvio Pendezza for their fundamental assistance.

References

1. Muller, Z.; Bercovitch, F.; Brand, R.; Brown, D.; Brown, M.; Bolger, D.; Carter, K.; Deacon, F.; Doherty, J.B.; Fennessy, J.; et al. Giraffa Camelopardalis. The IUCN Red List of Threatened Species 2016: e.T9194A51140239. Available online: https://www.iucnredlist.org/species/9194/109326950 (accessed on 17 April 2020).
2. Paul-Murphy, J.; Molter, C. Overview of Animal Welfare in Zoos. In *Fowler's Zoo and Wild Animal Medicine Current Therapy*; Miller, R.E., Lamberski, N., Calle, P.P., Eds.; Elsevier: St. Louis, MO, USA, 2019; Volume 9, pp. 64–72. ISBN 978-0-323-55228-8.
3. Hosey, G.; Melfi, V.; Pankhurst, S. Animal Welfare. In *Zoo Animals: Behaviour, Management, and Welfare*, 2nd ed.; Oxford University Press: Oxford, UK, 2013; pp. 212–250. ISBN 0199693528.
4. Normando, S.; Pollastri, I.; Florio, D.; Ferrante, L.; Macchi, E.; Isaja, V.; de Mori, B. Assessing animal welfare in animal-visitor interactions in zoos and other facilities. A pilot study involving giraffes. *Animals* **2018**, *8*, 153. [CrossRef] [PubMed]
5. Dagg, A.I. Physiology. In *Giraffe: Biology, Behaviour and Conservation*; Dagg, A.I., Ed.; Cambridge University Press: New York, NY, USA, 2014; pp. 117–134. ISBN 9781139542302.
6. Gage, L.J. Giraffe Husbandry and Welfare. In *Fowler's Zoo and Wild Animal Medicine Current Therapy*;

Miller, R.E., Lamberski, N., Calle, P.P., Eds.; Elsevier: St. Louis, MO, USA, 2019; Volume 9, pp. 619–622. ISBN 978-0-323-55228-8.

7. Glatston, A.R.; Smit, S. Analysis of the urine of the okapi (Okapia johnstoni). *Acta Zool. Pathol. Antverp.* **1980**, *75*, 49–58.
8. Zhang, Q. Hypertension and Counter-Hypertension Mechanisms in Giraffes. *Cardiovasc. Hematol. Disord. Targets* **2012**, *6*, 63–67. [CrossRef] [PubMed]
9. Hargens, A.R.; Millard, R.W.; Pettersson, K.; Johansen, K. Gravitational haemodynamics and oedema prevention in the giraffe. *Nature* **1987**, *329*, 59–60. [CrossRef]
10. Agaba, M.; Ishengoma, E.; Miller, W.C.; McGrath, B.C.; Hudson, C.N.; Bedoya Reina, O.C.; Ratan, A.; Burhans, R.; Chikhi, R.; Medvedev, P.; et al. Giraffe genome sequence reveals clues to its unique morphology and physiology. *Nat. Commun.* **2016**, *7*, 1–8. [CrossRef]
11. Patel, K.P.; Luo, F.J.G.; Plummer, N.S.; Hostetter, T.H.; Meyer, T.W. The production of p-Cresol sulfate and indoxyl sulfate in vegetarians versus omnivores. *Clin. J. Am. Soc. Nephrol.* **2012**, *7*, 982–988. [CrossRef]
12. Laghi, L.; Zhu, C.; Campagna, G.; Rossi, G.; Bazzano, M.; Laus, F. Probiotic supplementation in trained trotter horses: Effect on blood clinical pathology data and urine metabolomic assessed in field. *J. Appl. Physiol.* **2018**, *125*, 654–660. [CrossRef]
13. Barbara, G.; Scaioli, E.; Barbaro, M.R.; Biagi, E.; Laghi, L.; Cremon, C.; Marasco, G.; Colecchia, A.; Picone, G.; Salfi, N.; et al. Gut microbiota, metabolome and immune signatures in patients with uncomplicated diverticular disease. *Gut* **2017**, *66*, 1252–1261. [CrossRef]
14. Kok, D.E.G.; Rusli, F.; van der Lugt, B.; Lute, C.; Laghi, L.; Salvioli, S.; Picone, G.; Franceschi, C.; Smidt, H.; Vervoort, J.; et al. Lifelong calorie restriction affects indicators of colonic health in aging C57Bl/6J mice. *J. Nutr. Biochem.* **2018**, *56*, 152–164. [CrossRef]
15. Liao, Y.; Hu, R.; Wang, Z.; Peng, Q.; Dong, X.; Zhang, X.; Zou, H.; Pu, Q.; Xue, B.; Wang, L. Metabolomics Profiling of Serum and Urine in Three Beef Cattle Breeds Revealed Different Levels of Tolerance to Heat Stress. *J. Agric. Food Chem.* **2018**, *66*, 6926–6935. [CrossRef] [PubMed]
16. Sink, C.A.; Weinstein, N.M. Specimen procurement. In *Practical Veterinary Urinalysis*; John Wiley & Sons: Chichester, West Sussex, UK, 2012; pp. 9–18.
17. Zhu, C.; Li, C.; Wang, Y.; Laghi, L. Characterization of yak common biofluids metabolome by means of proton nuclear magnetic resonance spectroscopy. *Metabolites* **2019**, *9*, 41. [CrossRef] [PubMed]
18. Zhu, C.; Faillace, V.; Laus, F.; Bazzano, M.; Laghi, L. Characterization of trotter horses urine metabolome by means of proton nuclear magnetic resonance spectroscopy. *Metabolomics* **2018**, *14*, 106. [CrossRef] [PubMed]
19. Zhu, C.; Laghi, L.; Zhang, Z.; He, Y.; Wu, D.; Zhang, H.; Huang, Y.; Li, C.; Zou, L. First Steps toward the Giant Panda Metabolome Database: Untargeted Metabolomics of Feces, Urine, Serum, and Saliva by 1 H NMR. *J. Proteome Res.* **2020**, *19*, 1052–1059. [CrossRef]
20. Chen, X.; Kyle, D.; Orskov, E.; Hovell, F. Renal clearance of plasma allantoin in sheep. *Exp. Physiol.* **1991**, *76*, 59–65. [CrossRef]
21. Sheedy, J.R.; Gooley, P.R.; Nahid, A.; Tull, D.L.; McConville, M.J.; Kukuljan, S.; Nowson, C.A.; Daly, R.M.; Ebeling, P.R. 1H-NMR analysis of the human urinary metabolome in response to an 18-month multi-component exercise program and calcium–vitamin-D3 supplementation in older men. *Appl. Physiol. Nutr. Metab.* **2014**, *39*, 1294–1304. [CrossRef]
22. Mukherjee, K.; Edgett, B.A.; Burrows, H.W.; Castro, C.; Griffin, J.L.; Schwertani, A.G.; Gurd, B.J.; Funk, C.D. Whole blood transcriptomics and urinary metabolomics to define adaptive biochemical pathways of high-intensity exercise in 50-60 year old masters athletes. *PLoS ONE* **2014**, *9*, e92031. [CrossRef]
23. Enea, C.; Seguin, F.; Petitpas-Mulliez, J.; Boildieu, N.; Boisseau, N.; Delpech, N.; Diaz, V.; Eugène, M.; Dugué, B. 1H NMR-based metabolomics approach for exploring urinary metabolome modifications after acute and chronic physical exercise. *Anal. Bioanal. Chem.* **2010**, *396*, 1167–1176. [CrossRef]
24. Ginnett, T.F.; Demment, M.W. Sex differences in giraffe foraging behavior at two spatial scales. *Oecologia* **1997**, *110*, 291–300. [CrossRef]
25. Mayneris-Perxachs, J.; Bolick, D.T.; Leng, J.; Medlock, G.L.; Kolling, G.L.; Papin, J.A.; Swann, J.R.; Guerrant, R.L. Protein-and zinc-deficient diets modulate the murine microbiome and metabolic phenotype. *Am. J. Clin. Nutr.* **2016**, *104*, 1253–1262. [CrossRef]
26. Schüßler, D.; Greven, H. Quantitative aspects of the ruminating process in giraffes (Giraffa camelopardalis) fed with different diets. *Zoo Biol.* **2017**, *36*, 407–412. [CrossRef] [PubMed]

27. Schnackenberg, L.K.; Sun, J.; Espandiari, P.; Holland, R.D.; Hanig, J.; Beger, R.D. Metabonomics evaluations of age-related changes in the urinary compositions of male Sprague Dawley rats and effects of data normalization methods on statistical and quantitative analysis. In Proceedings of the BMC Bioinformatics, Hong Kong, China, 27–30 August 2007.
28. Slupsky, C.M.; Rankin, K.N.; Wagner, J.; Fu, H.; Chang, D.; Weljie, A.M.; Saude, E.J.; Lix, B.; Adamko, D.J.; Shah, S.; et al. Investigations of the effects of gender, diurnal variation, and age in human urinary metabolomic profiles. *Anal. Chem.* **2007**, *79*, 6995–7004. [CrossRef] [PubMed]
29. Theodorou, M.K.; France, J. Rumen microorganisms and their interactions. In *Quantitative Aspects of Ruminant Digestion and Metabolism*; Dijkstra, J., Forbes, J., France, J., Eds.; CAB International: Wallingford, Oxfordshire, UK, 2009; Volume 2, pp. 207–228.
30. Schmidt, J.M.; Henken, S.; Dowd, S.E.; McLaughlin, R.W. Analysis of the Microbial Diversity in the Fecal Material of Giraffes. *Curr. Microbiol.* **2018**, *75*, 323–327. [CrossRef] [PubMed]
31. Soupart, P. Urinary excretion of free amino acids in normal adult men and women. *Clin. Chim. Acta* **1959**, *4*, 265–271. [CrossRef]
32. Pechlivanis, A.; Kostidis, S.; Saraslanidis, P.; Petridou, A.; Tsalis, G.; Mougios, V.; Gika, H.G.; Mikros, E.; Theodoridis, G.A. 1H NMR-based metabonomic investigation of the effect of two different exercise sessions on the metabolic fingerprint of human urine. *J. Proteome Res.* **2010**, *9*, 6405–6416. [CrossRef] [PubMed]
33. Karen, A.; Szabados, K.; Reiczigel, J.; Beckers, J.F.; Szenci, O. Accuracy of transrectal ultrasonography for determination of pregnancy in sheep: Effect of fasting and handling of the animals. *Theriogenology* **2004**, *61*, 1291–1298. [CrossRef]
34. Sun, L.; Guo, Y.; Fan, Y.; Nie, H.; Wang, R.; Wang, F. Metabolic profiling of stages of healthy pregnancy in Hu sheep using nuclear magnetic resonance (NMR). *Theriogenology* **2017**, *92*, 121–128. [CrossRef]
35. Kenéz, Á.; Dänicke, S.; Rolle-Kampczyk, U.; von Bergen, M.; Huber, K. A metabolomics approach to characterize phenotypes of metabolic transition from late pregnancy to early lactation in dairy cows. *Metabolomics* **2016**, *12*, 165. [CrossRef]
36. Diaz, S.O.; Barros, A.S.; Goodfellow, B.J.; Duarte, I.F.; Carreira, I.M.; Galhano, E.; Pita, C.; Almeida, M.D.C.; Gil, A.M. Following healthy pregnancy by nuclear magnetic resonance (NMR) metabolic profiling of human urine. *J. Proteome Res.* **2013**, *12*, 969–979. [CrossRef]
37. Bouckenooghe, T.; Remacle, C.; Reusens, B. Is taurine a functional nutrient? *Curr. Opin. Clin. Nutr. Metab. Care* **2006**, *9*, 728–733. [CrossRef]
38. Bristow, A.W.; Whitehead, D.C.; Cockburn, J.E. Nitrogenous constituents in the urine of cattle, sheep and goats. *J. Sci. Food Agric.* **1992**, *59*, 387–394. [CrossRef]
39. Dasarathy, J.; Gruca, L.L.; Bennett, C.; Parimi, P.S.; Duenas, C.; Marczewski, S.; Fierro, J.L.; Kalhan, S.C. Methionine metabolism in human pregnancy. *Am. J. Clin. Nutr.* **2010**, *91*, 357–365. [CrossRef] [PubMed]
40. Sarosiek, I.; Schicho, R.; Blandon, P.; Bashashati, M. Urinary metabolites as noninvasive biomarkers of gastrointestinal diseases: A clinical review. *World J. Gastrointest. Oncol.* **2016**, *8*, 459. [CrossRef] [PubMed]
41. Edwards, S.M.; Cunningham, S.A.; Dunlop, A.L.; Corwin, E.J. The Maternal Gut Microbiome during Pregnancy. *MCN Am. J. Matern. Nurs.* **2017**, *42*, 310–316. [CrossRef] [PubMed]
42. Bercovitch, F.B.; Berry, P.S.M. Reproductive life history of Thornicroft's giraffe in Zambia. *Afr. J. Ecol.* **2010**, *48*, 535–538. [CrossRef]
43. Bercovitch, F.B.; Berry, P.S.M. Giraffe birth locations in the South Luangwa National Park, Zambia: Site fidelity or microhabitat selection? *Afr. J. Ecol.* **2015**, *53*, 206–213. [CrossRef]
44. Lee, D.E.; Bond, M.L.; Bolger, D.T. Season of birth affects juvenile survival of giraffe. *Popul. Ecol.* **2017**, *59*, 45–54. [CrossRef]
45. R Development Core Team. *R: A Language and Environment for Statistical Computing*; Vienna, Austria, 2011; Volume 1. Available online: http://www.r-project.org (accessed on 24 March 2020).
46. Kneen, M.A.; Annegarn, H.J. Algorithm for fitting XRF, SEM and PIXE X-ray spectra backgrounds. *Nucl. Instrum. Methods Phys. Res. Sect. B Beam Interact. Mater. Atoms* **1996**, *109–110*, 209–213. [CrossRef]
47. Liland, K.H.; Almøy, T.; Mevik, B.H. Optimal choice of baseline correction for multivariate calibration of spectra. *Appl. Spectrosc.* **2010**, *64*, 1007–1016. [CrossRef]
48. Dieterle, F.; Ross, A.; Schlotterbeck, G.; Senn, H. Probabilistic quotient normalization as robust method to account for dilution of complex biological mixtures. Application in 1H NMR metabonomics. *Anal. Chem.* **2006**, *78*, 4281–4290. [CrossRef]

49. Box, G.E.P.; Cox, D.R. An Analysis of Transformations. *J. R. Stat. Soc. Ser. B* **2018**, *26*, 211–243. [CrossRef]
50. Chambers, J.M.; Freeny, A.; Heiberger, R.M. Analysis of variance; designed experiments. In *Statistical Models in S*; Wadsworth and Brooks/Cole Advanced Books and Software; Springer: Pacific Grove, CA, USA, 1992; pp. 145–193.
51. Bazzano, M.; Laghi, L.; Zhu, C.; Magi, G.E.; Serri, E.; Spaterna, A.; Tesei, B.; Laus, F. Metabolomics of tracheal wash samples and exhaled breath condensates in healthy horses and horses affected by equine asthma. *J. Breath Res.* **2018**, *12*, 46015. [CrossRef] [PubMed]
52. Hubert, M.; Rousseeuw, P.J.; Vanden Branden, K. ROBPCA: A new approach to robust principal component analysis. *Technometrics* **2005**, *47*, 64–79. [CrossRef]

Gas Chromatography–Mass Spectrometry-Based Metabolomic Analysis of Wagyu and Holstein Beef

Tomoya Yamada *, Mituru Kamiya and Mikito Higuchi

Division of Livestock Feeding and Management, National Agriculture and Food Research Organization, Nasushiobara, Tochigi 329-2793, Japan; m_kamiya@affrc.go.jp (M.K.); mikito@affrc.go.jp (M.H.)
* Correspondence: toyamada@affrc.go.jp.

Abstract: Japanese Black cattle (Wagyu) beef is characterized by high intramuscular fat content and has a characteristic sweet taste. However, the chemical components for characterizing the sweet taste of Wagyu beef have been unclear. In this experiment, we conducted a metabolomic analysis of the longissimus muscle (sirloin) in Wagyu and Holstein cattle to determine the key components associated with beef taste using gas chromatography–mass spectrometry (GC-MS). Holstein sirloin beef was characterized by the abundance of components such as glutamine, ribose-5-phosphate, uric acid, inosine monophosphate, 5-oxoproline, and glycine. In contrast, Wagyu sirloin beef was characterized by the abundance of sugar components (maltose and xylitol). Dietary fat is known to increase the intensity of sweet taste. These results suggest that the sweet taste of Wagyu beef is due to the synergetic effects of higher sugar components and intramuscular fat.

Keywords: metabolome; beef; Wagyu; Holstein

1. Introduction

Food quality, especially the taste of food, is affected by numerous chemical components. Metabolomic analysis has been used to select biomarkers from numerous metabolites. Therefore, the relationship between metabolomic profiling and food quality has been investigated to identify quality-related components. Previous reports showed that the metabolomic profiles of foods such as fermented alcoholic beverages [1], soybeans [2], and tomatoes [3], strongly affect their taste.

Metabolomic analyses of meat have also been conducted to identify quality-related components. The relationships between metabolomic profiles and processing conditions of hams [4], the muscle type of pork [5], and the sensory perceptions of pork [6] have been reported. In addition, metabolome studies of beef have shown that metabolomic profiles were affected by geographical origin [7], breed [8,9], storage conditions [10], and aging periods [11].

Japanese Black cattle, also called Wagyu, are characterized by their great capacity for intramuscular adipose tissue accumulation [12,13]. The high intramuscular adipose tissue content of beef, called marbling, improves the texture, juiciness, and tenderness of Wagyu beef [13]. In sensory tests, Wagyu beef had significantly higher sensory characteristic scores than beef from other cattle breeds [14,15]. Interestingly, Wagyu beef has a characteristic sweet aroma and sweet taste that are not detected in other cattle breeds in sensory tests [15–17]. Previous reports indicated that the sweet aroma of Wagyu beef was affected by the lactone and decenal components [18,19]. However, the metabolomic biomarkers discerning the breed differences in beef, especially the characteristic sweet taste of Wagyu beef, have remained unclear. Holstein cattle are categorized as a dairy breed, and Holstein beef is characterized as lean meat [20,21]. Previous sensory test results have indicated breed differences between sensory characteristic scores of Wagyu and Holstein beef [14,15]. Therefore, to elucidate the breed differences in beef taste, a comparison of Wagyu and Holstein is thought to be the optimal model. In the present study, we conducted a metabolomic analysis of longissimus muscle (sirloin) samples from Wagyu and Holstein cattle to identify the metabolomic biomarkers characterizing the breed differences in beef taste.

2. Results

Gas chromatography–mass spectrometry (GC-MS) analysis detected 67 metabolites in the sirloin samples of Wagyu and Holstein. Full results are shown in Supplementary Table S1. The principal component analysis (PCA) score plots showed that the metabolomic profile was divided into Wagyu and Holstein groups (Figure 1). The heatmap of metabolites also showed a difference between Wagyu and Holstein groups (Figure 2). Metabolites contributing to cluster 7, which characterized the Holstein sample, were mainly composed of amino acids (proline and glycine), amino compounds (succinic acid, amino propanoic acid, creatinine, and pyruvic acid), and nucleic acid metabolites (inosine and ribose). In contrast, cluster 1, which characterized the Wagyu sample, was mainly composed of sugar components (maltose and xylitol) and fatty acids (stearic acid, palmitic acid and nonanoic acid). Table 1 shows the differences in the relative quantity of the main metabolite compounds in Wagyu and Holstein samples. The amount of glutamine, ribose-5-phosphate, uric acid, inosine monophosphate, 5-oxoproline, and glycine in Holstein samples was significantly higher than in Wagyu samples. In contrast, the amount of maltose and xylitol in Wagyu samples was significantly higher than that in Holstein samples.

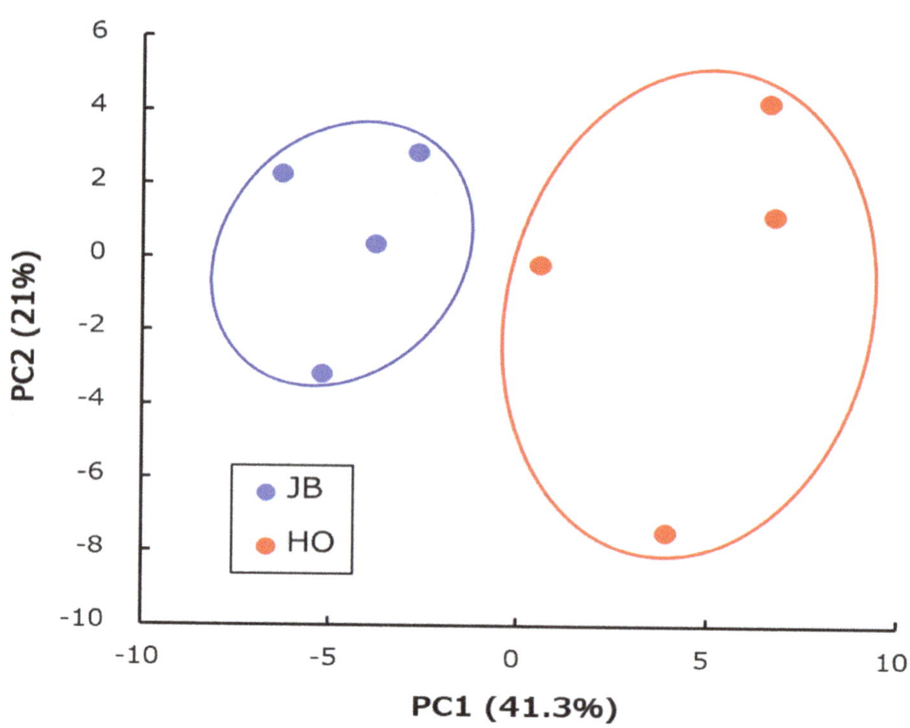

Figure 1. Principal component analysis (PCA) of metabolome data from Japanese Black Wagyu (JB) and Holstein (HO) sirloin samples. •: JB (n = 4), •: HO (n = 4).

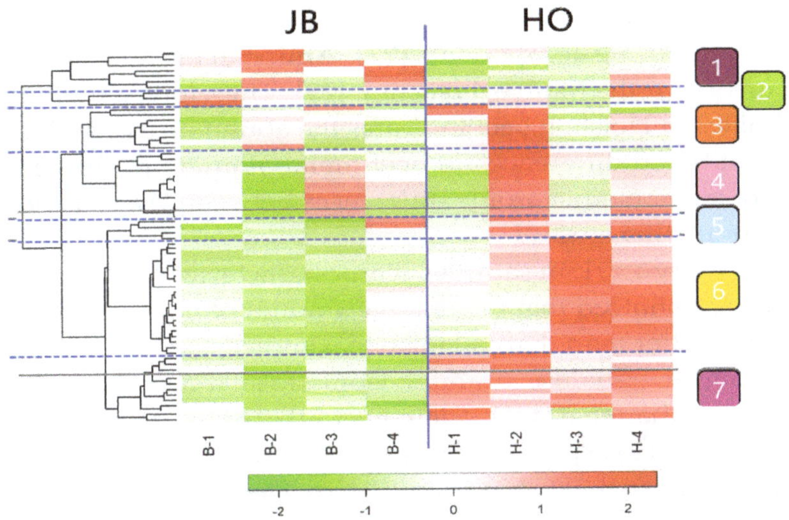

Figure 2. Heatmap of metabolites in Japanese Black Wagyu (JB) and Holstein (HO) sirloin samples. The upregulated metabolites are shown in red, and the downregulated metabolites are presented in green. Cluster 1, which characterized the Wagyu sample, was mainly composed of maltose, xylitol, stearic acid, palmitic acid, and nonanoic acid. Cluster 7, which characterized the Holstein sample, was mainly composed of proline, glycine, succinic acid, amino propanoic acid, creatinine, pyruvic acid, inosine, and ribose. B1-B4: Japanese Black Wagyu (n = 4); H1-H4: Holstein (n = 4).

Table 1. Main metabolite compounds in Japanese Black Wagyu and Holstein sirloin samples.

Change	Compound Name	Relative Area				Comparative Analysis	
		JB		Ho		JB/Ho	
		Mean	S.D.	Mean	S.D.	Ratio	p-Value
Increase	Maltose	7.1×10^4	1.4×10^4	4.0×10^4	1.5×10^4	1.8	0.022 *
	Xylitol	2.7×10^4	4.4×10^3	1.8×10^4	2.3×10^3	1.5	0.022 *
	Palmitic acid	8.8×10^4	3.6×10^4	6.0×10^4	8.8×10^3	1.5	0.221
	Stearic acid	6.7×10^4	4.0×10^4	5.2×10^4	1.3×10^4	1.3	0.513
	Ribose	4.3×10^5	1.2×10^5	3.5×10^5	5.6×10^4	1.2	0.264
	Sedoheptulose7-phosphate	7.9×10^4	3.2×10^4	6.5×10^4	2.3×10^4	1.2	0.513
	Mannose	1.1×10^6	2.7×10^5	8.9×10^5	2.8×10^5	1.2	0.406
	Glycerol 3 phosphate	2.1×10^4	7.9×10^3	2.1×10^4	3.3×10^3	1.1	0.623
Decrease	Glycine	1.6×10^6	1.1×10^5	2.6×10^6	2.4×10^5	0.6	0.001 **
	Ornithine	3.9×10^4	6.3×10^3	6.7×10^4	2.3×10^4	0.6	0.094
	5-Oxoproline	9.2×10^4	9.2×10^3	1.7×10^5	3.2×10^4	0.6	0.014 *
	Inosine monophosphate	1.8×10^5	7.9×10^4	3.3×10^5	1.4×10^4	0.5	0.030 *
	Uric acid	1.1×10^4	1.4×10^3	2.2×10^4	4.8×10^3	0.5	0.014 *
	2-Hydroxyglutaric acid	5.8×10^3	2.0×10^3	1.3×10^4	5.3×10^3	0.5	0.075
	Ribose 5-phosphate	3.2×10^4	5.4×10^3	6.9×10^4	1.2×10^4	0.5	0.004 **
	Glutamine	3.5×10^4	3.8×10^3	9.8×10^4	3.1×10^4	0.4	0.026 *

Values are expressed as means and S.D. ratio: fold intensity of metabolite compounds (JB/Ho). Annotation and relative quantification of metabolites was measured by each peak using the gas chromatography–mass spectrometry (GC-MS) solution (Shimadzu) and GC/MS Metabolite Database Ver. 2 (Shimadzu). Japanese Black Wagyu (JB, n = 4), Holstein (HO, n = 4) * $p < 0.05$, ** $p < 0.01$.

3. Discussion

In the present study, we showed that the Holstein sirloin samples were characterized by amino acids, amino compounds and nucleic-acid metabolites. We also showed that the relative amount of glutamine, ribose-5-phosphate, uric acid, inosine monophosphate, 5-oxoproline, and glycine in the Holstein sirloin samples was significantly higher than in the Wagyu samples. The abundance of amino acids, amino compounds, and nucleic-acid metabolites clearly reflects the lean meat content of Holstein beef. Previous studies also showed the relationship between metabolomic profiling and beef quality [7–11]. These results indicate that metabolomic analysis is an optimum approach to identify quality-related chemical components of beef. The cooking of meat forms a characteristic taste via numerous chemical reactions. The Maillard reaction is one of the major chemical reactions of cooked meat that produces many flavoring components [22]. Amino acids and nucleic-acid metabolites are main meat-flavor precursors for the Maillard reaction [23]. Therefore, a higher amount of amino acid and nucleic-acid metabolites in Holstein beef indicates the abundance of a Maillard reaction substrate. These results suggest that breed differences in beef metabolomic profiles affect the taste of cooked meats. The elucidation of metabolomic profile differences between Wagyu and Holstein cooked meat is an important subject for further study.

We showed that the Wagyu sirloin samples were characterized by sugar components and fatty acids. The present study also showed that the relative amount of maltose and xylitol in Wagyu sirloin samples was significantly higher than that in Holstein samples. Ueda et al. reported that the relative amount of malic acid, maltose, trehalose, arabitol, isomaltose, n-acetylserine, and inositol in Wagyu beef was significantly higher than that in Holstein beef [9]. The aging periods of beef affect the meat quality and metabolomic profile [11]. The difference between the results of our study and those of Ueda et al. would be attributed to meat aging conditions. The meat aging period in the present study was at 4 °C for 7 days after slaughter. In contrast, the aging in Ueda et al. was at 4 °C for 20 days [9]. On the other hand, results showing a higher amount of maltose in Wagyu beef than in Holstein beef were common to both experiments. These results suggest that a higher maltose concentration is a primary feature of Wagyu beef, independent of the aging period. The causative substance of the sweet taste of Wagyu beef remains unclear. Amino acids have different taste properties depending upon their chemical structure [24]. Glycine is categorized as a "sweet" amino acid [24]. The present results showed that glycine was abundant in the Holstein sirloin samples. Therefore, the possibility that amino acids contribute to the sweet taste of Wagyu beef would be excluded. In contrast, the present study showed that maltose and xylitol, categorized as sugar components, are abundant in Wagyu beef. Wagyu is characterized by higher intramuscular fat content than Holstein [12,13]. Previous reports indicated that dietary fat increased the intensity of sweet taste [25,26]. These results suggest that the sweet taste of Wagyu beef is affected by the synergetic interaction between higher sugar components and intramuscular fat. Threshold sweetness concentrations of maltose and xylitol have been reported using sensory test methods [27,28]. Kearsley et al. reported that the threshold sweetness concentration of maltose was 1.07% (w/v), and that of xylitol was 0.51% (w/v) [27]. In molar volumes, Birch et al. showed that the threshold sweetness concentration of maltose was 21.0 mM/l, and that of xylitol was also 21.0 mM/l [28]. However, the effect of dietary fat on the threshold sweetness concentration of sugar components has not been reported. The slaughter age of Wagyu (aged 29–30 months) and Holsteins (aged 21–22 months) in this study was defined in accordance with the commonly applied fattening periods of each breed in Japan. Ueda et al. also analyzed beef samples of Wagyu (aged 31–32 months) and Holsteins (aged 21 months) that were similar to the age of cattle used in this study [9]. Previous studies have shown that the slaughter age affects meat quality and sensory traits of beef [29,30]. To our knowledge, there are no previous studies examining the effects of slaughter age on metabolomic profiling of beef. Therefore, the differences between fattening periods of Wagyu and Holsteins in this study may affect the metabolomic profiling of beef. In addition, there is a possibility that other metabolites, which we could not detect in this study, might affect the sweetness of beef. Further studies are needed to clarify the effects of metabolites on the sweet taste of Wagyu beef.

4. Materials and Methods

4.1. Animals

Wagyu steers (aged 29–30 months, $n = 4$) and Holstein steers (aged 21–22 months, $n = 4$) were used in this study. They received a concentrate (78% total digestible nutrients and 13% crude protein) and orchard grass hay (56% total digestible nutrients and 8% crude protein) ad libitum from 10 months of age until they were slaughtered. Longissimus muscle (sirloin) samples were collected at slaughter. The sirloin samples (1.5 kg) were collected between the third and fourth lumbar vertebrae from the left side of the carcass. Samples were vacuum-packed and wet-aged at 4 °C for 7 days and then stored at −80 °C for later metabolome analysis. All animals received humane care as outlined in the Guide for the Care and Use of Experimental Animals (No.1631B004, National Agriculture and Food Research Organization).

4.2. GC-MS Analysis

Frozen sirloin samples were powdered with liquid nitrogen and weighed (100 mg) in the frozen state. Frozen samples were plunged into 80% methanol and homogenized using zirconia beads and an ultrasonic homogenizer for 5 min. The samples were centrifuged at 15,000 rpm for 5 min. The supernatant was filtered using a Mono-Spin C18 column (GL Science, Tokyo, Japan), and then the filtration (50 μL) was dried by a nitrogen gas flow. Methoxyamine hydrochloride solubilized with pyridine (20 mg/mL, 50 μL) was added to each sample, and oxime formation was achieved by reacting at 30 °C for 90 min. Trimethylsilyl-trifluoroacetamide (50 μL) was then added to each sample, and trimethylsilylation was carried out by reacting at 37 °C for 30 min. Analyses were performed on a gas chromatography–mass spectrometer (GC-MS, QP2010Ultra, Shimadzu, Kyoto, Japan) using a DB-5 column (Agilent Technologies, Santa Clara, CA, USA) at the Kazusa DNA Research Institute. The carrier gas was helium in a flow of 1.1 mL/min. The injection temperature was 280 °C, and the injection volume was 0.5 μL. The temperature program was isothermal for 4 min at 100 °C, then raised at a rate of 4 °C / min to 320 °C and held for 8 min. The ion source temperature and scan speed were set to 200 °C and 2500 u/sec, respectively. Sample peaks were recorded over the mass range of 45–600 *m/z*. The retention time correction of peaks was carried out based on the retention time of a standard alkane series mixture (C-7 to C-33). Annotation and relative quantification of metabolite was measured by each peak using the GC-MS solution (Shimadzu) and GC/MS Metabolite Database Ver. 2 (Shimadzu). The relative area was calculated using the peak area of each metabolite relative to the analyzed sample weight at Kazusa DNA Research Institute.

4.3. Statistical Analysis

Principal component analysis (PCA) was conducted using SampleStat software (Human Metabolome Technologies, Japan). Hierarchical cluster analysis (HCA) was performed and heatmap formation analyzed using PeakStat software (Human Metabolome Technologies). Differences between the relative quantity of metabolites in Wagyu and Holstein samples were evaluated using Welch's t-test. Results are presented as means, and S.D. values of $p < 0.05$ were considered significant.

Author Contributions: Conceptualization, T.Y. and M.H.; methodology, T.Y.; software, T.Y.; validation, T.Y., M.K., and M.H.; formal analysis, T.Y.; investigation, T.Y., M.K., and M.H.; resources, T.Y., M.K., and M.H.; data curation, T.Y., M.K., and M.H.; writing—original draft preparation, T.Y.; writing—review and editing, T.Y., M.K., and M.H.; visualization, T.Y.; supervision, T.Y.; project administration, T.Y.; funding acquisition, T.Y. All authors have read and agreed to the published version of the manuscript.

Acknowledgments: We would like to thank Kazuki Iguchi (Human Metabolome Technologies) for technical support.

References

1. Takahashi, K.; Kohno, H. Different polor metabolites and protein profiles between high- and low-quality Japanese Ginjo sake. *PLoS ONE* **2016**, *11*, e0150524. [CrossRef] [PubMed]
2. Sugimoto, M.; Goto, H.; Otomo, K.; Ito, M.; Onuma, H.; Suzuki, A.; Sugawara, M.; Abe, S.; Tomita, M.; Soga, T. Metabolomic profiles and sensory attributes of edamame under various storage duration and temperature conditions. *J. Agric. Food Chem.* **2010**, *58*, 8418–8425. [CrossRef] [PubMed]

3. Malmendal, A.; Amoresano, C.; Trotta, R.; Lauri, I.; De Tito, S.; Novellino, E.; Randazzo, A. NMR spectrometers as "magic tongues": Prediction of sensory descriptors in canned Tomatos. *J. Agric. Food Chem.* **2011**, *59*, 10831–10838. [CrossRef] [PubMed]
4. Sugimoto, M.; Obiya, S.; Kaneko, M.; Enomoto, A.; Honma, M.; Wakayama, M.; Soga, T.; Tomita, M. Metabolic profiling as a possible reverse engineering tool for estimating processing conditions of dry-cured hams. *J. Agric. Food Chem.* **2017**, *65*, 402–410. [CrossRef] [PubMed]
5. Muroya, S.; Oe, M.; Nakajima, I.; Ojima, K.; Chikuni, K. CE-TOFMS-based metabolomic profiling revealed characteristic metabolic pathways in postmortem porcine fast and slow type muscles. *Meat Sci.* **2014**, *98*, 726–735. [CrossRef] [PubMed]
6. Straadt, I.K.; Aaslyng, M.D.; Bertram, H.C. An NMR-based metabolomics study of pork from different crossbreeds and relation to sensory perception. *Meat Sci.* **2014**, *96*, 719–728. [CrossRef] [PubMed]
7. Jung, Y.; Lee, J.; Kwon, J.; Lee, K.S.; Ryu, D.H.; Hwang, G.S. Discrimination of the geographical origin of beef by H NMR-based metabolomics. *J. Agric. Food Chem.* **2010**, *58*, 10458–10466. [CrossRef]
8. Ritota, M.; Casciani, L.; Failla, S.; Valentin, M. HRMAS-NMR spectroscopy and multivariate analysis meat characterization. *Meat Sci.* **2012**, *92*, 754–761. [CrossRef]
9. Ueda, S.; Iwamoto, E.; Kato, Y.; Shinohara, M.; Shirai, Y.; Yamanoue, M. Comparative metabolomics of Japanese Black cattle beef and other meats using gas chromatography-mass spectrometry. *Biosci. Biotechnol. Biochem.* **2019**, *83*, 137–147. [CrossRef]
10. Argyri, A.A.; Mallouchos, A.; Panagou, E.Z.; Nychas, G.J.E. The dynamics of the HS/SPME-GC-MS as a tool to assess the spoilage of minced beef stored under different packaging and temperature conditions. *Int. J. Food Microbiol.* **2015**, *193*, 51–58. [CrossRef]
11. Muroya, S.; Oe, M.; Ojima, K.; Watanabe, A. Metabolomic approach to key metabolites characterizing postmortem aged loin muscle of Japanese Black (Wagyu) cattle. *Asian-Australas. J. Anim. Sci.* **2019**, *32*, 1172–1185. [CrossRef] [PubMed]
12. Zembayashi, M.; Lunt, D.K. Distribution of intramuscular lipid throughout M. longissimus thoracis et lumborum in Japanese Black, Japanese Shorthorn, Holstein and Japanese Black crossbreds. *Meat Sci.* **1995**, 211–216. [CrossRef]
13. Motoyama, M.; Sasaki, K.; Watanabe, A. Wagyu and the factors contributing to its beef quality: A Japanese industry overview. *Meat Sci.* **2016**, *120*, 10–18. [CrossRef] [PubMed]
14. Corbin, C.H.; O'Quinn, T.G.; Garmyn, A.J.; Legato, J.F.; Hunt, M.R.; Dinh, T.T.N.; Rathmann, R.J.; Brooks, J.C.; Miller, M.F. Sensory evaluation of tender beef strip loin steaks of varying marbling levels and quality treatments. *Meat Sci.* **2015**, *100*, 24–31. [CrossRef] [PubMed]
15. Sasaki, K.; Ooi, M.; Nagura, N.; Motoyama, M.; Narita, T.; Oe, M.; Nakajima, I.; Tatsuro, H.; Ojima, K.; Kobayashi, M.; et al. Classification and characterization of Japanese consumer's beef preferences by external preference mapping. *J. Sci. Food. Agric.* **2017**, *97*, 3453–3462. [CrossRef]
16. Watanabe, G.; Motoyama, M.; Orita, K.; Takita, K.; Aonuma, T.; Nakajima, I.; Tajima, A.; Abe, A.; Sasaki, K. Assessment of the dynamics of sensory perception of Wagyu beef strip loin prepared with different cooking methods and fattening periods using the temporal dominance of sensations. *Food Sci. Nutr.* **2019**, *7*, 3538–3548. [CrossRef]
17. Matsuishi, M.; Fujimori, M.; Okitani, A. Wagyu beef aroma in Wagyu (Japanese Black Cattle) beef preferred by the Japanese over imported beef. *Anim. Sci. J.* **2001**, *72*, 498–504. [CrossRef]
18. Matsuishi, M.; Kume, J.; Itou, Y.; Takahashi, M.; Arai, M.; Nagatomi, H.; Watanabe, K.; Hayase, F.; Okitani, A. Aroma components of Wagyu beef and imported beef. *Nihon Chikusan Gakkaiho* **2004**, *75*, 409–415. [CrossRef]
19. Inagaki, S.; Amano, Y.; Kumazawa, K. Identification and characterization of volatile components causing the characteristics flavor of Wagyu beef (Japanese Black cattle). *J. Agric. Food. Chem.* **2017**, *65*, 8691–8695. [CrossRef]
20. Matsuzaki, M.; Takizawa, S.; Ogawa, M. Plasma insulin metabolite concentrations and carcass characteristics of Japanese Black, Japanese Brown, and Holstein steers. *J. Anim. Sci.* **1997**, *75*, 3287–3293. [CrossRef]
21. Yamada, T.; Kawakami, S.; Nakanishi, N. Expression of adipogenic transcription factors in adipose tissue of fattening Wagyu and Holstein steers. *Meat Sci.* **2009**, *81*, 86–92. [CrossRef] [PubMed]
22. Kerth, C.R.; Miller, R.K. Beef flavor: A review from chemistry to consumer. *J. Sci. Food Agric.* **2015**, *95*, 2783–2798. [CrossRef] [PubMed]

23. Khan, M.I.; Jo, C.; Tariq, M.R. Meat flavor precursors and factors influencing flavor precursors-A systematic review. *Meat Sci.* **2015**, *110*, 278–284. [CrossRef] [PubMed]
24. Kawai, M.; Sekine-Hayakawa, Y.; Okiyama, A.; Ninomiya, Y. Gustatory sensation of L- and D- amino acids in humans. *Amino Acids.* **2012**, *43*, 2349–2358. [CrossRef] [PubMed]
25. Tuoria, H.; Sommardahl, C.; Hyvönen, L.; Leporanta, K.; Merimma, P. Does fat affect the timing of flavor perception? a case study with yoghurt. *Food Qual. Pref.* **1995**, *6*, 55–58. [CrossRef]
26. Metcalf, K.L.; Vickers, Z.M. Taste intensities of oil-in-water emulsions with varying fat content. *J. Sens. Stud.* **2002**, *17*, 379–390. [CrossRef]
27. Kearsley, M.W.; Dziedzic, S.Z.; Birch, G.G.; Smith, P.D. The production and properties of glucose syrups III. Sweetness of glucose syrups and related carbohydrates. *Starch* **1980**, *32*, 244–247. [CrossRef]
28. Birch, G.G.; Munton, S.L. Evidence for constant number of available sweet receptor sites at threshold concentration of sugars. *Experientia* **1981**, *37*, 839–840. [CrossRef]
29. Bouton, P.E.; Ford, P.V.; Harris, P.V.; Shorthose, W.R.; Ratcliff, D.; Morgan, J.H.L. Influence of animal age on the tenderness of beef: Muscle differences. *Meat Sci.* **1978**, *2*, 301–311. [CrossRef]
30. Okumura, T.; Saito, K.; Sowa, T.; Sakuma, H.; Ohhashi, F.; Tameoka, N.; Hirayama, M.; Nakayama, S.; Sato, S.; Gogami, T.; et al. Changes in beef sensory traits as somatic-cell-cloned Japanese black steers increased in age from 20 to 30 months. *Meat Sci.* **2012**, *90*, 159–163. [CrossRef]

PERMISSIONS

All chapters in this book were first published in MDPI; hereby published with permission under the Creative Commons Attribution License or equivalent. Every chapter published in this book has been scrutinized by our experts. Their significance has been extensively debated. The topics covered herein carry significant findings which will fuel the growth of the discipline. They may even be implemented as practical applications or may be referred to as a beginning point for another development.

The contributors of this book come from diverse backgrounds, making this book a truly international effort. This book will bring forth new frontiers with its revolutionizing research information and detailed analysis of the nascent developments around the world.

We would like to thank all the contributing authors for lending their expertise to make the book truly unique. They have played a crucial role in the development of this book. Without their invaluable contributions this book wouldn't have been possible. They have made vital efforts to compile up to date information on the varied aspects of this subject to make this book a valuable addition to the collection of many professionals and students.

This book was conceptualized with the vision of imparting up-to-date information and advanced data in this field. To ensure the same, a matchless editorial board was set up. Every individual on the board went through rigorous rounds of assessment to prove their worth. After which they invested a large part of their time researching and compiling the most relevant data for our readers.

The editorial board has been involved in producing this book since its inception. They have spent rigorous hours researching and exploring the diverse topics which have resulted in the successful publishing of this book. They have passed on their knowledge of decades through this book. To expedite this challenging task, the publisher supported the team at every step. A small team of assistant editors was also appointed to further simplify the editing procedure and attain best results for the readers.

Apart from the editorial board, the designing team has also invested a significant amount of their time in understanding the subject and creating the most relevant covers. They scrutinized every image to scout for the most suitable representation of the subject and create an appropriate cover for the book.

The publishing team has been an ardent support to the editorial, designing and production team. Their endless efforts to recruit the best for this project, has resulted in the accomplishment of this book. They are a veteran in the field of academics and their pool of knowledge is as vast as their experience in printing. Their expertise and guidance has proved useful at every step. Their uncompromising quality standards have made this book an exceptional effort. Their encouragement from time to time has been an inspiration for everyone.

The publisher and the editorial board hope that this book will prove to be a valuable piece of knowledge for researchers, students, practitioners and scholars across the globe.

LIST OF CONTRIBUTORS

Kaijie Yang, Jue Wang, Yabiao Luo, Qiguo Tang, Ximing Liu and Meiying Fang
Department of Animal Genetics and Breeding, National Engineering Laboratory for Animal Breeding, MOA Laboratory of Animal Genetics and Breeding, Beijing Key Laboratory for Animal Genetic Improvement, College of Animal Science and Technology, China Agricultural University, Beijing 100193, China

Kejun Wang
Department of Animal Genetics and Breeding, National Engineering Laboratory for Animal Breeding, MOA Laboratory of Animal Genetics and Breeding, Beijing Key Laboratory for Animal Genetic Improvement, College of Animal Science and Technology, China Agricultural University, Beijing 100193, China
College of Animal Science and Veterinary Medicine, Henan Agricultural University, Zhengzhou 450002, China

Maryam Gholizadeh and Jamal Fayazi
Department of Animal Science, Faculty of Animal Science and Food Technology, Agricultural Sciences and Natural Resources University of Khuzestan, Mollasani, Ahvaz 6341773637, Iran

Yazdan Asgari
Department of Medical Biotechnology, School of Advanced Technologies in Medicine, Tehran University of Medical Sciences, Tehran 1416753955, Iran

Hakimeh Zali
School of Advanced Technologies in Medicine, Shahid Beheshti University of Medical Sciences, Tehran 1416753955, Iran

Lars Kaderali
Institute of Bioinformatics, University Medicine Greifswald, Felix-Hausdor -Str. 8, 17475 Greifswald, Germany

Timothy D.W. Luke, Jennie E. Pryce and Simone J. Rochfort
Agriculture Victoria Research, AgriBio, Centre for AgriBioscience, Bundoora 3083, Australia
School of Applied Systems Biology, La Trobe University, Bundoora 3083, Australia

Aaron C. Elkins
Agriculture Victoria Research, AgriBio, Centre for AgriBioscience, Bundoora 3083, Australia

William J. Wales
Agriculture Victoria Research, Ellinbank Centre, Ellinbank 3821, Australia

Tatsuhiko Goto
Research Center for Global Agromedicine, Obihiro University of Agriculture and Veterinary Medicine, Obihiro, Hokkaido 080-8555, Japan
Department of Life and Food Sciences, Obihiro University of Agriculture and Veterinary Medicine, Obihiro, Hokkaido 080-8555, Japan

Hiroki Mori
Department of Life and Food Sciences, Obihiro University of Agriculture and Veterinary Medicine, Obihiro, Hokkaido 080-8555, Japan

Shunsuke Shiota
Graduate School of Agriculture, Kyoto University, Kyoto 606-8502, Japan

Susumu Muroya
NARO Institute of Livestock and Grassland Science, Tsukuba, Ibaraki 305-0901, Japan

Shuji Ueda
Graduate School of Agricultural Science, Kobe University, Hyogo 657-8501, Japan

Tomohiko Komatsu
Livestock Research Institute of Yamagata Integrated Research Center, Shinjo, Yamagata 996-0041, Japan

Takuya Miyakawa
Graduate School of Agricultural and Life Sciences, University of Tokyo, Bunkyo-ku, Tokyo 113-8657, Japan

Per Ertbjerg
Department of Food and Nutrition, University of Helsinki, 00014 Helsinki, Finland

Masaaki Taniguchi, Aisaku Arakawa, Motohide Nishio, Toshihiro Okamura and Kazuo Ishii
Institute of Livestock and Grassland Science, National Agriculture and Food Research Organization, Tsukuba Ibaraki 305-0901, Japan

Chika Ohnishi
Miyazaki Station, National Livestock Breeding Center, Kobayashi Miyazaki 886-0004, Japan

Kouen Kadowaki
Ibaraki Station, National Livestock Breeding Center, Chikusei Ibaraki 308-0112, Japan

Kimiko Kohira, Fumika Homma and Kazunori Matsumoto
National Livestock Breeding Center, Nishigo Fukushima 961-8511, Japan

Saliya Gurusinghe
Graham Center for Agricultural Innovation, Locked Bag 588, Wagga Wagga, NSW 2678, Australia

Sajid Latif
Graham Center for Agricultural Innovation, Locked Bag 588, Wagga Wagga, NSW 2678, Australia
School of Animal and Veterinary Sciences, Charles Sturt University, Wagga Wagga, NSW 2678, Australia

Paul A. Weston and Leslie A. Weston
Graham Center for Agricultural Innovation, Locked Bag 588, Wagga Wagga, NSW 2678, Australia
School of Agriculture and Wine Sciences, Charles Sturt University, Wagga Wagga, NSW 2678, Australia

Russell A. Barrow
Graham Center for Agricultural Innovation, Locked Bag 588, Wagga Wagga, NSW 2678, Australia
Plus 3 Australia Pty Ltd., Hawker, ACT 2614, Australia

John W. Piltz
New South Wales Department of Primary Industries, Wagga Wagga, NSW 2678, Australia

Marilena Bazzano, Enrica Lotito, Beniamino Tesei and Fulvio Laus
School of Biosciences and Veterinary Medicine, University of Camerino, Via Circonvallazione, 93/95, 62024 Matelica, Italy

Stefano Sgariglia
Practitioner, 63900 Fermo, Italy

Kamil Stastny, Kristina Putecova, Lenka Leva, Milan Franek and Martin Faldyna
Veterinary Research Institute, Hudcova 296/70, 62100 Brno, Czech Republic

Petr Dvorak
Faculty of Veterinary Hygiene and Ecology, University of Veterinary and Pharmaceutical Sciences Brno, Palackeho tr. 1946/1, 61242 Brno, Czech Republic

Satoko Naoe
Laboratory for Metabolomics, RIKEN Center for Integrative Medical Sciences, Yokohama 230-0045, Japan

Hiroshi Tsugawa
Laboratory for Metabolomics, RIKEN Center for Integrative Medical Sciences, Yokohama 230-0045, Japan
Metabolome informatics research team, RIKEN Center for Sustainable Resource Science, Yokohama 230-0045, Japan

Mikiko Takahashi
Metabolome informatics research team, RIKEN Center for Sustainable Resource Science, Yokohama 230-0045, Japan

Kazutaka Ikeda
Laboratory for Metabolomics, RIKEN Center for Integrative Medical Sciences, Yokohama 230-0045, Japan
Cellular and Molecular Epigenetics Laboratory, Graduate School of Medical Life Science, Yokohama City University, Tsurumi, Yokohama 230-0045, Japan

Makoto Arita
Laboratory for Metabolomics, RIKEN Center for Integrative Medical Sciences, Yokohama 230-0045, Japan
Cellular and Molecular Epigenetics Laboratory, Graduate School of Medical Life Science, Yokohama City University, Tsurumi, Yokohama 230-0045, Japan
Division of Physiological Chemistry and Metabolism, Graduate School of Pharmaceutical Sciences, Keio University, Minato-ku, Tokyo 105-8512, Japan

Kiriko Nakamura, Akira Ohtsuka and Daichi Ijiri
Department of Agricultural Sciences and Natural Resources, Kagoshima University, Korimoto, Kagoshima 890-0065, Japan

Saki Shimamoto
Department of Agricultural Sciences and Natural Resources, Kagoshima University, Korimoto, Kagoshima 890-0065, Japan
Graduate School of Science and Technology, Niigata University, Nishi-ku, Niigata 950-2181, Japan

Shozo Tomonaga
Division of Applied Biosciences, Graduate School of Agriculture, Kyoto University, Sakyo-ku, Kyoto 606-8502, Japan
Graduate School of Agriculture, Kyoto University, Kyoto 606-8502, Japan

Satoru Furukawa
Furukawa Research Office Co. Ltd., Setagaya-ku, Tokyo 157-0066, Japan

Priyanka Banerjee, Victor Adriano Okstoft Carmelo and Haja N. Kadarmideen
Quantitative Genomics, Bioinformatics and Computational Biology Group, Department of Applied Mathematics and Computer Science, Technical University of Denmark, 2800 Kongens Lyngby, Denmark

Chenglin Zhu and Luca Laghi
Department of Agro-Food Science and Technology, University of Bologna, 47521 Cesena, Italy

Sabrina Fasoli and Gloria Isani
Department of Veterinary Medical Sciences, University of Bologna, Ozzano Emilia, 40064 Bologna, Italy

Tomoya Yamada, Mituru Kamiya and Mikito Higuchi
Division of Livestock Feeding and Management, National Agriculture and Food Research Organization, Nasushiobara, Tochigi 329-2793, Japan

Index

www.ingramcontent.com/pod-product-compliance
Lightning Source LLC
Chambersburg PA
CBHW081216180926
51613CB00003B/253
* 9 7 8 1 6 4 7 4 0 4 1 0 9 *